Nanocharacterization of Cementitious Composites

Wengui Li · Hanbing Zhao · Kejin Wang

Nanocharacterization of Cementitious Composites

Unveiling Properties at Multiscale

Springer

Wengui Li
Centre for Infrastructure Engineering
and Safety, School of Civil
and Environmental Engineering
The University of New South Wales
Sydney, NSW, Australia

Hanbing Zhao
Centre for Infrastructure Engineering
and Safety, School of Civil
and Environmental Engineering
The University of New South Wales
Sydney, NSW, Australia

Kejin Wang
Department of Civil, Construction,
and Environmental Engineering
Iowa State University
Ames, IA, USA

ISBN 978-981-96-2916-9 ISBN 978-981-96-2917-6 (eBook)
https://doi.org/10.1007/978-981-96-2917-6

This Springer imprint is published by the registered company Springer Nature Singapore Pte Ltd.
The registered company address is: 152 Beach Road, #21-01/04 Gateway East, Singapore 189721, Singapore

Competing Interests The authors have no competing interests to declare that are relevant to the content of this manuscript.

Contents

Chapter 1
Introduction

1.1 Ordinary Portland Cement

The earliest application of cementitious materials in construction can be traced back to ancient Rome and Egypt. With the continuous exploration of people, cementitious materials have developed from the lime, gypsum and volcanic ash to the mature Portland cement materials (see Fig. 1.1). Ordinary Portland cement remains today the most important building material due to its low cost, wide raw material availability, easy to form various sizes and shapes, and high mechanical properties, required for civil infrastructures. The raw materials mined from quarry, such as limestones, clay, slates and some shales, are crushed, ground, and then calcined to form cement clinker, which is finally ground with 3–5% of gypsum to produce ordinary Portland cement [1]. The classification of cement materials according to the mainstream national standards is shown in Table 1.1.

1.1.1 C–S–H in Cement Paste

The main components of cement are tricalcium silicate ($3CaO{\cdot}SiO_2$), dicalcium silicate ($2CaO{\cdot}SiO_2$), tricalcium aluminate ($3CaO{\cdot}Al_2O_3$), tetra-calcium aluminoferrite ($4CaO{\cdot}Al_2O_3Fe_2O_3$), and gypsum ($CaSO_4{\cdot}0.5H_2O$), which are usually abbreviated as C_3S, C_2S, C_3A, C_4AF, and $C\overline{S}$ respectively [8, 9]. When Portland cement is mixed with an appropriate proportion of water, cement particles start to react with water to form hydration products, provide cementing properties that enable to bond solid particles together. With hydration time, the cement paste will gain stiffness, changing from a flowable, plastic state to a solid and hardened state, enabling to carry loads, especially compression [8].

The cementing properties (like adhesion) of a cement paste mainly comes from the reaction of C_2S and C_3S with water to form calcium silicate hydrate

W. Li et al., *Nanocharacterization of Cementitious Composites*,
https://doi.org/10.1007/978-981-96-2917-6_1

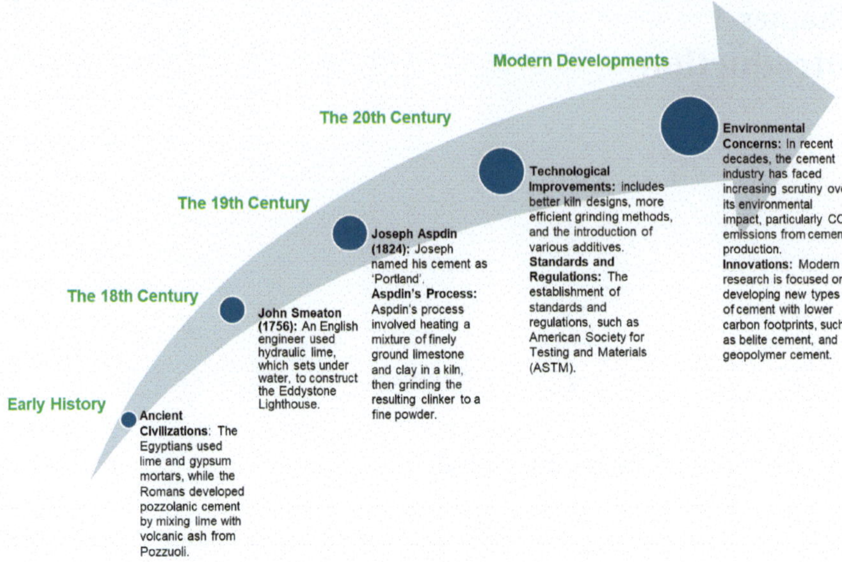

Fig. 1.1 The development history of cementitious materials

($3CaO{\cdot}2SiO_2{\cdot}3H_2O$), which can be abbreviated as C–S–H, in a gel form [10]. Figure 1.2 shows the scanning electron microscope (SEM) image of cement paste at 4000 times of magnification. The large area of amorphous phases in the image is mainly C–S–H gel. The hydration reaction formula of C_2S and C_3S can be simply written as:

$$C_2S + H_2O \rightarrow CSH + CH \tag{1.1}$$

$$C_3S + H_2O \rightarrow CSH + CH \tag{1.2}$$

where CH is calcium hydroxide ($Ca(OH)_2$).

As seen in Fig. 1.2, C–S–H exhibits a dense amorphous colloidal structure. In recent years, with the development of microscopic characterization technology, the exploration of the C–S–H structure has gradually moved towards the nano-level. Trimethylsilylation (TMS) and solid-state Nuclear magnetic resonance spectrometer (NMR) studies have revealed that C–S–H is consisted of numerous dimeric silicate chains in the early stages of formation [12, 13]. Each dimer contains two basic tetrahedral unit ($(SiO_4)^{4-}$), which share one identical Si^{2+}, Ca^{2+} and the surrounding O^{2-} form an octahedral unit. As the hydration reaction proceeds, more complex silicon chains are formed. Calcium chains and silicon chains present a highly disordered layer structure with an average chain length of $3n - 1$, where $n = 1, 2, 3$, etc., as shown in the schematic diagram in Fig. 1.2.

Table 1.1 Cement types in standards

Type	Description	Standard ID
General Purpose (GP) Cement	• Most common type used in general construction • Composed primarily of Portland cement clinker and gypsum	AS 3972 Portland and Blended Cements [2]
General Purpose Blended (GB) Cement	• Contains Portland cement and one or more types of supplementary cementitious materials (SCMs) such as fly ash, slag, or silica fume	
High Early Strength (HE) Cement	• Develops strength rapidly, ideal for situations requiring early removal of formwork	
Low Heat (LH) Cement	• Generates lower heat during hydration, suitable for large concrete structures like dams	
Shrinkage Limited (SL) Cement	• Designed to limit drying shrinkage in concrete	
Portland Cements (CEM I)	• Composed predominantly of Portland cement clinker with a small percentage of gypsum (5% or less)	EN 197-1 Cement Composition, Specifications and Conformity Criteria for Common Cements [3]
Portland-Composite Cement (CEM II)	• CEM II/A: 6–20% of other constituents (fly ash, slag, silica fume, limestone, etc.) • CEM II/B: 21–35% of other constituents	
Blast Furnace Cements (CEM III)	• CEM III/A: 36–65% slag • CEM III/B: 66–80% slag • CEM III/C: 81–95% slag	
Pozzolanic Cements (CEM IV)	• CEM IV/A: 11–35% pozzolana • CEM IV/B: 36–55% pozzolana	
Composite Cements (CEM V)	• CEM V/A: 18–30% blast furnace slag and 18–30% pozzolana or fly ash • CEM V/B: 31–50% blast furnace slag and 31–50% pozzolana or fly ash	
Masonry Cement (MC)	• Contains various proportions of clinker, limestone, and other additives, intended for masonry work	

(continued)

Table 1.1 (continued)

Type	Description	Standard ID
High Alumina Cement (HAC)	• Contains a high proportions of alumina, suitable for refractory and high-temperature applications	
Type I-General Purpose	• Suitable for general construction when there are no special requirements	ASTM C150 Standard Specification for Portland Cement
Type II-Moderate Sulfate Resistance	• Used where moderate sulfate resistance is needed, often in drainage structures and in environments with moderate sulfate levels in soil or water	
Type III-High Early Strength	• Provides high early strength, useful for rapid construction and early form removal	
Type IV-Low Heat of Hydration	• Generates less heat during hydration, suitable for large concrete structures like dams to minimize thermal cracking	
Type V-High Sulfate Resistance	• Offers high sulfate resistance, ideal for structures exposed to high sulfate environments, such as soils and groundwater with high sulfate content	
Type IP-Portland-Pozzolan Cement	• Contains a mixture of Portland cement and pozzolans, improving durability and reducing permeability	ASTM C595 Standard Specification for Blended Hydraulic Cements [4]
Type IS-Portland-Slag Cement	• Contains a mixture of Portland cement and slag, enhancing durability and reducing heat of hydration	
Type IL-Portland-Limestone Cement	• Contains a mixture of Portland cement and limestone, improving workability and sustainability	
Type IT-Ternary Blended Cement	• Contains a combination of Portland cement, slag, and pozzolan or other SCMs	
Type GU-General Use	• Suitable for general construction where no special properties are required	ASTM C1157 Standard Performance Specification for Hydraulic Cement [5]
Type HE-High Early Strength	• Provides high early strength, similar to ASTM C150 Type III	

(continued)

Table 1.1 (continued)

Type	Description	Standard ID
Type MS-Moderate Sulfate Resistance	• Offers moderate sulfate resistance, similar to ASTM C150 Type II	
Type HS-High Sulfate Resistance	• Offers high sulfate resistance, similar to ASTM C150 Type V	
Type MH-Moderate Heat of Hydration	• Generates moderate heat during hydration, useful for mass concrete structures	
Type LH-Low Heat of Hydration	• Generates low heat during hydration, similar to ASTM C150 Type IV	
Ordinary Portland Cement (P.O)	• Similar to the ASTM Type I, used for general construction purpose	GB 175 Standard for Portland Cement [6]
Portland Cement (P.C)	• General-purpose cement used widely in construction	
Sulphate Resistant Portland Cement (P.SR)	• Used in environments with high sulfate content, similar to ASTM Type V	
Portland Blast Furnace Slag Cement (P.S.A and P.S.B)	• Contains a mix of Portland cement and ground granulated blast furnace slag (GGBFS) • P.S.A: Higher content of slag ($\geq$ 20% but $\leq$ 50%) • P.S.B: Lower content of slag ($\geq$ 20% but $\leq$ 40%)	GB 1344 Standard for Special Cements [7]
Portland Pozzolana Cement (P.P)	• Contains Portland cement and pozzolanic materials like fly ash • Improves durability and reduces permeability	
Portland Fly-Ash Cement (P.F)	• Contains Portland cement and fly ash • Enhances durability and reduces permeability	
Portland Limestone Cement (P.L)	• Contains Portland cement and limestone • Improves workability and sustainability	
Composite Portland Cement (P.C)	• Contains Portland cement and a combination of different SCMs like fly ash, slag and limestone	

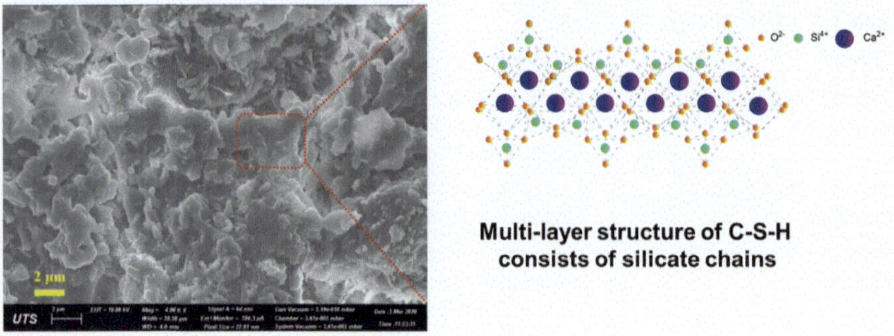

Fig. 1.2 Portland cement paste under 4000× magnification [11]

The solid-state ^{29}Si NMR has also been used to identify the length of the silica chain, and thus the hydration degree of cement, because more basic $(SiO_4)^{4-}$ connected to the central Si atom increases the average electron density around the Si atom, which results in a more negative chemical shift in ^{29}Si NMR spectra [14]. Figure 1.3 shows a typical ^{29}Si NMR spectrum of cement paste and the corresponding deconvolution results. The first peak Q^0 located at -70 ppm represents basic $(SiO_4)^{4-}$ unit [15]. The superscript Q represents the number of basic $(SiO_4)^{4-}$ connecting the central Si atom [15]. Therefore, the appearance of deconvolution peaks with larger superscripts represents the formation of more complex silicon chains and the deeper degree of hydration.

Fig. 1.3 ^{29}Si NMR spectrum deconvolution results of cement paste [16]

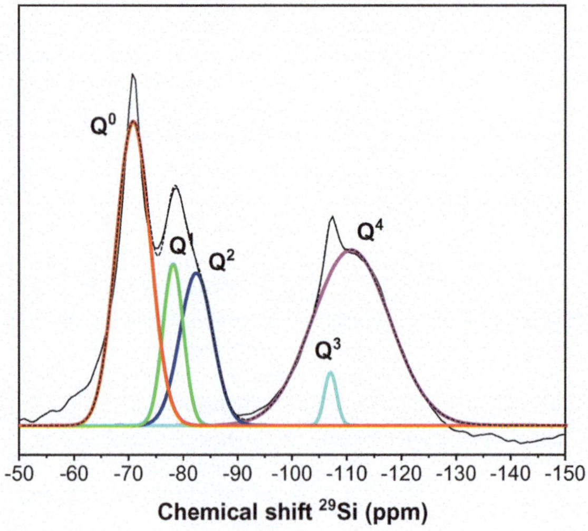

1.1.2 CH in Cement Paste

CH, also called portlandite, is another important hydration product of cement hydration, usually in the form of massive crystals mixed with amorphous C–S–H gel [17]. It is also a layered mineral. Each hydroxyl group is connected to three calcium atoms in the same layer and is surrounded by three hydroxyl groups in the adjacent layer, as shown in Fig. 1.4. Larger portlandite crystals can be generated in a cement paste with a high water-to-cement ratio [18]. High porosity and weak bond between crystals and C–S–H gel induces a loose structure leading to the poor performance of cement paste.

CH is also responsible for the alkalinity of cement-based materials. It is easy to dissolve in water and acidic solutions, causing leaching, and easy to react with other chemicals, such as acids and salts.

Portlandite in cement could react with dissolved SiO_2 and Al_2O_3 to produce hydrated gels of C–S–H and C–A–H [19]. This process is called the pozzolanic reaction and has been focused on in numerous studies on the hydration reaction mechanism of cement-based materials. Natural volcanic ash contains large amounts of SiO_2 and Al_2O_3 [20]. The early Romans and Greeks collected volcanic ash around their residences and mixed it with lime and water as a cementitious building material [21]. Therefore, in academia, people refer to the powders whose main components are active SiO_2 and Al_2O_3 as pozzolanic materials. The process of CH reacting with water and pozzolanic materials as pozzolanic reaction.

The pozzolanic materials that are widely used today are industrial or municipal by-products, such as fly ash produced by burning coal, blast furnace slag in steel-making furnaces, silica fume produced in industrial silicon, and waste glass powder in cities [22]. Since cement hydration will inevitably generate a part of calcium hydroxide, which are detrimental to the bonding performance of cement paste, the cement manufacturer chooses to add around 20–60% of supplementary cementitious materials (SCMs) to the cement [23]. They intend to consume the CH in cement paste and produces more C–S–H gel through pozzolanic reaction. The most typical products are fly ash cement and slag cement [24]. Generally, pozzolanic reaction activity

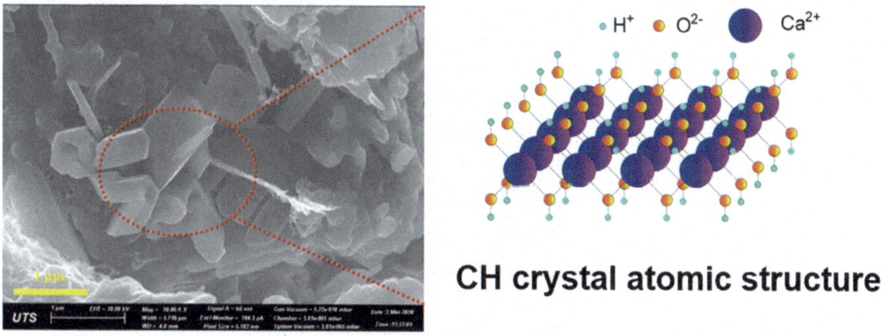

Fig. 1.4 CH crystals under 20,000× magnification

is much lower than that of ordinary Portland cement hydration reaction, especially under a room and lower temperature [25]. Due to the dilution effect of cement mixed with a large amount of SCMs, the early strength growth is slower than that of concrete made of ordinary Portland cement [26]. However, as the pozzolanic reaction proceeds, the mechanical properties of concrete containing pozzolanic materials such as fly ash and slag will be better than ordinary concrete [27].

1.1.3 AFt and AFm in Cement Paste

In addition to the main hydration products (C–S–H and CH) in cement paste, calcium-aluminum sulfate products such as ettringite (AFt) and monosulfate (AFm) are also worthy of attention, especially when the concrete is in a sulfate-rich environment [28]. Ettringite is a needle-shaped trigonal crystal in the SEM image of cement paste, as shown in Fig. 1.5. Its Mohs hardness is between 2 and 2.5 and its specific gravity is 1.77 [29]. In an environment with a high content of sulfate ions, a large amount of ettringite is generated in cement paste, which can easily lead to cracking of the paste matrix due to crystal expansion [30]. Its influence on the overall bonding strength of cement paste exceeds that of CH mentioned above. As the sulfate ion concentration of the pore solution in the cement paste decreases, ettringite becomes unstable and gradually transforms into AFm-formed phase at a slow rate [31]. The difference between AFt and AFm is mainly reflected in bond water and charged anions [32]. AFt has the general constitutional formula $[Ca_3(Al, Fe)(OH)_6 \cdot 12H_2O]_2 \cdot X_3 \cdot xH_2O$ where $x \leq 2$ and X represents one formula unit of a doubly charged anion [33]. The general formula of AFm is slightly different from AFt. The general formula of AFm is $[Ca_2(Al, Fe)(OH)_6] \cdot X \cdot xH_2O$, where $x \leq 2$ and X represents one formula unit of a singly charged anion, or a half formula unit of a doubly charged anion [34]. For cement paste, X is generally OH^- and $SO_4{}^{2-}$.

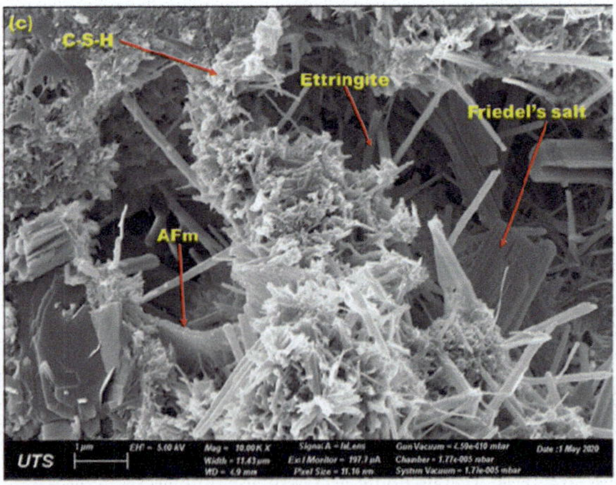

Fig. 1.5 AFt and AFm in cement paste under SEM images [35]

Normally, the concentration of the Portland cement with deionized water paste was mainly distributed in two ranges, divided by the time when gypsum was completely dissolved [36]. Gypsum is an essential component in the commercial cement. Gypsum is added to cement to slow down the setting time. Without gypsum, the hydration of C_3A in cement can lead to a rapid setting, often referred to as flash set. The main function of gypsum is to control this process, allowing for more manageable workability and placement of the cement. In addition, gypsum can react with C_3A to form ettringite, which is beneficial as it occurs during the early stages of hydration and helps maintain a balance in the hydration process. Before the gypsum depletion, the composition of the pore solution was dominated by K^+, Na^+, SO_4^{2-}, Ca^{2+}, and OH^- [37]. The continuous dissolution of gypsum kept the high concentration level of Ca^{2+} and SO_4^{2-}, forming AFt and CH as the major hydrates at this stage [38]. The other main constituents of the cement, such as SiO_4^{4-} and $[Al(OH)_6]^{3-}$, were initially very low but increased gradually along with the hydration [39].

After the gypsum was depleted, a notable change in various ions concentration can be observed between 5 and 24 h. Due to the lack of direct supply from gypsum dissolution, Ca^{2+} and SO_4^{2-} were significantly decreased by AFt formation and started to convert, or form directly, to AFm phases instead. Meanwhile, the concentration of OH^- had to rise to maintain the charge balance of the solution; SiO_4^{4-} and $[Al(OH)_6]^{3-}$ were then slightly increased with the increasing pH according to the solubility required. With the increasing degree of hydration accompanied by the consuming liquid volume, alkali was released continuously into the gradually concentrated pore solution. Therefore at the late stage, the growing pH increased the SiO_4^{4-} and $[Al(OH)_6]^{3-}$ and decreased the Ca^{2+} based on the solubility [28].

In Portland cement paste, both chloride ions and hydroxyl ions can be chemically bound by the principal layer of AFm structure to balance charges as shown in Fig. 1.6. This combination product includes calcium chloroaluminate, commonly known as Friedel's salt: $3CaO{\cdot}Al_2O_3{\cdot}CaCl_2{\cdot}10H_2O$. During the process, the adsorbed OH^- ions are more easily displaced by Cl^- ions, due to the fact that OH^- hardly competes with Cl^- in balancing charge [40]. Meanwhile, the free OH^- ions are retained in the pore solution, leading to an increase of pH. Therefore, the increase of alkalinity in pore solution is due to the significant substitution of adsorbed OH^- ions by Cl^- ions, and it also means that the metastable hydroxy-AFm hydrates hardly coexist with aqueous Cl^- [41]. At temperature of 25 °C and the standard atmosphere (100,000 Pa), monocarboaluminate (CO_3-AFm) is more stable than monosulfoaluminate (SO_3-AFm) in Portland cement, which promotes a progressive substitution of CO_3^{2-} in SO_3-AFm to convert into hemi carboaluminate or, more likely, CO_3-AFm [42]. Similarly to CO_3^{2-}, Cl^- can also displace sulfate in SO_3-AFm, converting to Friedel's salt as the more stable AFm hydrates [43].

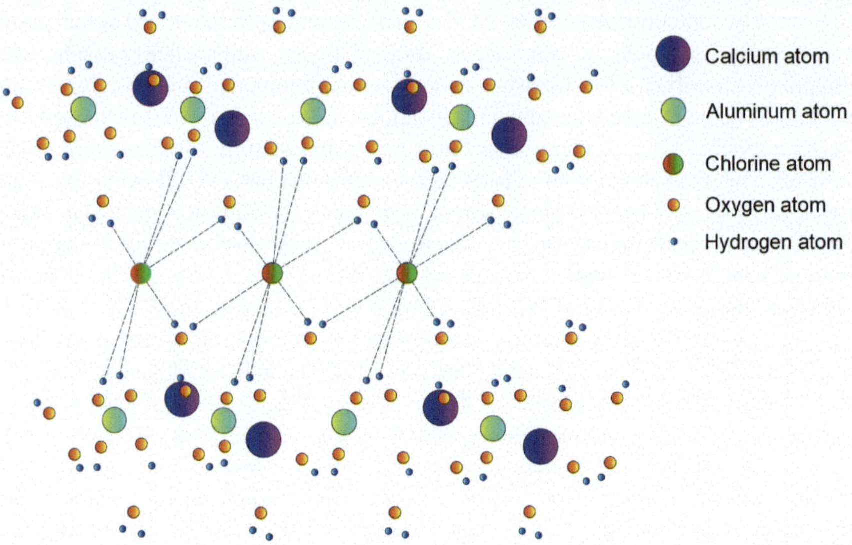

Fig. 1.6 Structural representations of AFm phase: hydrogen bond network within the interlayer region

1.2 Alkali Activated Cementitious Materials

Alkali activated cementitious materials (AACMs) are made with precursors and activators, and they can be classified as two groups: (1) alkali activated cement (AAC), mainly by activating precursors that are rich in calcium, such as Class C fly ash and blast furnace slag and (2) geopolymer, primarily by activating precursors that are rich in aluminum and silica, such as Class F fly ash and metakaolin. Both ACC and geopolymers are clinker-free cements, and they are well known with benefits of reusing industrial by-products and reducing carbon dioxide emission, which have attracted extensive studies [44–47]. Reactions in AAC primarily involves the formation of C–S–H phases, while it may also form some calcium aluminosilicate hydrate (C–A–S–H) gels, depending on the specific mix. ACC generally has properties similar to ordinary Portland cement in terms of setting, hardening, and durability. In geopolymers, the activation of aluminosilicates in an alkaline solution initiates the dissolution of Si and Al species, and these Si and Al species then re-polymerize to form an amorphous sodium aluminosilicate hydrate (N–A–S–H) gel, or geopolymer gel, a three-dimensional network with Si–O–Al linkages, leading to superior thermal and chemical resistances. C–S–H gels may coexist within the geopolymer matrix when precursors contain calcium.

Replacing industrial by-products such as fly ash and slag for Portland cement can not only bring environmental benefits, but also improve the long-term strength and durability of concrete. However, fly ash and slag as supplementary cementitious materials are just used to partially replace Portland cement with replacing ratio generally lower than 30–40% [48]. A large amount of fly and slag still have not been properly treated. AACMs is a promising construction material that can maximize the use of industrial by-products and develop concrete that is truly clinker-free [47]. Proper design of AACMs can make their mechanical properties and durability comparable to Portland cement materials [49]. Despite the great advantages and potential of geopolymer, the understanding and application of AACMs are still far less than that of Portland cement concrete. In addition to general engineering properties, great efforts are needed in basic research to gain more scientific insights into AACMs.

Research has revealed that at the microscopic scale, AACMs paste is mainly composed of N–A–S–H gel, unreacted and partially reacted precursor materials, crystalline phases and defects [50]. The N–A–S–H gel is the most crucial component linked to the performance of AACMs [51]. The basic properties, nanoscale reinforcement of N–A–S–H gel and related mechanisms can be characterized by the rapidly developing and popular cutting-edge instrument technology. Figure 1.7 shows the microstructure of a typical geopolymer matrix. The gel phase is the most crucial component of cementitious binders. Unlike the extensively studied C–S–H gel in Portland and blended cement pastes, research on the N–A–S–H gel in geopolymer, despite some nanoindentation studies on its elastic modulus, hardness, creep, and fracture toughness, remains limited in comparison [52]. The insufficient studies not only leave gaps but also lead to inconsistent results [53]. Typically, based on the investigation on heat-cured alkali-activated fly ash (AAFA), ambient cured AAFA and heat-cured alkali-activated metakaolin, study found that the mature N–A–S–H gels (6 months) show almost the constant intrinsic Young's modulus of 17–18 GPa, which is independent of the precursor materials and the temperature curing regime [54]. In other studies [55–57], the elastic modulus of N–A–S–H gel was reported to be 16.3 GPa, 4.44–16.78 GPa (after 28-day) and 11–25 GPa (at 28 days). Thus, the results are different in these studies and also varied in a large range in the same study. This phenomenon is similar to nanomechanical investigation for Portland cement paste. There are a plenty number of studies indicating that the elastic modulus of C–S–H is one of intrinsic material properties and does not depend on water to cement ratio, type of cement, admixtures, heat curing, etc. [58, 59]. Besides, when comparing cement paste with different curing age, tests at 28 d reveal similar modulus results as those at longer time of curing of 5 months and even 1 year [60]. However, some other studies [61] found that the elastic modulus of cement paste is influenced by the water-to-cement ratio.

The different phenomenon above for geopolymer may be attributed to the difference between mature and immature gel [63]. However, it shall also be noticed that different testing, analysis methods and different parameters were adopted in the above studies. For the statistical analysis of geopolymer, the least-square estimation (LSE) method is normally adopted while different bin sizes such as 2 GPa and 1 GPa and different phase numbers of 3 or 4 were used [64]. The different parameters also possibly affected the results. Therefore, further studies are necessary on the microstructure composition and micromechanical properties of geopolymer paste, and the nanoindentation testing and data analysis method themselves. After all, the

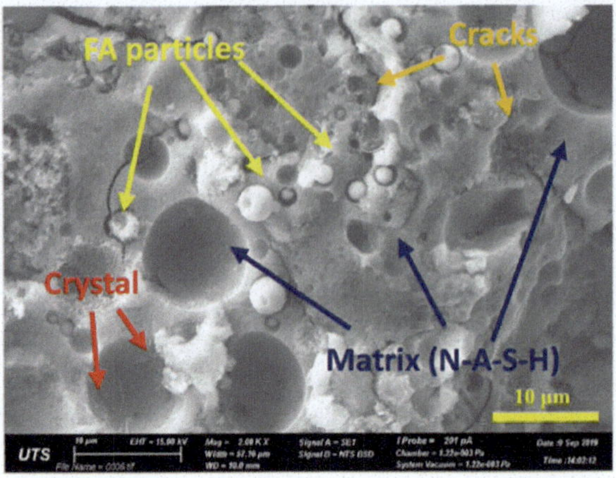

Fig. 1.7 Microstructure of a typical alkali activate fly ash geopolymer matrix [62]

correct understanding of the elastic modulus of N–A–S–H should be based on the premise that the test and data deconvolution analysis method are consistent and reliable to eliminate errors caused by the analysis process [65].

For Portland cement paste, mixing of C–S–H gel with nanoscale CH crystal or AFt and AFm is the most common form of the mixed phase detected in the microscopic observation [62]. CH crystal does not exist in the geopolymer, and nanoscale crystal is not considered in any current nanoindentation investigation of geopolymer paste [66]. However, the AAFA has crystal phases, and some micron-sized crystals can be observed directly by SEM as in Fig. 1.8.

Some of those crystals would also exit in smaller sizes like nano and sub-micron scale [67]. In addition, based on an analysis of extensive existing results, Provis et al. [68] proposed that the geopolymer binders are formed by agglomerating of nanocrystalline zeolites and amorphous gel phase, and the nanocrystalline zeolites

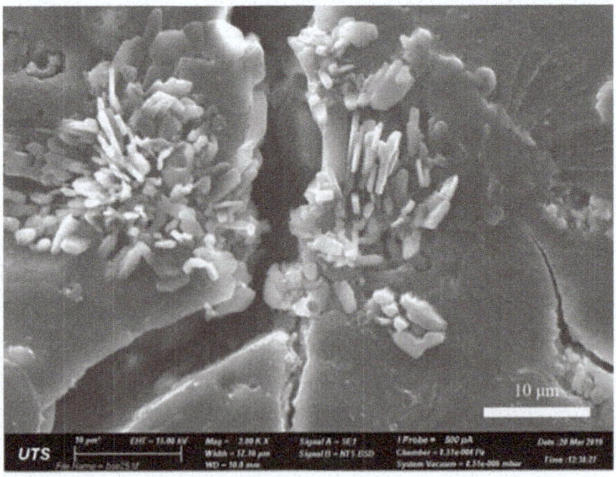

Fig. 1.8 Morphology of micron-sized crystals in AAFA [65]

phase cannot be detected by X-ray diffraction (XRD) due to the small size. The existence of crystals will also be reflected on micromechanical properties of geopolymer pastes [69]. However, similar to CH in cement paste, crystals in geopolymer would not provide any cohesion strength for the interface with N–A–S–H gel [70].

The investigation of the effects of design parameters such as silica modulus on the properties of N–A–S–H gels is vital since it is closely linked to the engineering performance of geopolymer [71]. In addition to optimizing the general design parameters, some studies modified geopolymer with nano-engineering approaches [72]. Typically, geopolymer strength was improved by the addition of nano-SiO_2, nano-Al_2O_3, nano-TiO_2, nano-clay and carbon nanotubes [64]. Higher durability such as enhanced sulphuric acid attack resistance, and lower water absorption were realized by incorporating nano-SiO_2 and nano-clay, respectively [73]. However, there is no nanomechanical testing to help understand the changes in the micro- and nanoscale properties of the modified geopolymer.

1.3 Interfacial Transition Zone

From the perspective of raw materials, concrete is generally made with aggregates, Portland cement and water. After the reaction, the latter two become cement paste, which bonds aggregate and form concrete, which has three phases of aggregate, paste matrix and interfacial transition zone (ITZ) between paste and aggregate particles. Paste matrix and ITZ, as shown in Fig. 1.9, are the dominant phases that control the properties of concrete [74]. ITZ is the weakest phase in normal strength concrete, being often considered as pre-existing microcracks in pristine concrete and predominantly responsible for the initial development of microcracks in concrete under loading [75]. The ITZ in PC concrete is characterized by high porosity and a large number of crystalline phases, which has an important impact on mechanical properties, transport properties and durability of concrete [76]. It is therefore considered to be the key to develop high-performance concrete. Although the typical width of ITZ in PC concrete is between 15 and 20 μm, they still account for 20 to 30% of the total paste volume in concrete, which highlights the importance of understanding and enhancing the ITZ [77].

1.3.1 ITZs in Recycled Aggregate Concrete

The mechanical properties and durability of recycled aggregate concrete (RAC) are generally lower than natural aggregate concrete (NAC), and the relatively poor performance of RAC mainly comes from the attached old mortar on the surface of recycled aggregates. The performance of the attached old mortar highly depends on the sources of waste concrete, which makes the performance of recycled aggregate various [78].

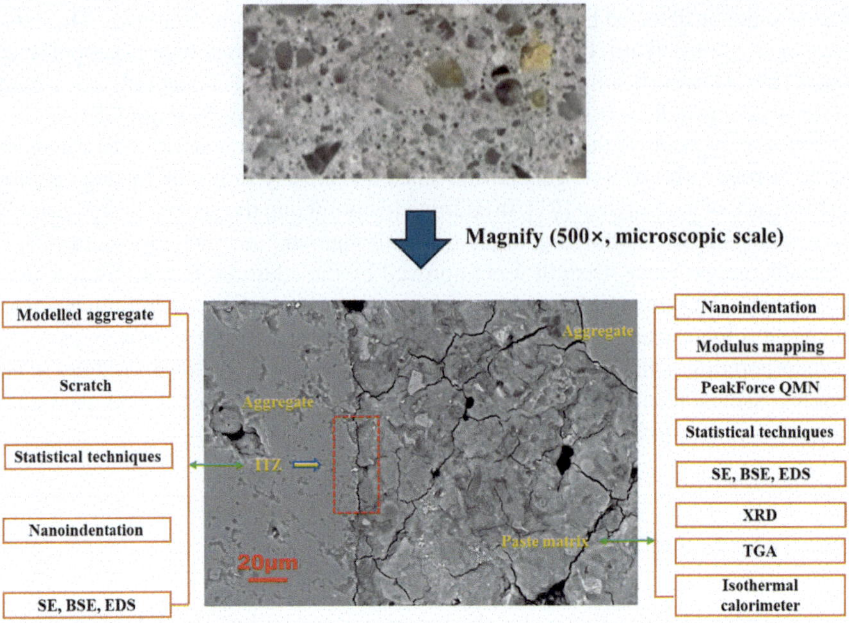

Fig. 1.9 Structure of concrete and research techniques

Besides, the high mortar content of the recycled aggregate reveals some unfavourable features such as high porosity, high water absorption, and high crushing index [78].

As shown in Fig. 1.10, compared to conventional concrete, RAC contains more types of ITZs. When recycled aggregate is applied and mixed in concrete, the attached mortar brings another ITZ i.e., the new ITZ, while ITZ is generally considered to be the weak zone in concrete.

In order to improve the properties of RAC, many studies have been conducted. The key to performance enhancement of RAC is to improve the cohesion strength of ITZ between recycled aggregate and new cement paste [80]. Some researchers have tried modifying the surface of recycled aggregate, such as acidic solutions pre-soaking of recycled aggregates, carbonization of recycled aggregates and pre-coating the recycled aggregate with sulfoaluminate cement and mineral additives, while others have attempted to densify ITZs of RAC by additions, for instance, adding mineral admixtures to RAC for mechanical properties and durability improvement [81]. Although some progress has been made by these efforts, there is still a great distance for engineering application of the RAC and techniques still need to be improved to find more feasible ways.

The rapid development of nanotechnology has been typically accompanied by the optimization of properties and the reduction of cost of nanomaterials, as well as the growing popularity of versatile instruments for characterization and understanding of materials at the micro- and nanoscale [82]. Thus, it is expected that nanotechnology shall be used more and more widely. In recent years, there are increasing

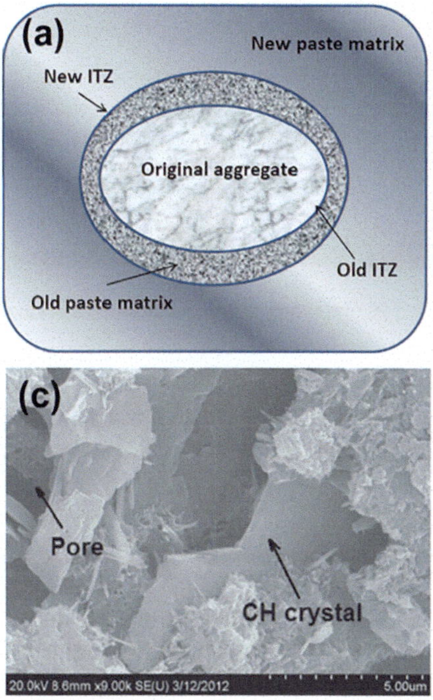

Fig. 1.10 Characteristics of ITZs in RAC [79]: **a** schematic of RAC; **b** old ITZ; **c** new ITZ

number of studies have introduced nanomaterials and nanoscale characterization into the field of cement-based materials research, which also promotes the interest of using nanotechnology to study and modify of RAC [83]. In theory, the ultrafine size of nanomaterials enables itself to repair or fill the pores and cracks within ITZs at the scale, which is inaccessible by other traditional materials, while the ultra-high resolution characterization allows researchers to better understand the link between microstructure and macroscopic properties and then improve macroscopic properties of RAC by nanoscale manipulation [83]. The existing research results also demonstrate that nanotechnology displays huge potential for promoting the application of RAC [84]. Li et al. [79] successfully studied the influence of different mixing methods on the ITZ nanomechanical properties and microstructure characteristics of RAC by nanoindentation and SEM. Mukharjee et al. [85] studied the ITZ parameters like porosity of nanomodified RACs by energy dispersive X-ray spectroscopy (EDS) and backscattered electron (BSE), then the relationships between macroscopic mechanical properties and ITZ parameters were established. Hosseini et al. [86] modified ITZs in RAC with NS and found that the RAC with 3% NS and 100% recycled aggregate replacement ratio had slightly higher compressive strength than the corresponding NAC. The application of NS modified ITZs of RAC into real project by Zhang et al. [87] also confirmed the NS can improve the mechanical properties of RAC and the crack resistance of RAC beams.

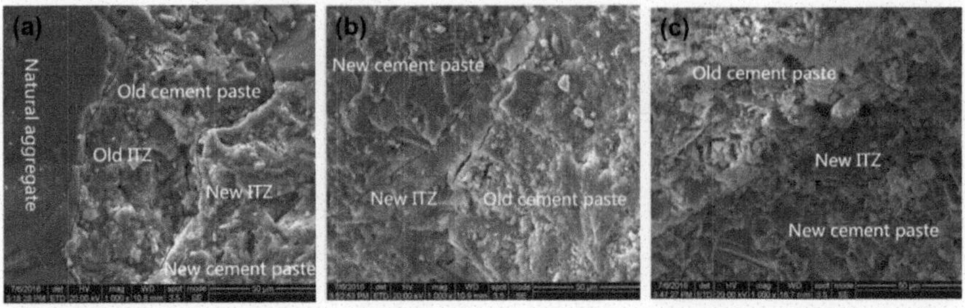

Fig. 1.11 Microstructure of old and new ITZ in RAC incorporating nanoparticles [78]: **a** normal ITZ; **b** ITZ with NL; **c** ITZ with NS

There are also studies on the microstructure of ITZs of RAC after incorporating of nanoparticles. Hosseini et al. [86] observed the NAC, RAC, and NS modified ITZs of RAC by SEM. Results revealed that RAC had more porosity in ITZ than that of NAC, but the ITZ of RAC could be improved by nanosilica (NS). After adding NS into RAC, obviously more uniform and denser ITZ was observed. ITZ microstructure characterization conducted by Mukharjee et al. [85] revealed that the microhardness of ITZ was enhanced and the porosity of ITZ was reduced after NS incorporation. Then the relationships between different mechanical properties and ITZ parameters were established by Mukharjee et al. [85]. Zhang et al. [84, 87] conducted nanoindentation test on RAC with recycled aggregate treated by different nano slurries such as NS and nano-limestone (NL) slurries. They found that the micromechanical properties of the old mortar and new ITZs of RAC had been improved by both these nano slurries, but which did not work on the old ITZ. As shown in Fig. 1.11, SEM observation, water absorption measurement and mercury intrusion porosimetry (MIP) test conducted by Li et al. [78] revealed that the appropriate dosages of NS could reduce porosity and improve the microstructure of RAC, but no obvious positive effect was found for adding of NC into RAC due to the agglomeration of nanoparticles. Ying et al. [88] investigated pore structure of old ITZ and new ITZ in RAC after adding NS and nano-TiO_2 (NT) particles by the MIP test. It was found that the pore structure was refined by both nanoparticles while NT was slightly more effective than NS for pore refinement. Shaikh et al. [89] compared the influence of different strengthening methods, namely, pre-soaking of aggregate by NS and modifying the RAC directly by NS, and found that both the MIP test and SEM observation supported that the way of pre-soaking aggregate in NS brought more significant positive effect on the cohesion of ITZs in RAC.

1.3.2 ITZs in Geopolymer Concrete

The research on the ITZ between aggregate and paste of geopolymer is limited. The limited studies seem to indicate that the ITZ in geopolymer concrete is not a distinct

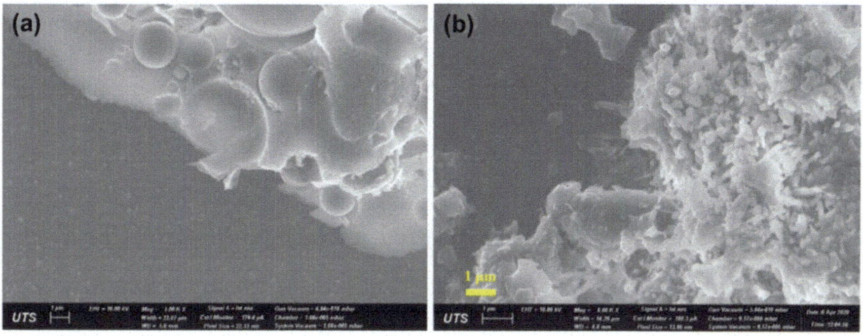

Fig. 1.12 The interface between aggregate and paste [11]. **a** Geopolymer paste; **b** Portland cement paste

weak region as the ITZ in Portland cement concrete [11]. A pull-out study showed that for most fibers, the bond strength in fly ash based-geopolymer mortars is stronger than that in Portland cement mortar [90]. Similarly, fly ash based-geopolymer paste and metakaolin/GGBFS-based geopolymer paste were found to have better bond with aggregate than that of Portland cement paste [91]. At the micro scale, SEM observation revealed that no significant ITZ was identified between the siliceous aggregates and fly ash-based geopolymer [92]. The strong interface could be attributed to the alkali solution used, especially the presence of soluble silicates [93]. By using SEM, the ITZ in sodium silicate-activated slag was also reported to be more homogeneous and have less porosity than that in Portland cement mortars as shown in Fig. 1.12.

One of the most important aspects to understand ITZ is its mechanical properties. Owing to the small width of ITZ, micro- and nanomechanical test techniques are important tools to quantitatively evaluate interfacial properties [94]. At present, micro-indentation and nanoindentation have been used to investigate the properties of ITZ in geopolymer. Nanoindentation studies on fly ash-based geopolymeric recycled aggregate concrete show that the new ITZ formed between geopolymer and old cement paste is not an obvious weak region [95]. Geopolymer even makes the pre-existing old ITZ achieve similar micromechanical properties as surrounded paste [95]. However, the ITZ in metakaolin/GGBFS-based geopolymer was found to have lower microhardness than paste [96]. For alkali-activated slag, a study indicated that the average Vickers micro-hardness of ITZ was 76.8% of bulk paste, while another study found that, the microhardness of ITZ could be either higher or lower than paste, which depends on the alkali solution used [97].

There are comparison studies between ITZs in geopolymer and Portland cement concretes by using grid nanoindentation [89]. Grid nanoindentation tests revealed that for concrete made with nature aggregate, the well-bonded ITZ in PC paste does not significantly differ in elastic modulus and hardness compared to the bulk paste [84]. The modulus and hardness of ITZ in geopolymer were more heterogeneous than that in Portland cement concrete. When recycled aggregate was used, one study indicated that the ITZ between new paste and old mortar coated recycled aggregate in Portland cement concrete was about 20 μm according to the micromechanical distribution, but no obvious weak ITZs were found in geopolymer concrete, whereas another study found no distinct new ITZ in both Portland cement concrete and geopolymer concrete [98].

1.3.3 Modelled Aggregate Concrete and ITZs

When applying grid nanoindentation in the study of ITZ, the ITZ sample is generally randomly selected from one or several local areas of a real concrete sample. It is known that the ITZ is highly heterogeneous in Portland cement concrete. In Portland cement concrete, ITZs are different depending on the aggregate investigated and are also heterogeneous around the same aggregate [74]. For instance, internal bleeding beneath an aggregate would lead to different water to cement ratio and then different properties of ITZ when compared with ITZ in other locations surrounding the aggregate [77]. Moreover, the degree of internal bleeding is different for aggregates in PC concrete with different shape and size [77]. Mixture process would also lead to heterogeneous of ITZs. The complex nature of ITZ causes great difficulties in quantitative testing.

It is a very time-consuming and challenging task to determine an 'average' property of ITZ even by doing BSE image statistical analysis in a real concrete [75]. Therefore, most of quantitative ITZ research is conducted using fine aggregate with small size in mortar samples. Even for mortars, when the whole ITZ surrounding a sand particle was characterized by BSE analysis, the number of sand grains used for test would be few due to the large test workload. Alternatively, if too many sand particles were selected for analysis, only few regions around the sand particle with certain long could be taken for BSE analysis. The workload levels increase dramatically for a similar study of ITZ in coarse aggregate by BSE. Therefore, comparative studies of ITZ between different concrete samples by BSE would be either not reliable enough or involved too much workload. Different from BSE technique where an image captures information from a several hundred microns by several hundred microns area, the indentation test obtains just one point from a several minutes test. It is impossible for nanoindentation to face the challenge that cannot even be solved by BSE [82].

To solve the above problems, Shah and Winter [99] are the first to propose a novel modelled concrete with cylindrical aggregate for analysing the inelastic mechanical properties of concrete. Subsequently, Xiao et al. [100] extended the study of this concrete to the microscopic level. In recent years, Luo et al. [101], Fu et al. [102], and Zhan et al. [103] used a similar concept to simulate ITZ by aggregate blocks and obtained satisfactory microstructure, chemical composition and micromechanical performance test results. It has to be admitted that the ITZ simulated by modelled concrete idealises the actual situation and ignores the randomness and complexity of the real ITZ. However, with the help of characteristics of modelled concrete, the bonding mechanism of aggregate and mortar can be more effectively explored.

The method to prepare modelled ITZs is displayed in Fig. 1.13. In Fig. 1.13a, an aggregate with a polished surface embedded in the mortar of modelled sample, and the Fig. 1.13b displays the cross section. The magnified details of the interface are shown in Fig. 1.14c. The modelled interface can allow for avoiding the issues for the grid nanoindentation relating to uneven interfaces in Fig. 1.14a, b. In addition, the interference resulting from factors (e.g. different types, shapes, size, relative positions, local mix condition of aggregates, different locations in ITZs from an aggregate) can be avoided, enabling more accurate comparison of ITZ with an acceptable statistical workload. The modelled interface is also convenient for other characterizations like EDS mapping of the pure ITZ. The origin of ITZ is mainly ascribed to the 'wall effect' [104]. Therefore, the modelled interface would not have a significant difference compared to the real, well-formed ITZs in the real concrete, but would be not able to reflect the heterogeneity and complexity.

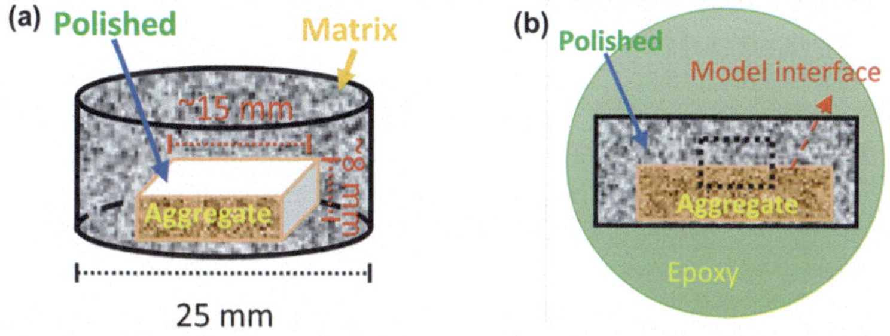

Fig. 1.13 Schematic diagram of modelled sample preparation [11]: **a** modelled aggregate concrete sample; **b** surface for testing

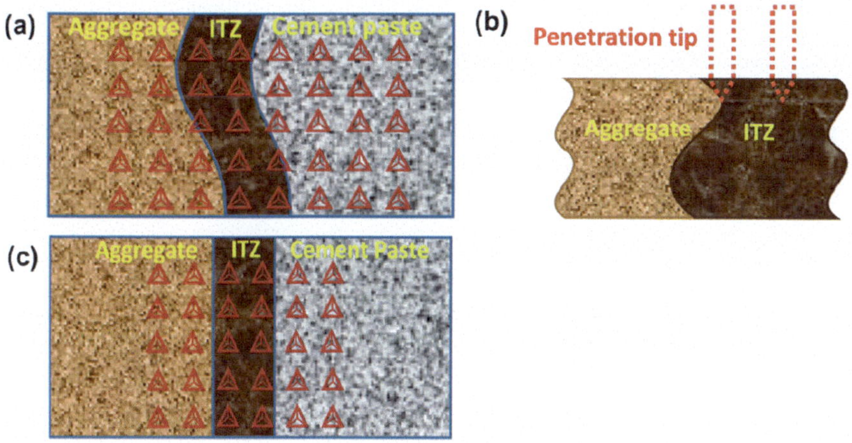

Fig. 1.14 Grid nanoindentation tests on different types of ITZs [11]: **a** the real interface in concrete; **b** penetration in ITZ; **c** test on the modelled interface

1.4 Summary

After more than 300 years of development, Portland cement has become the most important cementitious material in engineering construction. Researchers have also shifted their focus from the macro- to the micro-level and are looking for more environmentally friendly cementitious materials. The micromechanical properties, microstructure and atomic structure at nanoscale of constituents such as C–S–H, CH, AFt and AFm in Portland cement paste have been fully studied. However, there is a lack of research on the microscopic and nanoscopic aspects of N–A–S–H and other crystalline components. In addition, ITZ is generally considered to be the weakest area in concrete and is the key to determining the mechanical properties and permeability resistance of concrete. However, researchers obtained different observations between studies of ordinary cement concrete and geopolymer concrete. As geopolymer gel has a three-dimensional network of alumino-silicate bonds (Si–O–Al) with a dense microstructure, there is no obvious difference in the microscopic performance between ITZ and paste matrix in geopolymer concrete. When recycled aggregate is used in concrete, old mortar makes bonding at the interface more complicated. Therefore, researchers have also proposed many modification methods. In order to minimize the impact of ITZ inhomogeneity on micro- and nano-level test results, researchers designed modelled aggregate concrete, which can be widely applied on the investigation of ITZs in ordinary cement concrete and geopolymer concrete.

References

1. H. Pöllmann, *Cementitious materials: composition, properties, application* (De Gruyter, Berlin, Germany, Boston, Massachusetts, 2017)
2. A. Standard, Portland and Blended Cements, AS 3972. Sydney, Australia (2010)
3. B.S. Institution, Cement Composition, Specifications and Conformity Criteria for Common Cements, EN 197-1, London (2011)
4. ASTM, Standard Specification for Blended Hydraulic Cements, ASTM C595-08. ASTM, West Conshohocken, PA (2008)
5. ASTM, Standard Performance Specification for Hydraulic Cement, ASTM G1157-08. ASTM, West Conshohocken, PA (2008)
6. C. Standard, Standard for Portland Cement, GB 175-2007. AQSIQ; SAC (2007)
7. C. Standard, Standard for Special Cements, GB 1344-1999. State Quality and Technical Supervision Bureau (1999)
8. J.W. Bullard, H.M. Jennings, R.A. Livingston, A. Nonat, G.W. Scherer, J.S. Schweitzer, K.L. Scrivener, J.J. Thomas, Mechanisms of cement hydration. Cem. Concr. Res. **41**, 1208–1223 (2011)
9. A.A. Hamami, P. Turcry, A. Aït-Mokhtar, Influence of mix proportions on microstructure and gas permeability of cement pastes and mortars. Cem. Concr. Res. **42**, 490–498 (2012)
10. E. John, B. Lothenbach, Cement hydration mechanisms through time—a review. J. Mater. Sci. **58**, 9805–9833 (2023)
11. Z. Luo, W. Li, K. Wang, A. Castel, S.P. Shah, Comparison on the properties of ITZs in fly ash-based geopolymer and Portland cement concretes with equivalent flowability. Cem. Concr. Res. **143**, 106392 (2021)

12. P. Li, W. Li, Z. Sun, L. Shen, D. Sheng, Development of sustainable concrete incorporating seawater: a critical review on cement hydration, microstructure and mechanical strength. Cement Concr. Compos. **121**, 104100 (2021)
13. B. Lothenbach, F. Winnefeld, C. Alder, E. Wieland, P. Lunk, Effect of temperature on the pore solution, microstructure and hydration products of Portland cement pastes. Cem. Concr. Res. **37**, 483–491 (2007)
14. B. Lothenbach, K. Scrivener, R. Snellings, *A practical guide to microstructural analysis of cementitious materials* (CRC Press, Boca Raton, Florida, 2016)
15. D.C. MacLaren, M.A. White, Cement: its chemistry and properties. J. Chem. Educ. **80**(6), 623–635 (2003)
16. H. Zhao, W. Li, Y. Gan, K. Wang, Z. Luo, Nano/microcharacterization and image analysis on bonding behaviour of ITZs in recycled concrete enhanced with waste glass powder. Constr. Build. Mater. **392**, 131904 (2023)
17. I.G. Richardson, The nature of the hydration products in hardened cement pastes. Cement Concr. Compos. **22**, 97–113 (2000)
18. K.L. Scrivener, P. Juilland, P.J.M. Monteiro, Advances in understanding hydration of Portland cement. Cem. Concr. Res. **78**, 38–56 (2015)
19. J. Xiao, W. Li, Y. Fan, X. Huang, An overview of study on recycled aggregate concrete in China (1996–2011). Constr. Build. Mater. **31**, 364–383 (2012)
20. J. Xiao, W. Li, C.S. Poon, Recent studies on mechanical properties of recycled aggregate concrete in China—a review. Sci. China Tech. Sci. **55**(6), 1463–1480 (2012)
21. B. Zhan, J.d. Brito, C.S. Poon, *New Trends in Recycled Aggregate Concrete.* MDPI (2019)
22. J. Zhang, G.W. Scherer, Comparison of methods for arresting hydration of cement. Cem. Concr. Res. **41**, 1024–1036 (2011)
23. D. Jóźwiak-Niedźwiedzka, A.P. Fantilli, *Supplementary Cementitious Materials in Concrete.* Basel, Switzerland MDPI—Multidisciplinary Digital Publishing Institute (2021)
24. J. Skibsted, R. Snellings, Reactivity of supplementary cementitious materials (SCMs) in cement blends. Cem. Concr. Res. **124**, 105799 (2019)
25. Y. Jeong, S.-H. Kang, M.O. Kim, J. Moon, Acceleration of cement hydration from supplementary cementitious materials: performance comparison between silica fume and hydrophobic silica. Cement Concr. Compos. **112**, 103688 (2020)
26. G. Yao, Q. Wang, Z. Wang, J. Wang, X. Lyu, Activation ofhydration properties ofiron ore tailings and their application as supplementary cementitious materials in cement. Powder Technol. **360**, 863–871 (2020)
27. K. Wu, H. Han, C. Roßler, L. Xu, H.M. Ludwig, Rice hush ash as supplementary cementitious material for calcium aluminate cement—effects on strength and hydration. Constr. Build. Mater. **302**, 124198 (2021)
28. P. Li, W. Li, K. Wang, J.L. Zhou, A. Castel, S. Zhang, S.P. Shah, Hydration of Portland cement with seawater toward concrete sustainability: phase evolution and thermodynamic modelling. Cement Concr. Compos. **138**, 105007 (2023)
29. P. Li, W. Li, T. Yu, F. Qu, V.W.Y. Tam, Investigation on early-age hydration, mechanical properties and microstructure of seawater sea sand cement mortar. Constr. Build. Mater. **249**, 118776 (2020)
30. K.L. Scrivener, J.F. Young, *Mechanisms of chemical degradation of cement-based systems* (CRC Press, an imprint of Taylor and Francis, Boca Raton, FL, 2014)
31. I. Galan, H. Beltagui, M. García Maté, F.P. Glasser, M.S. Imbabi, Impact of drying on pore structures in ettringite-rich cements. Cem. Concr. Res. **84**, 85–94 (2016)
32. R. Talero, Performance of metakaolin and Portland cements in ettringite formation as determined by ASTM C 452–68: kinetic and morphological differences. Cem. Concr. Res. **35**, 1269–1284 (2005)
33. L.P. Pinto, C.A.d. Luz, R.D. Hooton, Ettringite instability in supersulfated cement: investigation under different curing regimes. Mater. Struct. **57**(55) (2024)
34. L.G. Baquerizo, T. Matschei, K.L. Scrivener, M. Saeidpour, L. Wadsö, Hydration states of AFm cement phases. Cem. Concr. Res. **73**, 143–157 (2015)

35. F. Qu, W. Li, K. Wang, V.W.Y. Tam, S. Zhang, Effects of seawater and undesalted sea sand on the hydration products, mechanical properties and microstructures of cement mortar. Constr. Build. Mater. **310**, 125229 (2021)
36. S. Mohammed, O. Safiullah, Optimization of the SO_3 content of an Algerian Portland cement: study on the effect of various amounts of gypsum on cement properties. Constr. Build. Mater. **164**, 362–370 (2018)
37. K.A. Snyder, X. Feng, B.D. Keen, T.O. Mason, Estimating the electrical conductivity of cement paste pore solutions from OH-, K+ and Na+ concentrations. Cem. Concr. Res. **33**, 793–798 (2003)
38. M. Balonis, M. Medala, F.P. Glasser, Influence of calcium nitrate and nitrite on the constitution of AFm and AFt cement hydrates. Adv. Cem. Res. **23**(3), 129–143 (2011)
39. Y. Zhang, P. Yu, F. Pan, Y. He, The synergistic effect of AFt enhancement and expansion in Portland cement-aluminate cement-FGD gypsum composite cementitious system. Constr. Build. Mater. **190**, 985–994 (2018)
40. G. Paul, E. Boccaleri, L. Buzzi, F. Canonico, D. Gastaldi, Friedel's salt formation in sulfoaluminate cements: a combined XRD and ^{27}Al MAS NMR study. Cem. Concr. Res. **67**, 93–102 (2015)
41. Y. Zhang, J. Chang, J. Ji, AH3 phase in the hydration product system of AFt-AFm-AH3 in calcium sulfoaluminate cements: a microstructural study. Constr. Build. Mater. **167**, 587–596 (2018)
42. M. Balonis, B. Lothenbach, G.L. Saout, F.P. Glasser, Impact of chloride on the mineralogy of hydrated Portland cement systems. Cem. Concr. Res. **40**, 1009–1022 (2010)
43. N.C.M. Marty, S. Grangeon, E. Elkaïm, C. Tournassat, C. Fauchet, F. Claret, Thermodynamic and crystallographic model for anion uptake by hydrated calcium aluminate (AFm): an example of molybdenum. Scient. Rep. **8** (2018)
44. A. Palomo, M. Palacios, Alkali-activated cementitious materials: Alternative matrices for the immobilisation of hazardous wastes Part II. Stabilisation of chromium and lead. Cement Concr. Res. **33**, 289–295 (2003)
45. T. Liu, Y. Chen, B. Yuan, W. Zhuang, H.J.H. Brouwers, Q. Yu, Sodium aluminate activated waste glass: reduced efflorescence behavior by C(N)−A−S−H transformation. Cem. Concr. Res. **181**, 107527 (2024)
46. Y. Liu, C. Shi, Z. Zhang, N. Li, An overview on the reuse of waste glasses in alkali-activated materials. Resour. Conserv. Recycl. **144**, 297–309 (2019)
47. S. Luhar, T.-W. Cheng, D. Nicolaides, I. Luhar, D. Panias, K. Sakkas, Valorisation of glass waste for development of Geopolymer composites—mechanical properties and rheological characteristics: a review. Constr. Build. Mater. **220**, 547–564 (2019)
48. J.M.J.M. Bijen, R.V. Selst, *Fly ash as addition to concrete* (Balkema for CUR, Rotterdam, Netherlands, 1992)
49. T. Luukkonen, Z. Abdollahnejad, J. Yliniemi, P. Kinnunen, M. Illikainen, One-part alkali-activated materials: a review. Cem. Concr. Res. **103**, 21–34 (2018)
50. R. Xiao, X. Dai, J. Zhong, Y. Ma, X. Jiang, J. He, Y. Wang, B. Huang, Toward waste glass upcycling: preparation and characterization of high-volume waste glass geopolymer composites. Sustain. Mater. Technol. **40**, e00890 (2024)
51. P. Zhang, K. Wang, Q. Li, J. Wang, Y. Ling, Fabrication and engineering properties of concretes based on geopolymers/alkali-activated binders—a review. J. Clean. Prod. **258**, 120896 (2020)
52. E.T. Rodriguez, K. Garbev, D. Merz, L. Black, I.G. Richardson, Thermal stability of C-S-H phases and applicability of Richardson and Groves' and Richardson C-(A)-S-H(I) models to synthetic C-S-H. Cem. Concr. Res. **93**, 45–56 (2017)
53. I. Garcia-Lodeiro, A. Palomo, A. Fernández-Jiménez, D.E. Macphee, Compatibility studies between N-A-S-H and C–A-S-H gels. Study in the ternary diagram Na_2O–CaO–Al_2O_3–SiO_2–H_2O. Cement Concr. Res. **41**, 923–931 (2011)
54. J. Nemecek, V. Kralík, V. Smilauer, L. Polívka, A. Jager, Tensile strength of hydrated cement paste phases assessed by microbending tests and nanoindentation. Cement Concr. Compos. **73**, 164–173 (2016)

55. S. Das, P. Yang, S.S. Singh, J.C.E. Mertens, X. Xiao, N. Chawla, N. Neithalath, Effective properties of a fly ash geopolymer: synergistic application of X-ray synchrotron tomography, nanoindentation, and homogenization models. Cem. Concr. Res. **78**, 252–262 (2015)
56. H. Lee, V. Vimonsatit, P. Chindaprasirt, Mechanical and micromechanical properties of alkali activated fly-ash cement based on nano-indentation. Constr. Build. Mater. **107**, 95–102 (2016)
57. Y. Ma, G. Ye, J. Hu, Micro-mechanical properties of alkali-activated fly ash evaluated by nanoindentation. Constr. Build. Mater. **147**, 407–416 (2017)
58. O. Bernard, F.-J. Ulm, E. Lemarchand, A multiscale micromechanics-hydration model for the early-age elastic properties of cement-based materials. Cem. Concr. Res. **33**, 1293–1309 (2003)
59. G. Constantinides, F.-J. Ulm, The effect of two types of C-S-H on the elasticity of cement-based materials: results from nanoindentation and micromechanical modeling. Cem. Concr. Res. **34**, 67–80 (2004)
60. M.J. DeJong, F.-J. Ulm, The nanogranular behavior of C-S-H at elevated temperatures (up to 700 °C). Cem. Concr. Res. **37**, 1–12 (2007)
61. H.M. Jennings, J.J. Thomas, J.S. Gevrenov, G. Constantinides, F.-J. Ulm, A multi-technique investigation of the nanoporosity of cement paste. Cem. Concr. Res. **37**, 329–336 (2007)
62. Z. Luo, W. Li, Y. Gan, K. Mendu, S.P. Shah, Maximum likelihood estimation for nanoindentation on sodium aluminosilicate hydrate gel of geopolymer under different silica modulus and curing conditions. Compos. B **198**, 108185 (2020)
63. L. Yun-Ming, H. Cheng-Yong, L.-Y. Li, N.A. Jaya, M.M.A.B. Abdullah, T.S. Jin, K. Hussin, Formation of one-part-mixing geopolymers and geopolymer ceramics from geopolymer powder. Constr. Build. Mater. **156**, 9–18 (2017)
64. Z. Luo, W. Li, Y. Gan, X. He, A. Castel, D. Sheng, Nanoindentation on micromechanical properties and microstructure of geopolymer with nano-SiO_2 and nano-TiO_2. Cement Concr. Compos. **117**, 103883 (2021)
65. Z. Luo, W. Li, Y. Gan, K. Mendu, S.P. Shah, Applying grid nanoindentation and maximum likelihood estimation for N-A-S-H gel in geopolymer paste: investigation and discussion. Cem. Concr. Res. **135**, 106112 (2020)
66. Z. Luo, W. Li, P. Li, K. Wang, S.P. Shah, Investigation on effect of nanosilica dispersion on the properties and microstructures of fly ash-based geopolymer composite. Constr. Build. Mater. **282**, 122690 (2021)
67. Y. Ling, K. Wang, X. Wang, S. Hua, Effects of mix design parameters on heat of geopolymerization, set time, and compressive strength of high calcium fly ash geopolymer. Constr. Build. Mater. **228**, 116763 (2019)
68. J.L. Provis, S.A. Bernal, Geopolymers and related alkali-activated materials. Annu. Rev. Mater. Res. **44**, 299–327 (2014)
69. P. Duxson, J.L. Provis, G.C. Lukey, S.W. Mallicoat, W.M. Kriven, J.S.J.v. Deventer, Understanding the relationship between geopolymer composition, microstructure and mechanical properties. Coll. Surf. A: Physicochem. Eng. Aspects **269**, 47–58 (2005)
70. P. Zhang, Y. Zheng, K. Wang, J. Zhang, A review on properties of fresh and hardened geopolymer mortar. Compos. B **152**, 79–95 (2018)
71. Y. Ling, K. Wang, C. Fu, Shrinkage behavior of fly ash based geopolymer pastes with and without shrinkage reducing admixture. Cement Concr. Compos. **98**, 74–82 (2019)
72. M. Zhang, T. El-Korchi, G. Zhang, J. Liang, M. Tao, Synthesis factors affecting mechanical properties, microstructure, and chemical composition of red mud–fly ash based geopolymers. Fuel **134**, 315–325 (2014)
73. F. Qu, W. Li, K. Wang, S. Zhang, D. Sheng, Performance deterioration of fly ash/slag-based geopolymer composites subjected to coupled cyclic preloading and sulfuric acid attack. J. Clean. Prod. **321**, 128942 (2021)
74. M. Alexander, G. Arliguie, A. Bentur, A. Katz, *Interfacial Transition Zone in Cementitious Composites* (CRC Press, Boca Raton, FL, 2014)
75. Y. Gao, H. Jing, Z. Yu, L. Li, J. Wu, W. Chen, Particle size distribution of aggregate effects on the reinforcing roles of carbon nanotubes in enhancing concrete ITZ. Constr. Build. Mater. **327**, 126964 (2022)

76. G. Mazzucco, V.A. Salomoni, C. Majorana, A cohesive contact algorithm to evaluate the mechanical behaviour of concrete ITZ at different roughness conditions. Constr. Build. Mater. **294**, 123479 (2021)
77. S. Diamond, J. Huang, The ITZ in concrete-a different view based on image analysis and SEM observations. Cement Concr. Compos. **23**, 179–188 (2001)
78. W. Li, C. Long, V.W.Y. Tam, C.-S. Poon, W.H. Duan, Effects of nano-particles on failure process and microstructural properties of recycled aggregate concrete. Constr. Build. Mater. **142**, 42–50 (2017)
79. W. Li, J. Xiao, Z. Sun, S. Kawashima, S.P. Shah, Interfacial transition zones in recycled aggregate concrete with different mixing approaches. Constr. Build. Mater. **35**, 1045–1055 (2012)
80. J. Xiao, W. Li, Z. Sun, D.A. Lange, S.P. Shah, Properties of interfacial transition zones in recycled aggregate concrete tested by nanoindentation. Cement Concr. Compos. **37**, 276–292 (2013)
81. H. Sasanipour, F. Aslani, J. Taherinezhad, Chloride ion permeability improvement of recycled aggregate concrete using pretreated recycled aggregates by silica fume slurry. Constr. Build. Mater. **270**, 121498 (2021)
82. J. Liu, K. Ma, J. Shen, J. Zhu, G. Long, Y. Xie, B. Liu, Influence of recycled concrete aggregate enhancement methods on the change of microstructure of ITZs in recycled aggregate concrete. Constr. Build. Mater. **371**, 130772 (2023)
83. Z. Luo, W. Li, V.W.Y. Tam, J. Xiao, S.P. Shah, Current progress on nanotechnology application in recycled aggregate concrete. J. Sustain. Cement-Based Mater. **8**(2), 79–96 (2019)
84. H. Zhang, X. Xu, W. Liu, B. Zhao, Q. Wang, Influence of the moisture states of aggregate recycled from waste concrete on the performance of the prepared recycled aggregate concrete (RAC)—a review. Constr. Build. Mater. **326**, 126891 (2022)
85. B.B. Mukharjee, S.V. Barai, Mechanical and microstructural characterization of recycled aggregate concrete containing silica nanoparticles. J. Sustain. Cement-Based Mater. **6**(1), 37–53 (2017)
86. P. Hosseini, A. Booshehrian, A. Madari, Developing concrete recycling strategies by utilization of nano-SiO_2 particles. Waste Biomass Valor **2**, 347–355 (2011)
87. H. Zhang, Y. Zhao, T. Meng, S.P. Shah, The modification effects of a nano-silica slurry on microstructure, strength, and strain development of recycled aggregate concrete applied in an enlarged structural test. Constr. Build. Mater. **95**, 721–735 (2015)
88. J. Ying, B. Zhou, J. Xiao, Pore structure and chloride diffusivity of recycled aggregate concrete with nano-SiO_2 and nano-TiO_2. Constr. Build. Mater. **150**, 49–55 (2017)
89. F. Shaikh, V. Chavda, N. Minhaj, H.S. Arel, Effect of mixing methods of nano silica on properties of recycled aggregate concrete. Struct. Concr. **19**, 387–399 (2018)
90. M.H. Al-Majidi, A. Lampropoulos, A.B. Cundy, Steel fibre reinforced geopolymer concrete (SFRGC) with improved microstructure and enhanced fibre-matrix interfacial properties. Constr. Build. Mater. **139**, 286–307 (2017)
91. A. Bhutta, M. Farooq, C. Zanotti, N. Banthia, Pull-out behavior of different fibers in geopolymer mortars: effects of alkaline solution concentration and curing. Mater. Struct. **50**, 80 (2017)
92. W.K.W. Lee, J.S.J.v. Deventer, The interface between natural siliceous aggregates and geopolymers. Cement Concr. Res. **34**, 195–206 (2004)
93. W.K.W. Lee, J.S.J.v. Deventer, Chemical interactions between siliceous aggregates and low-Ca alkali-activated cements. Cement Concr. Res. **37**, 844–855 (2007)
94. R.S. Nicolas, S.A. Bernal, R.M.d. Gutiérrez, J.S.J.v. Deventer, J. L.Provis, Distinctive microstructural features of aged sodium silicate-activated slag concretes. Cement Concr. Res. **65**, 41–51 (2014)
95. P.G. Allison, C.A.W. jr., R.D. Moser, A.J. Diaz, O.G. Rivera, S.S. Holton, Nanoindentation and SEM/EDX characterization of the geopolymerto-steel interfacial transition zone for a reactive porcelain enamel coating. Compos. Part B **78**, 131–137 (2015)

96. H. Peng, C. Cui, C.S. Cai, Y. Liu, Z. Liu, Microstructure and microhardness property of the interface between a metakaolin/GGBFS-based geopolymer paste and granite aggregate. Constr. Build. Mater. **221**, 263–273 (2019)
97. Y. Ding, J. Dai, C. Shi, Fracture properties of alkali-activated slag and ordinary Portland cement concrete and mortar. Constr. Build. Mater. **165**, 310–320 (2018)
98. Z. Liu, C.S. Cai, H. Peng, F. Fan, Experimental Study of the Geopolymeric Recycled Aggregate Concrete. J. Mater. Civ. Eng. **28**(9), 04016077 (2016)
99. S.P. Shah, G. Winter, Inelastic behavior and fracture of concrete. ACI Mater. J. **63**(9), 925–930 (1966)
100. J. Xiao, W. Li, D.J. Corr, S.P. Shah, Effects of interfacial transition zones on the stress–strain behavior of modeled recycled aggregate concrete. Cem. Concr. Res. **52**, 82–99 (2013)
101. Z. Luo, W. Li, K. Wang, S.P. Shah, D. Sheng, Nano/micromechanical characterisation and image analysis on the properties and heterogeneity of ITZs in geopolymer concrete. Cem. Concr. Res. **152**, 106677 (2022)
102. C. Fu, Y. Ling, K. Wang, An innovation study on chloride and oxygen diffusions in simulated interfacial transition zone of cementitious material. Cement Concr. Compos. **110**, 103585 (2020)
103. B. Zhan, D. Xuan, C.S. Poon, K.L. Scrivener, Characterization of interfacial transition zone in concrete prepared with carbonated modeled recycled concrete aggregates. Cem. Concr. Res. **136**, 106175 (2020)
104. W. Li, Z. Luo, Y. Gan, K. Wang, S.P. Shah, Nanoscratch on mechanical properties of interfacial transition zones (ITZs) in fly ash-based geopolymer composites. Compos. Sci. Technol. **214**, 109001 (2021)

Chapter 2
Nano-Characterization Techniques for Cementitious Materials

2.1 Introduction

At present, advanced characterization and simulation techniques allow the study of materials from small scales such as micro and nanoscales. The introduction of these techniques [1–4] has greatly promoted the understanding of cementitious materials. Micro-characterization facilities are crucial in studying cementitious materials, enabling researchers to delve into the microstructure, micromechanical properties and chemical composition of materials. These facilities typically incorporate a range of advanced imaging, spectroscopy, and analytical techniques, which allow for comprehensive analysis, promoting advancements in material optimization for low-carbon, high-performance concrete and smart concrete sensors. Micro-characterization supports the development of cementitious composites with enhanced durability, sustainability, and suitability for modern infrastructure applications. In this chapter, the advancements and applications of nano-characterization techniques for cementitious materials are introduced.

2.2 Nanoindentation Technique

2.2.1 *Basis of Nanoindentation Technique*

Nanoindentation is the most commonly used technique to measure nano/microscale elastic modulus and hardness of materials. It can also be used for other investigations such as creep [5], fracture [6–8] and tensile strength [9]. The instrument and testing area are shown in Fig. 2.1.

$$H = \frac{P_{max}}{A} \tag{2.1}$$

W. Li et al., *Nanocharacterization of Cementitious Composites*,
https://doi.org/10.1007/978-981-96-2917-6_2

$$S = \left.\frac{dP}{dh}\right|_{h=h_{\max}} = \frac{2}{\sqrt{\pi}} E_r \sqrt{A} \tag{2.2}$$

$$\frac{1}{E_r} = \frac{1-v^2}{E} + \frac{1-v_i^2}{E_i} \tag{2.3}$$

$$A(h_c) = m_0 h_c^2 + m_1 h_c + m_2 m_c^{1/2} + \ldots + m_n h_c^{(1/2^{(n-1)})} \tag{2.4}$$

$$h_c = h - (\varepsilon P)/S \tag{2.5}$$

where P and h are the indentation load and indentation depth, respectively. A is the projected contact area. S is the initial unloading stiffness. E_r is the reduced elastic modulus, which contains the elastic response of both indenter tip and tested material. E_i and v_i are the elastic modulus and Poisson's ratio of the indenter, respectively. E and v are the corresponding parameters for sample. The projected contact area A is determined by the contact depth h_c and tip area function. For Berkovich tip, ε is 0.75. Coefficient m_0, $m_1 \ldots m_n$ can be obtained by doing calibration on reference material such as fused silica.

Nanoindentation test can be performed in different models such as load control model and depth control model. Load control is generally adopted for the test of cementitious materials. In a test, an indenter with known geometry and mechanical parameters is pressed into the tested material under a continuously increasing load. After reaching the pre-set value, the load is generally kept constant for a while to eliminate creep and then removed gradually. A diagram of load depth curve during a test is shown in Fig. 2.2a. Based on it, the elastic modulus E and hardness H of tested material can be determined by Eqs. 2.1–2.3 [10]. Although this technique is called as 'nanoindentation', for cement-based material research, the test depth, tip radius, and involved range is sometimes more than 100 nm. A typical radius of curvature for the commonly used Berkovich probe is 150 nm (COOPER 2007).

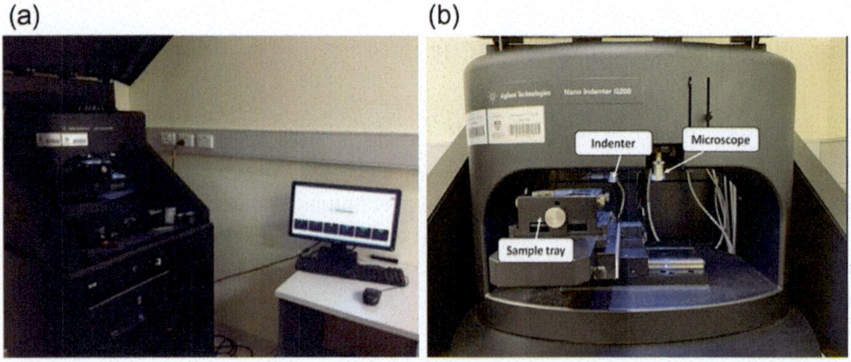

Fig. 2.1 Nano indenter: **a** instrument; **b** testing area

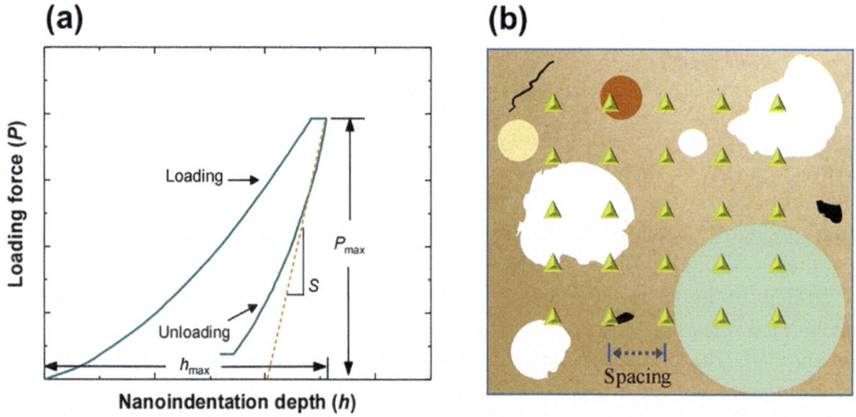

Fig. 2.2 Nanoindentation test technique: **a** load-depth curve in a test; **b** grid nanoindentation

Tip with large size like 600 nm was also reported for the study of Portland cement paste [11]. The suitable maximum indentation depth used for studying C–S–H of cement-based materials frequently ranges from 100 to 300 nm [12]. Research has also shown that the actual interaction range is 3–4 times the penetration depth [13], and the typical interaction volume would be up to 1 μm^3 [14]. Nevertheless, this term can help distinguish it from the microindentation technique. Because of these factors, different terms such as 'nanomechanical properties', 'micromechanical properties' and 'nano/micromechanical properties' are found to describe the properties of gel phase tested by nanoindentation.

Initially, the application of nanoindentation was mainly for homogeneous materials. However, lots of materials are composed of more than one component. There are some challenges in applying this advanced technique directly to obtain the mechanical properties of each micro-constituent. When the microstructure of the target components has a large characteristic length size, the investigation could be conducted on each specified component with the aid of microscope. However, when the material is still heterogeneity at the scale of several microns, it becomes very difficult to identify and locate the components. To solve the problem, the method of using grid nanoindentation shown in Fig. 2.2b coupled with statistical deconvolution techniques has been developed and widely used for many heterogeneous materials such as concrete, bone, shale, and ceramic composites [15]. It is called as statistical indentation technique (SNT). During the course of development, some studies clarified the principles of SNT and established test criteria [16]. These works can be understood from two parts: the test part and the deconvolution part.

2.2.2 Principles and Criteria for Statistical Nanoindentation Techniques

The nanoindentation testing theory is based on the infinite half-space model and assumes that the surface is perfectly flat and smooth. Surface roughness criteria were proposed for the nanoindentation study of cement-based materials [17]. Another study pointed out that there would be intrinsic porosity, so a low limit should be specified to avoid the filling of pores caused by over-polishing artefacts [18].

To determine the critical indentation depth that could generate a homogeneous mechanical response of a component, studies were conducted on a binary layered system [19]. The thin films-substrates model was considered to represent the most uncomplicated but most severe geometric condition the nanoindentation test would experience in a two-component material. Depending on the test parameters, the test results on this system may come from the corresponding single-phase response, or from different degrees of composite response. As given in Eq. 2.6 [12], the composite modulus E_{eff} is related to E_{f} and E_{s}, the elastic modulus of the film and substrate, respectively. The weight coefficient Ψ reflects the contribution of different components of the binary layered system. When Ψ is 0, it means that the elastic modulus obtained is totally tested from the film. The increase of the weight coefficient indicates that the role of the substrate becomes more and more significant. The influence of film is negligible for the case that the coefficient is close to 1.

The weight coefficient is considered to be highly relevant to test and material parameters such as the size and shape of indenter, the test depth, the thickness of the film and the stiffness difference between film and substrate. The quantitative relationship of them was established and developed by many researchers [20, 21]. The typical evaluation equations are given in Eqs. 2.7–2.8 [12, 20], where the allowed modulus ratio $E_{\mathrm{s}}/E_{\mathrm{f}}$ for analysis is in the wide range of 10^{-2} to 10^{2}, and the empirical constant n is around 1.27. It is clear that for a given coated substrate material, the weight coefficient Ψ decreases with the increase of the ratio of film thickness t to contact radius a. The summary study [12] indicated that these studies could draw a consistent conclusion: if the modulus ratio $E_{\mathrm{s}}/E_{\mathrm{f}}$ is in the range of 0.2–5, indentation tests with a depth to film thickness ratio h/t less than 10% can ensure that the results come from the individual response of the film without being affected by the substrate. Thus, a scale separability condition shown in Eq. 2.9 was proposed [12]. An indentation test was considered to be able to assess the homogeneous mechanical properties of a micro-component once the contact depth satisfies this rule of thumb. One thing to note is that this widely used criterion is based on depth h. At the same depth, indenter with different sizes and shapes may result in different responses.

$$E_{eff} = E_f + \Psi\left(E_s - E_f\right) \tag{2.6}$$

$$\Psi = \left[1 + \left(\frac{kt}{a}\right)^n\right]^{-1} \tag{2.7}$$

$$\log(k) = -0.093 + 0.792\log\left(\frac{E_s}{E_f}\right) + 0.05\left(\log\left(\frac{E_s}{E_f}\right)\right)^2 \tag{2.8}$$

$$d \ll h_{\max} < D/10 \tag{2.9}$$

where D and d are the characteristic length size of the micro-constituents, and the largest microstructural heterogeneity within the micro-constituents, respectively.

The above principles and criteria were for a single test point. Due to the heterogeneous of composites materials, nanoindentation testing is usually performed in the form of grids to collect massive test points, as shown in Fig. 2.2b. Therefore, it is necessary to specify a proper grid spacing. Based on the analysis of two-component materials, the spacing was suggested to be greater than the characteristic size of the residual indentation mark to avoid the interference of adjacent test points [3, 22]. Besides, in order to reflect the spatial distribution of components, the spacing s was specified to be much larger than the characteristic size of the two components ($s\sqrt{N} \gg D$) [3, 22]. In fact, it is better to specify the spacing more strictly than these criteria. For the microindentation hardness test (ASTM E384-2017), the minimum spacing recommended for Knoop and Vickers indentations are 2 or 2.5 times of the impression diagonals. Although the degree of plastic damage may decrease for the smaller scale nanoindentation test, it cannot be ignored that the residual impression is much smaller than the contact region and the contact region is much smaller than the involved range. The involved range was reported to be around 3–4 times the test depth [23]. In addition, it is a fact that the properties of the micro-constituents themselves sometimes show local heterogeneous, which also calls for large spacing. Thus, it is recommended that for indentation test, especially the large depth test, the minimum spacing could be determined by considering both the 2–2.5 times of the impression diagonals and 3–4 times of depth criteria. Besides, for comparative study, sufficient test data set in the form of different random grids is suggested to avoid the influence of local heterogeneity of the material.

For the deconvolution part, the initial studies [4, 24, 25] assumed the mechanical property distribution (e.g. elastic modulus) of the heterogeneous material tested from SNT a one-dimensional Gaussian mixture model as given in Eq. 2.10. The experiment data were constructed into frequency density histogram with a specified bin size, where each bar provides the frequency observation of the range of property value in the bin. The unknown parameters of theoretical probability density function (PDF), namely, the mean value, standard deviation and proportion of each component can be obtained by minimizing the sum of squared errors between the observed values and the theoretical values. However, the results may vary with the bin size specified. Then, this method was developed to use the cumulative distribution function (CDF) given in Eqs. 2.12–2.13 [26]. Nevertheless, the PDF based method can intuitively present the peak of components, making it the most used deconvolution technique. The same point of these two methods is that the least-square estimation (LSE) method is used for fitting.

$$P(x) = \sum_{j=1}^{n} f_j P_j(x) \tag{2.10}$$

$$\min \sum_{i=1}^{m} \frac{\left(P^i - P(x_i)\right)^2}{m} \tag{2.11}$$

$$D_X(X_i) = \frac{i}{N} - \frac{1}{2N} \quad for\ i \in [1, N], X = (M, H) \tag{2.12}$$

$$\min \sum_{i=1}^{N} \sum_{X=(M,H)} \left(D_X(X_i) - \sum_{j=1}^{n} f_j D_j(X_i) \right)^2 \tag{2.13}$$

$$\sum_{j=1}^{n} f_j = 1 \tag{2.14}$$

where n and f is the number of the components of the material and their proportion, respectively. P_j (x) and D_j (X_i) are the normal distribution function to describe the mechanical distribution of the j-th component. P^i is the frequency value of the i-th bin. D_X (X_i) indicates the experimental cumulative distribution function for the sorted test data X_i. m is the number of bins specified in the frequency histogram, while N is the number of the indentation test points. Details can be found in references [27, 28].

The above studies are a major breakthrough in the application of nanomechanical testing in heterogeneous materials. These techniques and related approaches such as micromechanics [29–31] have promoted the understanding of the mechanical properties and structure of micro-constituents in materials. However, limitations were sometimes found for the application of the statistical nanoindentation technique.

2.2.3 Examinations on Statistical Nanoindentation Technique

Based on the proposed scale separability criterion and deconvolution technique, tests were conducted on Ti-TiB composites to verify the validity of the SNT method [32]. The elastic modulus of components obtained by SNT was found to agree fairly well with the values reported in the literature. The packing density of C–S–H determined by SNT and micromechanics was also in line with the results calculated based on the mass density and particle density values measured by experiment [33]. However, some studies questioned the validity of using of SNT in Portland cement paste, especially for the identification of both low-density (LD) C–S–H and high-density (HD) C–S–H from the peaks of probability density function [4, 34, 35]. These works were conducted by virtual statistical indentation experiments, as displayed in Fig. 2.3. According to the results [14], the main problems of SNT lies in the characteristic of

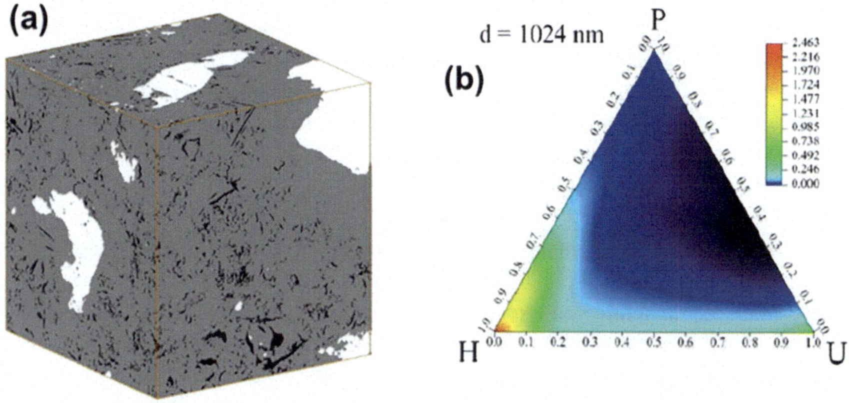

Fig. 2.3 Virtual experiment for statistical indentation on Portland cement paste [14]: **a** a FIB-nt sample of cement paste with a size of 14.0 × 14.2 × 13.1 μm^3; **b** ternary plot of Gaussian-filtered phase distribution densities

microstructure and the numerical instability of the LSE method to fit the experimental curve. Figure 2.3b provides the statistical results for distribution densities of phases detected by 1 μm^3 involved volume, which indicate that not all of the phases have an apparent peak. Besides, for deconvolution technique, the global optimum is hard to be found while significantly different results would be generated from different local optimum; the fewer number of phases of 2 and 3 are even found to better fit the frequency distribution histogram than 4 in Portland cement paste. These issues could also affect the investigation of geopolymer.

2.2.4 Progress of Nanoindentation Technique

In summary, the disadvantages of nanoindentation technique mainly include the time-consuming testing procedure, large involved volume, and numerical instability when applied in the form of SNT. Some efforts have been made to improve the nanoindentation technique. It was reported that conventionally, typical one point indentation test could cost 7–8 min to complete [36]. Some sophisticated nanoindentation equipment has developed the 'Express test', where a single point test takes only about one seconds to complete [37–39]. New breakthrough, accelerated property mapping (XPM) was reported to achieve an ultrahigh speed of six measurements per second. Several preliminary attempts for the ultra-high-speed nanoindentation have been found since 2017 [4, 23, 27]. This technique is also recently introduced for the cement-based materials research [4]. Despite the significant progress, there are some concerns. The high speed is realized by ignoring thermal drifts and creep, and the maximum scan area was reported to be only 90 μm × 90 μm [4]. Besides, the test needs to consider the strain rate effect [39, 40].

In addition to statistical analysis, efforts have been made on finding new methods for distinguishing different phases in a composite material. A 'manual' method was adopted by Davydov et al. [41], to identify specified phases using a BSE detector and then to navigate and select testing points on the specified phases using atomic force microscopy (AFM). Currently, there are increasing interests to combine nanoindentation with other research techniques like BSE [2, 3, 13], EDS [42–44], and XRD [7, 45, 46].

For the relatively large indenter tip and interaction volume of nanoindentation, it would make the results not pure individual phases and limit some delicate research at very small scale. While some researchers were trying to solve these problems by using advanced analysis methods, such as micromechanics, to evaluate the influence of embedded phases on the modulus of C–S–H at the indentation scale [3, 4, 33], others have introduced new research techniques, such as modulus mapping [47–49], PeakForce QNM [50–52] and nanoscratch test [8, 46, 53].

2.3 Modulus Mapping Technique

The modulus mapping technique incorporates the nanoscale dynamical mechanical analysis (nanoDMA) with in-situ scanning probe microscope (SPM) imaging, which can quantitatively map the nanomechanical properties of a material surface. In a modulus mapping test, the indenter scans the surface of the tested materials with a tiny high frequency sinusoidal oscillation, which is superimposed by a preset sinusoidal force and quasi-static force. Two key parameters, storage modulus E' and loss modulus E'', can be obtained by analyzing the recorded displacement amplitude and phase lag in the scanning process. The storage modulus and loss modulus represent the materials' capacity to store energy (elastic portion) and dissipate energy as heat (viscous portion) of the viscoelastic material, respectively [54]. The mechanism of modulus mapping was summarized in the literature [25, 30]. This technique is available on Hysitron Triboindenter when the Modulus Mapping module is equipped. As shown in some typical studies, the tip radius of the indenter used for a modulus mapping test is often large (e. g., 400 nm, 600 nm [21, 46]). Even if the same size of indenter tip is used for nanoindentation and modulus mapping tests, a significant increase in resolution has been observed in modulus mapping test [55, 56], due to the smaller force and more precise tip control. For some materials, only several nanometres of contact depth [7] and nanoscale influence radius of tip [57] were achieved, which allows a nearly non-destructive testing with very small test spacing. Each image of the modulus mapping has 256 × 256 test points. These massive test points can be obtained in just 10 min [32]. However, large scanning area would have increased time consumption [31, 58], and the maximum scanning area of this technique is only 50 μm × 50 μm [31, 58].

$$E' = \frac{K_S\sqrt{\pi}}{2\sqrt{A_c}} \tag{2.15}$$

$$E'' = \frac{\omega D_S \sqrt{\pi}}{2\sqrt{A_c}} \tag{2.16}$$

where A_c is the contact area; K_s and D_s are the contact stiffness and damping coefficient, respectively; ω is the angular frequency.

2.4 Atomic Force Microscopy-Based Nanomechanical Mapping Techniques

The typical AFM-based nanomechanical mapping technique is PeakForce quantitative nanomechanical mapping (PeakForce QNM). The AFM mainly has three operating modes: contact mode, tapping mode and non-contact mode, in which tapping mode is a very widely used mode for study in ambient conditions or aqueous medium [4, 59]. Presently, some new operating modes have been developed, for instance, the PeakForce Tapping mode. Different from the conventional tapping mode, where the vibration amplitude of the cantilever is kept constant and force curve is not available during the scanning process, the PeakForce QNM operates with PeakForce Tapping, which is characterized by accurately controlling of the maximum force during the acquisition of the force curves [2]. This control method can protect the tip and sample from damage, and minimize the contact area between them. Therefore, a noticeable improvement in resolution is achieved. By this technique, materials properties such as elastic modulus, adhesion, deformation and dissipation can be mapped with nanoscale resolution. Besides, high-resolution sample topography can be provided simultaneously.

The force versus tip-sample separation curve for a single cycle PeakForce tapping test is shown in Fig. 2.4a. Mechanical parameters can be extracted from this curve [11, 60]. The Derjaguin-Muller-Toporov (DMT) model given in Eq. 2.17 is used to fit the retract curve (blue line) to obtain the reduced modulus E^*. Then, the elastic modulus of the sample E_s can be acquired by Eq. 2.18. The typical elastic modulus contour map obtained by PeakForce QNM is revealed in Fig. 2.4b.

$$F_{tip} - F_{adh} = \frac{4}{3} E^* \sqrt{Rd^3} \tag{2.17}$$

$$E_s = \left(1 - v_S^2\right) \left[\frac{1}{E^*} - \frac{1 - v_{tip}^2}{E_{tip}} \right]^{-1} \tag{2.18}$$

where F_{tip} is the force on the AFM tip; the F_{adh} is the adhesion force between the tip and sample; R is the tip radius and d is sample deformation; E_{s} and v_{s} are the Young's modulus and Poisson's ratio of the sample, respectively; E_{tip} and v_{tip} are the corresponding parameters of the tip, respectively.

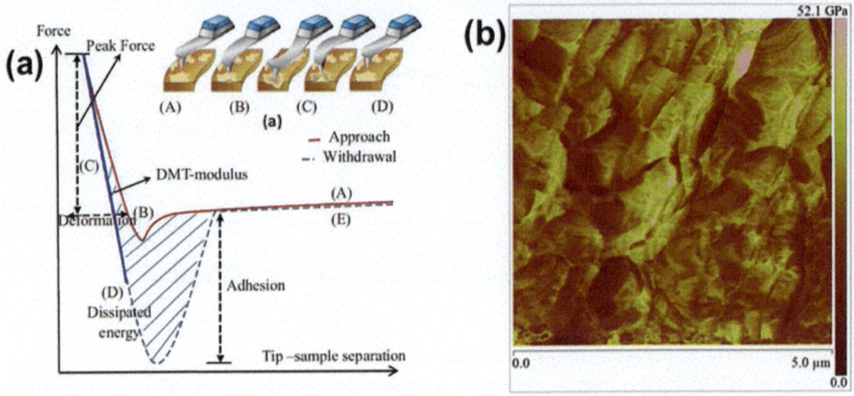

Fig. 2.4 Basic principle and results of the PeakForce QNM technique: **a** approach and withdrawal of AFM-tip in a single tapping cycle [61]; **b** elastic modulus contour map [62]

The spring constant for available cantilevers in PeakForce QNM range from 0.3 to 300 N/m, which enables force control from pN to μN [63]. The detectable elastic modulus of the testing materials was in the range of 700 kPa to 70 GPa or 1.0 kPa to 100 GPa depending on instruments [63]. In terms of indenter tip, the tip adopted in PeakForce QNM technique is smaller than that used in traditional nanoindentation and modulus mapping, which helps to realize unprecedented high-resolution. Tip radius of 8 nm, 5 nm, and even 2 nm were reported [64]. Typical contact depth in cement-paste materials research was only 2–10 nm [65, 66]. A test for 512 × 512 points on a 20 μm × 20 μm area was reported to take 17 min [67]. The test area is possible to reach 90 μm × 90 μm (5000 × 5000 points), but the test time would also rise to around one day [67].

2.5 Nanoscratch Technique

Nanoscratch has become an increasingly used technique for nanoscale analysis and characterization of thin films, coatings and bulk materials, which can be used to detect the adhesion strength, scratch hardness and wear resistant, etc. [33, 39, 68]. Similar to indentation test, scratch is a traditional test method, but it is not frequently used for cement-based materials, especially for nanoscale scratch. The parameters specified in a scratch test mainly include the maximum normal force, lateral scratch length and speed. As shown in Fig. 2.5, during a nanoscratch test, an indenter is pressed into the specimen. Then it is pulled straight over the surface of the tested specimen at a given speed. There are two transducers in the scratch testing system. One acquires control force (normal force P) and displacement in the press-in direction, and the other records the force (lateral force F_{scr}) and the displacement in the movement direction [4]. Prescan and postscan under low loads are usually combined with the scratch test

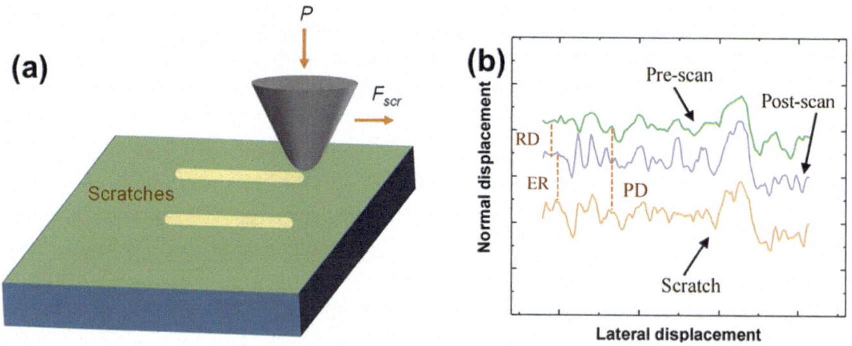

Fig. 2.5 Basic principle and results of the nanoscratch technique (constant load model): **a** scratch tests; **b** scratch displacement

to provide surface topography before and after the scratch [46, 69, 70]. The penetration depth value (PD) can be calculated by subtracting the prescan displacement from the scratch displacement, which consists of the residual depth (RD) part and the elastic recovery (ER) part, as displayed in Fig. 2.5b.

Based on the displacement, force and residue impressions acquired, scratch test can provide various information to reflect the elastic, plastic, fracture, etc. of materials, as described in below standards and literature [39, 57]. In a test, there are different modes to choose from. The main modes contain constant load scratch tests, ramped load scratch tests and wear tests. Progressively increasing normal force is generally used for the study of coatings (ASTM C1624-05, ASTM D7187-20) and fracture toughness of bulk materials, while constant force model can be used for the test of scratch hardness of both coated surfaces and some bulk materials (ASTM G171-03). In the above three standards, the tip's geometry is specified as spherical or hemispherical, different from the three-sided pyramid of the generally used Berkovich tip in the indentation test. According to ASTM G171-03, the hardness HS_P and stylus drag coefficient D_{sc} can be determined by Eqs. 2.19 and 2.20. Currently, the scratch study of fracture toughness of cement-based materials are almost all microscopic scratch [31, 35, 71], where the tip radius is 200 μm. The lack of nanoscratch tests may be due to the increased heterogeneity. Components would affect the nanoscale fracture results significantly. Besides, nanoscratch tests may sometimes not be severe enough to cause a fracture of some materials/phases. However, the tiny tip of nanoscratch technique makes it possible to test very small size objects, such as ITZs.

$$HS_p = \frac{kP}{w^2} \tag{2.19}$$

$$D_{sc} = \frac{F_{scr}}{P} \tag{2.20}$$

The current nanoscratch studies of the cement-based materials are mainly using the constant load nanoscratch, with the load applied commonly ranges from 2 to 8 mN and a scratch length ranges from 10 to 200 μm [31, 72, 73]. The nanoscratch technique is characterized by continuous testing. Therefore, it can be expected to acquire more intensive data than discrete point measurement methods. For instance, the data acquisition rate was reported to achieve as high as 192 kHz in NST^3 nano scratch tester. Nanoscratch was found to have some advantages over AFM on measurement stability and less tip wear on hard materials [72]. It acts as a powerful tool in simulating single asperity contact in tribological experiments [72]. However, similar to the nanoindentation, there are usually large applied force and penetration depth in nanoscratch test. For instance, a 600 nm radius Berkovich indenter was reported for nanoscratch test [23, 74]. Under the normal force of 4 mN, the average penetration depth in the C–S–H phase achieved about 466 nm. Thus, the large depth also leads to a large spacing between each scratch test. The spacing for microscope scratch hardness testing is no less than five scratch widths (ASTM G171-03).

2.6 Applications and Comparisons

Nanoindentation has been widely used for the mechanical properties investigation of micro-constituents in Portland cement pastes, blended cementitious composites and alkali-activated cementitious materials, as well as properties investigation of ITZs in aggregate and fibre reinforced cementitious composites [32]. Numerical model, micromechanical models and corresponding theories have also been developed based on the nanoindentation mechanical tests [44, 56, 58]. In addition to the usual elastic properties, this versatile technique has also been used for other investigations such as strength, fracture performance and creep [2, 39, 68]. Compared with the great achievements acquired by nanoindentation, the research results obtained by the other techniques such as modulus mapping, PeakForce QNM and nanoscratch are far less. In fact, these powerful techniques have their own advantages, which can be displayed by comparing their testing characteristics and current applications.

The significant difference of these techniques shows in testing speed, testing spatial resolution, testing area, material damage, etc. For the testing speed, XPM improves the nanoindentation testing speed from the conventional one testing point in a few minutes to six testing points in a second. However, modulus mapping can achieve 109 points per second (256×256 points/10 min), and PeakForce QNM can achieve 257 points per second (512×512 points/17 min) [4, 29]. The testing speed in nanoscratch depends on the specified scratch speed and data acquisition rate. Typically, for Agilent G200 Nano Indenter, the allowed maximum scratch speed is 2.5 mm/s, even if the data density is just 1 point/micron, it could reach 250 points per second. Thus, the maximum test speed of nanoscratch would be even far higher than PeakForce QNM. The test speed for this comparison is only the speed at which the test officially started. Other factors such as waiting time of thermal drift would also significantly increase the real testing time. Besides, small depth tests result in high requirements on surface roughness, parameters settings and tip calibration, which was reported to causes high preparation time for of modulus mapping and PeakForce QNM [54, 60].

For the testing spatial resolution (spacing between test points), compared with nanoindentation, modulus mapping technique decrease the applied force and depth, which results in high resolution. In addition to the applied force, PeakForce QNM also decreases the tip radius to several nanometres, which realizes further improvement in resolution. Any discrete points testing method above would have a spacing, while scratch test transformed the test into a continuous manner. For an instrument with high data acquisition rate, scratch testing could obtain more intensive data than the above techniques. However, it should be noted that the scratch technique only achieves high spatial resolution in the scratch direction, whereas modulus mapping and PeakForce QNM allow very small spacing in both x and y directions. Besides, the spacing in the discrete testing method is to avoid the influence of adjacent test points. There may be mutual interference between adjacent points in the continuous scratch tests.

For the testing area, there is almost no restriction for the grid size of conventional nanoindentation. For scratch, typically, the scratch length in Agilent G200 Nano Indenter can also reach up to 500 mm. However, the maximum scan area was decreased to 90 μm × 90 μm for XPM and PeakForce QNM, and 50 μm × 50 μm for modulus mapping [33, 75, 76]. Compared with XPM, modulus mapping and especially PeakForce QNM have smaller test depth, much denser data. As a consequence, the abundant test data significantly increased testing time for a given scan area. Therefore, modulus mapping and PeakForce QNM are mainly for very detailed local investigation, which are non-destructive test methods. As shown in Table 2.1, the test parameters set in some current studies can also clearly reveal the test characteristics of different techniques.

The current applications of these techniques were briefly summarized in Table 2.2. More details can be found in the author's published paper [79]. The different advantages of techniques can be displayed by comparing their applications in the research of ITZ. It is evident that the aggregate, fibre and reinforcing steel/mortar ITZs investigated by nanoindentation all have a large size, for instance, 50 μm for aggregate/mortar interfaces and approximately 200 μm for RPEC coated steel/mortar ITZs [80]. The modulus mapping and nanoscratch techniques were applied for small scale ITZ, the ITZ between unhydrated cement grain and C–S–H, which was detected to have sizes in the range of 155 nm to 3.3 μm [2, 24, 55, 69]. The PeakForce QNM was used to test the ITZ between C–S–H colonies in early stages of cement hydration, which was found to have a width less than 10 nm [28, 81]. It is obvious that, due to the ability to set large grid, nanoindentation has advantages in properties investigation of large ITZs, while modulus mapping and PeakForce QNM techniques can reveal very detailed local mechanical information at decreased scales. Nanoscratch shows similar advantages as modulus mapping in testing small scale objects. In fact, it can also investigate large size ITZ as that of nanoindentation.

Table 2.1 Typical parameters set in nanomechanical testing techniques

Literatures	Applied force (μm)	Pixel spacing (μm)	Testing area/length (μm)	Testing methods
[62, 77]	1200	3 × 3	100 × 100	NI
[36]	1000	3 × 3	87 × 87	NI
[73]	1200	3 × 5	30 × 10	NI
[70]	4 ± 3.5	0.137 × 0.059	35 × 35	MM
[73]	8 ± 3.5	0.059 × 0.059	15 × 15	MM
[70]	4 ± 3.5	0.028 × 0.028	7.0 × 7.0	MM
[18]	3.0	0.039 × 0.039	20 × 20	PF QNM
[36]	–	0.010 × 0.020	5.0 × 5.0	PF QNM
[62]	0.05	0.010 × 0.010	5.0 × 5.0	PF QNM
[18]	2000	–	10	NS
[73]	4000	–	12	NS
[78]	8000	–	200	NS

Note NI Nanoindentation; *MM* Modulus mapping; *PF QNM* PeakForce QNM; *NS* Nanoscratch, and the same below

The advantages of these mechanical mapping techniques have also been played in other aspects. Due to the feature of non-destructive test with dense test points, modulus mapping has been used for identification and then size measurement of phases, such as for inner product and outer product C–S–H in cement paste [82]. Similarly, PeakForce QNM also has the potential to identify phases [83]. Besides, the ultra-high-speed scanning capability of PeakForce QNM was used to test the early age properties of cement paste, where the topography and Young's Modulus of cement paste after 10 min, 14 min and 18 min of hydration were investigated. Considering the different advantages, these techniques can be combined used so as to extend the information to be obtained from the nanomechanical characterization. For instance, modulus mapping was helped to identify different phases, which was followed by traditional nanoindentation test on the target individual phase [70]. In addition to using multiple nanocharacterization techniques, there is a growing trend for the combination of nanoscience and nano-engineering approaches [79]. A typical case is the application of nanomechanical characterization to the study of nanomaterial modified cement-based materials. The combination could lead to potential breakthroughs for developing next generation of multifunctional and high performance cementitious materials.

Table 2.2 Summary of the advanced nanomechanical characterization techniques for the study of cement-based materials

Techniques	Commonly used testing parameters	Current common applications	Main advantages/concerns
NI	Young's modulus; Hardness; Indentation load; Indentation depth	Mechanical study of phases; Various tests at micro- and nano-scale (fracture, etc.) ITZs: aggregate, fibre and reinforcing steel/mortar	Measure in a large area; Relatively mature theory for testing cement paste Multiphase interaction; Large spacing restricts precise measurement
MM	Storage modulus; Loss Modulus; Contact force; Displacement amplitude	Identification and then mechanical/size measurement of phases; ITZs: unhydrated cement grain/C–S–H	Non-destructive, quick and high resolution mapping; Phase identification and relevant small scale research Local test; Test results may be more easily affected by the sample surface, tip and microstructural heterogeneity
PF QNM	Young's modulus; Adhesion force; Energy dissipation; Maximum deformation	Properties investigation of phases and small scale research like mechanical/size measurement of C–S–H colonies; ITZs: aggregate/cement paste; C–S–H colonies/C–S–H colonies	Non-destructive; Higher resolution and faster mapping than nanoindentation and modulus mapping; Phase identification; Extremely small scale research; Local test; Limited modulus test range; Test results may be more easily affected by the sample surface, tip and microstructural heterogeneity
NS	Deformation; Friction coefficient; Loads/critical loads; Displacement	Mechanical properties study of phases; ITZs: unhydrated cement grain/C–S–H; bond strength	Quick and continuous test; Adhesion strength/wear resistant/fracture No direct elastic modulus; Multiphase interaction

2.7 Summary

Although there are new techniques such as modulus mapping and PeakForce quantitative nanomechanical mapping that can avoid the large interaction volume of nanoindentation, these techniques are only available on some specific instruments. Besides, these techniques with very shallow penetration depth have high requirement on sample preparation and tip calibration. Typically, the sample for PeakForce quantitative nanomechanical mapping testing is not prepared by the regular polishing method, but by advanced focused ion beam (FIB) milling [18]. In fact, the FIB technique is also concerned for re-deposition of sputtered materials, Ga^+ ion damage and heat damage of microstructure, which may affect test results, especially for surface nanomechanical testing. The small depth would also increase the risk of interference from the microstructural heterogeneity of the constituent (*d* criterion in Eq. 2.9).

Additionally, the local information provided by these mapping techniques is not as good as the large size grid nanoindentation to represent the overall information of the materials for a comparative study. Thus, there is no perfect technique to determine the nano/micromechanical properties of highly heterogeneous materials. The regular statistical nanoindentation technique is currently still the most widely used method.

References

1. A.T. Akono, J. Chen, S. Kaewunruen, Friction and fracture characteristics of engineered crumb-rubber concrete at microscopic lengthscale. Constr. Build. Mater. **175**, 735–745 (2018)
2. Y. Ding, J.G. Dai, C.J. Shi, Fracture properties of alkali-activated slag and ordinary Portland cement concrete and mortar. Constr. Build. Mater. **165**, 310–320 (2018)
3. J. Liu, Q. Zeng, S. Xu, Is scratch test proper to characterize microstructure and mechanical properties of cement-based materials? The effects of loading level and routine. Cem. Concr. Res. **133** (2020)
4. Z. Luo, W. Li, Y. Gan, K. Mendu, S.P. Shah, Maximum likelihood estimation for nanoindentation on sodium aluminosilicate hydrate gel of geopolymer under different silica modulus and curing conditions. Compos. Part B: Eng. **198** (2020)
5. J. Němeček, Creep effects in nanoindentation of hydrated phases of cement pastes. Mater Charact **60**(9), 1028–1034 (2009)
6. A.T. Akono, N.X. Randall, F.J. Ulm, Experimental determination of the fracture toughness via microscratch tests: application to polymers, ceramics, and metals. J. Mater. Res. **27**(2), 485–493 (2012)
7. Z. Luo, N. Zhang, L. Zhao, J. Zeng, P. Liu, N. Li, Interaction of a hydraulic fracture with a hole in poroelasticity medium based on extended finite element method. Eng. Anal. Boundary Elem. **115**, 108–119 (2020)
8. X.H. Wang, S. Jacobsen, S.F. Lee, J.Y. He, Z.L. Zhang, Effect of silica fume, steel fiber and ITZ on the strength and fracture behavior of mortar. Mater. Struct. **43** (2010)
9. J. Nemecek, V. Kralík, V. Smilauer, L. Polívka, A. Jager, Tensile strength of hydrated cement paste phases assessed by microbending tests and nanoindentation. Cement Concr. Compos. **73**, 164–173 (2016)
10. W.C. Oliver, G.M. Pharr, An improved technique for determining hardness and elastic modulus using load and displacement sensing indentation experiments. J. Mater. Res. **7**(6), 1564–1583 (1992)
11. Q. Li, Y. Li, L. Zhou, Nanoscale evaluation of multi-layer interfacial mechanical properties of sisal fiber reinforced composites by nanoindentation technique. Compos. Sci. Technol. **152**, 211–221 (2017)
12. G. Constantinides, F.J. Ulm, The nanogranular nature of C-S-H. J. Mech. Phys. Solids **55**(1), 64–90 (2007)
13. J.J. Chen, L. Sorelli, M. Vandamme, F.J. Ulm, G. Chanvillard, A coupled nanoindentation/SEM-EDS study on low water/cement ratio portland cement paste: evidence for C-S-H/$Ca(OH)_2$ nanocomposites. J. Am. Ceram. Soc. **93**(5), 1484–1493 (2010)
14. P. Lura, P. Trtik, B. Münch, Validity of recent approaches for statistical nanoindentation of cement pastes. Cement Concr. Compos. **33**(4), 457–465 (2011)
15. M.J. DeJong, F.-J. Ulm, The nanogranular behavior of C-S-H at elevated temperatures (up to 700 °C). Cem. Concr. Res. **37**, 1–12 (2007)
16. L. Sorelli, G. Constantinides, F.J. Ulm, F. Toutlemonde, The nano-mechanical signature of ultra high performance concrete by statistical nanoindentation techniques. Cem. Concr. Res. **38**(12), 1447–1456 (2008)

17. M. Miller, C. Bobko, M. Vandamme, F.J. Ulm, Surface roughness criteria for cement paste nanoindentation. Cem. Concr. Res. **38** (2008)
18. P. Trtik, J. Kaufmann, U. Volz, On the use of peak-force tapping atomic force microscopy for quantification of the local elastic modulus in hardened cement paste. Cem. Concr. Res. **42** (2012)
19. M. Sumesh, U.J. Alengaram, M.Z. Jumaat, K.H. Mo, M.F. Alnahhal, Incorporation of nano-materials in cement composite and geopolymer based paste and mortar—a review. Constr. Build. Mater. **148**, 62–84 (2017)
20. B. Zhan, D. Xuan, C.S. Poon, K.L. Scrivener, Characterization of interfacial transition zone in concrete prepared with carbonated modeled recycled concrete aggregates. Cem. Concr. Res. **136**, 106175 (2020)
21. Z. Zhu, J.L. Provis, H. Chen, Quantification of the influences of aggregate shape and sampling method on the overestimation of ITZ thickness in cementitious materials. Powder Technol. **326**, 168–180 (2018)
22. J. Hu, Z. Qian, D. Wang, M. Oeser, Influence of aggregate particles on mastic and air-voids in asphalt concrete. Constr. Build. Mater. **93**, 1–9 (2015)
23. R. San Nicolas, S.A. Bernal, R. Mejía De Gutiérrez, J.S.J. Van Deventer, J.L. Provis, Distinctive microstructural features of aged sodium silicate-activated slag concretes. Cem. Concr. Res. **65**, 41–51 (2014)
24. T. Ji, Q. Gao, W. Zheng, X. Lin, H.-C. Wu, Interfacial transition zone of alkali-activated slag concrete. ACI Mater. J. **114**(3), 347 (2017)
25. M. Kuroda, T. Watanabe, N. Terashi, Increase of bond strength at interfacial transition zone by the use of fly ash. Cem. Concr. Res. **30**(2), 253–258 (2000)
26. C. Hu, Z. Li, A review on the mechanical properties of cement-based materials measured by nanoindentation. Constr. Build. Mater. **90**, 80–90 (2015)
27. J. Liu, Q. Zeng, S. Xu, The state-of-art in characterizing the micro/nano-structure and mechanical properties of cement-based materials via scratch test. Constr. Build. Mater. **254** (2020)
28. A. Ureña, J. Rams, M.D. Escalera, M. Sánchez, Characterization of interfacial mechanical properties in carbon fiber/aluminium matrix composites by the nanoindentation technique. Compos. Sci. Technol. **65**(13), 2025–2038 (2005)
29. J.L. Provis, Geopolymers and other alkali activated materials: Why, how, and what? Mater. Struct./Materiaux et Constructions **47**(1–2), 11–25 (2014)
30. X. Wang, P. Xu, R. Han, J. Ren, L. Li, N. Han, F. Xing, J. Zhu, A review on the mechanical properties for thin film and block structure characterised by using nanoscratch test. Nanotechnol. Rev. **8**(1), 628–644 (2019)
31. J. Xu, D.J. Corr, S.P. Shah, Nanoscratch study of the modification effects of nanoSiO_2 on C-S-H gel/cement grain interfaces. J. Mater. Civ. Eng. **29**(9) (2017)
32. M. Schöneich, M. Zamanzade, M. Stommel, Fiber-matrix interphase in applied short glass fiber composites determined by a nano-scratch method. Compos. Sci. Technol. **119**, 100–107 (2015)
33. H. Peng, C. Cui, C.S. Cai, Y. Liu, Z. Liu, Microstructure and microhardness property of the interface between a metakaolin/GGBFS-based geopolymer paste and granite aggregate. Constr. Build. Mater. **221**, 263–273 (2019)
34. K.L. Scrivener, A.K. Crumbie, P. Laugesen, The interfacial transition zone (ITZ) between cement paste and aggregate in concrete. Interface Sci. **12**(4), 411–421 (2004)
35. Y. Yan, S. Chang, T. Wang, Y. Geng, Scratch on polymer materials using AFM tip-based approach: a review. Polymers **11**(10) (2019)
36. T. Howind, J. Hughes, W. Zhu, Mapping of mechanical properties of cement-based materials at micro/nano-scale. J. Innov. Eng. **2**(1) (2014)
37. G. Fang, M. Zhang, The evolution of interfacial transition zone in alkali-activated fly ash-slag concrete. Cem. Concr. Res. **129** (2020)
38. M. Husem, The effects of bond strengths between lightweight and ordinary aggregate-mortar, aggregate-cement paste on the mechanical properties of concrete. Mater. Sci. Eng. A **363**(1–2), 152–158 (2003)

39. J.A. King, I. Miskioglu, D.D. Wright-Charlesworth, C.D. Van Karsen, Nanoscratch testing to assess the fiber adhesion of short-carbon-fiber composites. J. Appl. Polym. Sci. **103**(1), 328–335 (2007)
40. G. Mondal, D.C. Rai, Performance of harbour structures in Andaman Islands during 2004 Sumatra earthquake. Eng. Struct. **30**(1), 174–182 (2008)
41. D. Davydov, M. Jirásek, L. Kopecký, Critical aspects of nano-indentation technique in application to hardened cement paste. Cem. Concr. Res. **41** (2011)
42. J.L. Provis, G.C. Lukey, J.S.J. Van Deventer, Do geopolymers actually contain nanocrystalline zeolites? a reexamination of existing results. Chem. Mater. **17**(12), 3075–3085 (2005)
43. R. San Nicolas, J.L. Provis, The interfacial transition zone in alkali-activated slag mortars. Front. Mater. **2** (2015)
44. C. Shi, P. Xie, Interface between cement paste and quartz sand in alkali-activated slag mortars. Cem. Concr. Res. **28**(6), 887–896 (1998)
45. K.L. Scrivener, Backscattered electron imaging of cementitious microstructures: understanding and quantification. Cement Concr. Compos. **26**(8), 935–945 (2004)
46. M. Vandamme, F.J. Ulm, P. Fonollosa, Nanogranular packing of C-S-H at substochiometric conditions. Cem. Concr. Res. **40**(1), 14–26 (2010)
47. F. Sanchez, K. Sobolev, Nanotechnology in concrete—a review. Constr. Build. Mater. **24** (2010)
48. K.L. Scrivener, R.J. Kirkpatrick, Innovation in use and research on cementitious material. Cem. Concr. Res. **38** (2008)
49. S.A. Syed Asif, K.J. Wahl, R.J. Colton, O.L. Warren, Quantitative imaging of nanoscale mechanical properties using hybrid nanoindentation and force modulation. J. Appl. Phys. **90** (2001)
50. W.G. Li, J.Z. Xiao, S. Kawashma, G.S. Shekhawat, S.P. Shah, Experimental investigation on quantitative nanomechanical properties of cement paste. ACI Mater. J. **111** (2014)
51. J. Němeček, V. Králík, J. Vondřejc, Micromechanical analysis of heterogeneous structural materials. Cem. Concr. Compos. **36** (2013)
52. A. Peled, J. Castro, J. Weiss, Atomic force microscopy examinations of mortar made by using water-filled lightweight aggregate. J. Transp. Res. Board **2141** (2010)
53. M.R. Vanlandingham, S.H. McKnight, G.R. Palmese, J.R. Elings, X. Huang, T.A. Bogetti, R.F. Eduljee, J. Gillespie, Nanoscale indentation of polymer systems using the atomic force microscope. J. Adhes. **64** (1997)
54. A. Hajimohammadi, T. Ngo, J.L. Provis, T. Kim, J. Vongsvivut, High strength/density ratio in a syntactic foam made from one-part mix geopolymer and cenospheres. Compos. Part B: Eng. **173** (2019)
55. P.R. Rangaraju, J. Olek, S. Diamond, An investigation into the influence of inter-aggregate spacing and the extent of the ITZ on properties of Portland cement concretes. Cem. Concr. Res. **40**(11), 1601–1608 (2010)
56. Y. Wei, W. Kong, Y. Wang, A. Sha, Multifunctional application of nanoscratch technique to characterize cementitious materials. Cement Concr. Res. **140** (2021)
57. Z. Luo, W. Li, Y. Gan, K. Mendu, S.P. Shah, Applying grid nanoindentation and maximum likelihood estimation for N-A-S-H gel in geopolymer paste: investigation and discussion. Cement Concr. Res. **135** (2020)
58. Z. Zhang, J.L. Provis, A. Reid, H. Wang, Fly ash-based geopolymers: the relationship between composition, pore structure and efflorescence. Cem. Concr. Res. **64**, 30–41 (2014)
59. C.G. Hoover, F.J. Ulm, Experimental chemo-mechanics of early-age fracture properties of cement paste. Cem. Concr. Res. **75**, 42–52 (2015)
60. M. Khedmati, Y.R. Kim, J.A. Turner, H. Alanazi, C. Nguyen, An integrated microstructural-nanomechanical-chemical approach to examine material-specific characteristics of cementitious interphase regions. Mater Charact **138**, 154–164 (2018)
61. M. Wang, L. Liu, Investigation of microscale aging behavior of asphalt binders using atomic force microscopy. Constr. Build. Mater. **135**, 411–419 (2017)
62. W. Li, S. Kawashima, J. Xiao, D.J. Corr, C. Shi, S.P. Shah, Comparative investigation on nanomechanical properties of hardened cement paste. Mater. Struct./Materiaux et Constructions **49**(5), 1591–1604 (2016)

63. H. Lee, V. Vimonsatit, P. Chindaprasirt, Mechanical and micromechanical properties of alkali activated fly-ash cement based on nano-indentation. Constr. Build. Mater. **107**, 95–102 (2016)
64. T. Li, J. Xiao, C. Zhu, Z. Zhong, Experimental study on mechanical behaviors of concrete with large-size recycled coarse aggregate. Constr. Build. Mater. **120**, 321–328 (2016)
65. J.F. Lamond, R.L. Campbell Sr, A. Giraldi, N. Jenkins, T.R. Campbell, W. Halczak, R. Miller, J.A. Cazares, H.C. Hale Jr, P.T. Seabrook, Removal and reuse of hardened concrete. American Concrete Institute: Farmington Hills, MI, USA **26** (2001)
66. G. Li, Y. Zhao, S.S. Pang, Four-phase sphere modeling of effective bulk modulus of concrete. Cem. Concr. Res. **29**(6), 839–845 (1999)
67. H. Lee, V. Vimonsatit, P. Chindaprasirt, T. Ngo, P. Mendis, Creep properties of cement and alkali activated fly ash materials using nanoindentation technique. Constr. Build. Mater. **168**, 547–555 (2018)
68. C. Hu, Z. Li, Property investigation of individual phases in cementitious composites containing silica fume and fly ash. Cement Concr. Compos. **57**, 17–26 (2015)
69. Z. Luo, W. Li, K. Wang, A. Castel, S.P. Shah, Comparison on the properties of ITZs in fly ash-based geopolymer and Portland cement concretes with equivalent flowability. Cement Concr. Res. **143** (2021)
70. Y. Wei, X. Gao, S. Liang, A combined SPM/NI/EDS method to quantify properties of inner and outer C-S-H in OPC and slag-blended cement pastes. Cement Concr. Compos. **85**, 56–66 (2018)
71. M. Zhou, Y. Jiang, Y. Sui, Microstructure and properties of interfacial transition zone in ZTA particle–reinforced iron composites. Appl. Phys. A: Mater. Sci. Process. **125**(2) (2019)
72. Z.-Z. Wang, P. Gu, X.-P. Wu, H. Zhang, Z. Zhang, M.Y.M. Chiang, Micro/nano-wear studies on epoxy/silica nanocomposites. Compos. Sci. Technol. **79**, 49–57 (2013)
73. J. Xu, D.J. Corr, S.P. Shah, Nanomechanical investigation of the effects of nanoSiO_2 on C-S-H gel/cement grain interfaces. Cement Concr. Compos. **61**, 7–17 (2015)
74. W. Li, Z. Luo, Y. Gan, K. Wang, S.P. Shah, Nanoscratch on mechanical properties of interfacial transition zones (ITZs) in fly ash-based geopolymer composites. Compos. Sci. Technol. **214** (2021)
75. Z. Liu, C.S. Cai, H. Peng, F. Fan, Experimental study of the geopolymeric recycled aggregate concrete. J. Mater. Civ. Eng. **28**(9) (2016)
76. G. Mallikarjunachari, P. Ghosh, Analysis of strength and response of polymer nano thin film interfaces applying nanoindentation and nanoscratch techniques. Polymer **90**, 53–66 (2016)
77. W. Li, Z. Huang, T. Zu, C. Shi, W.H. Duan, S.P. Shah, Influence of nanolimestone on the hydration, mechanical strength, and autogenous shrinkage of ultrahigh-performance concrete. J. Mater. Civ. Eng. **28**(1) (2016)
78. J. Xu, W. Yao, Nano-scratch as a new tool for assessing the nano-tribological behavior of cement composite. Mater. Struct./Materiaux et Constructions **44**(9), 1703–1711 (2011)
79. Z. Luo, W. Li, K. Wang, S.P. Shah, Research progress in advanced nanomechanical characterization of cement-based materials. Cement Concr. Compos. **94**, 277–295 (2018)
80. P.G. Allison, C.A. Weiss, R.D. Moser, A.J. Diaz, O.G. Rivera, S.S. Holton, Nanoindentation and SEM/EDX characterization of the geopolymer-to-steel interfacial transition zone for a reactive porcelain enamel coating. Compos. B Eng. **78**, 131–137 (2015)
81. B. Nematollahi, J. Qiu, E.H. Yang, J. Sanjayan, Microscale investigation of fiber-matrix interface properties of strain-hardening geopolymer composite. Ceram. Int. **43**(17), 15616–15625 (2017)
82. J. Liu, R. Jiang, J. Sun, P. Shi, Y. Yang, Concrete damage evolution and three-dimensional reconstruction by integrating CT test and fractal theory. J. Mater. Civ. Eng. **29**(9) (2017)
83. A. Karimzadeh, M.R. Ayatollahi, Investigation of mechanical and tribological properties of bone cement by nano-indentation and nano-scratch experiments. Polym. Testing **31**(6), 828–833 (2012)

Chapter 3
Sample Preparation and Requirements for Micromechanical Property Tests

3.1 Introduction

The concept of nanoindentation testing is based on infinite half-space model and assumes that the surface is perfectly flat and smooth [1]. Some researchers have revealed that sample preparation has significant effects on the nanomechanical properties obtained from nanoindentation tests. Besides, the negative influence of high surface roughness on the results obtained by other new techniques like modulus mapping, PeakForce QNM and nanoscratch was also found. Therefore, the roughness of the sample surface determines whether accurate and meaningful data can be obtained in micromechanical property tests. However, the challenge faced by researchers is how to achieve a tolerable level of surface roughness for complex heterogeneous concrete and determine the corresponding relationship between different roughness and the degree of influence on the test results.

3.2 Sample Surface Processing Procedure

Along with the increasing use of nanoindentation technique, many researchers have explored the proper sample preparation methods. Miller et al. [2] put forward that the major objectives of sample preparation are to make the surface of sample as flat as possible, to minimize sample perturbation, and to acquire reliable and repeatable results. Then, Miller et al. [2] adopted a three-step sample preparation procedure to achieve these objectives. As shown in Fig. 3.1, these three steps were: (1) to slice the sample to the suitable size by a diamond saw; (2) to make the two ends of the sample parallel to each other by grinding the sample on 120 grit ZirMet (Buehler)\paper; and (3) to polish the sample surface flat and smooth with a hard, perforated pad charged with 1.0 μm oil-based diamond suspension for 8 h with a polishing speed of 18–25 cm/s. During the surface preparation procedure, the sample was cleaned

W. Li et al., *Nanocharacterization of Cementitious Composites*,
https://doi.org/10.1007/978-981-96-2917-6_3

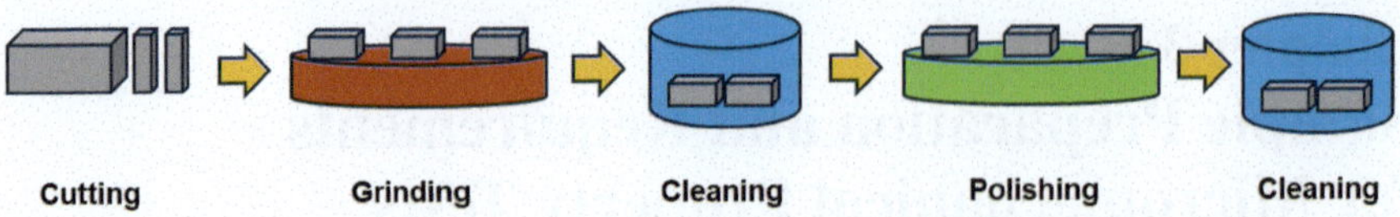

Fig. 3.1 Cementitious material sample preparation procedure for micromechanical properties tests

with ultrasonic bath in n-decane for 5 min. Similar sample preparation methods were widely used in other studies.

With the development of instrument technology, the processing of cement-based samples has become more automated, effectively saving time and labour costs. The photos of main facilities for sample surface processing are shown in Fig. 3.2. The precision diamond cutter and grinder can maintain a reasonable force and speed to process sample evenly. Moreover, they can continuously spray water onto the sample surface during operation to ensure that the gels are not destroyed by high temperatures. Epoxy resin mounting equipment is commonly used for embedding cement-based materials in resin for preservation, sectioning, or preparation for microscopic analysis. Epoxy resin provides a durable and stable mount, ensuring the sample remains intact during polishing or cutting. The mounting equipment usually possess a vacuum embedding system, which is used to ensure that epoxy resin penetrates all voids and cracks within the cement-based materials. This is crucial when preparing samples for microscopy or other detailed analyses. Ultrasonic bath can be timed to clean the sample surface from non-adherent particulate matter.

In recent years, many attempts have been made to improve these sample preparation procedures. Sakulich et al. [3] used a series of diamond suspensions with graded sizes (15, 9, 6, 3, 1 and 0.5 μm) for fine polishing. Others polished their samples using even smaller size range of diamond suspensions, such as 3–0.25 μm, 6–0.1 μm and 6–0.05 μm [4, 5]. Furthermore, prior to grinding and polishing, Xiao et al. [6] used vacuum impregnation of concrete slice with epoxy resin as shown in Fig. 3.3 to fill the pores and avoid the damage to microstructure in the subsequent processes. After cured in an oven at 70 °C for 12 h, the epoxy resin impregnated sample was mounted on a metal holder, and then the metal holder was placed in an annular metal collar to prevent wobbling and maintain parallel top and bottom surfaces. After being ground by Buehler-Met II paper disks with gradation of grits 280, 800, 1200 and 1500, the sample was further polished by an auto-polisher with ethanol-based diamond suspensions, with a grit size from 9, 6–3 μm. Then, a manual and semi-automated polish by diamond lapping films with a grit size of 3, 1 and 0.5 μm was also adopted in this sample preparation. Satisfactory results were subsequently obtained as shown in Fig. 3.4a, b. Epoxy resin is often used to protect microstructure during polishing of samples [7]. However, it could also stiffen the microstructure. In addition, applying of vacuum may also result in some undesirable changes in the microstructure of the samples studied.

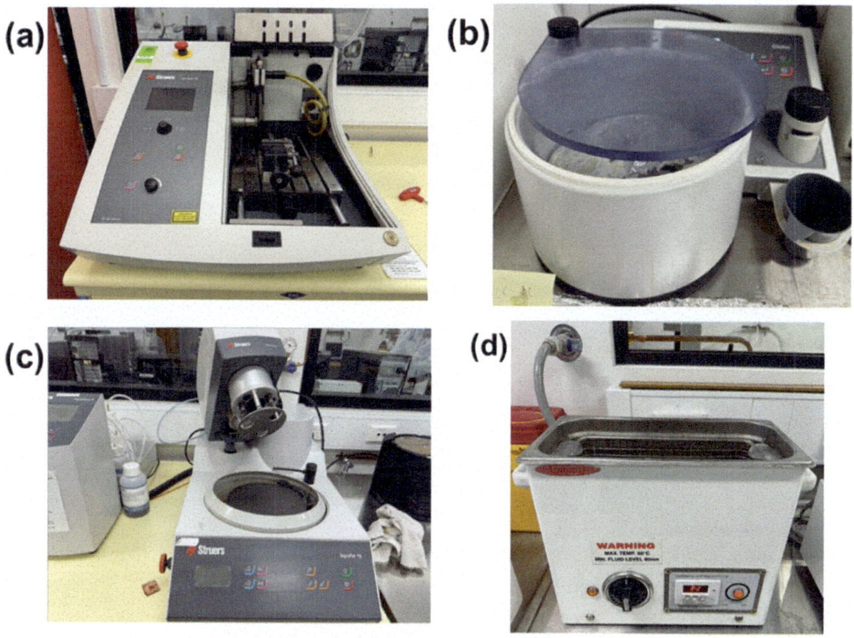

Fig. 3.2 Photos of sample processing facilities: **a** diamond cutter; **b** mounting equipment; **c** grinding and polishing machine; **d** ultrasonic bath

Fig. 3.3 Concrete sample impregnated with epoxy resin [6]

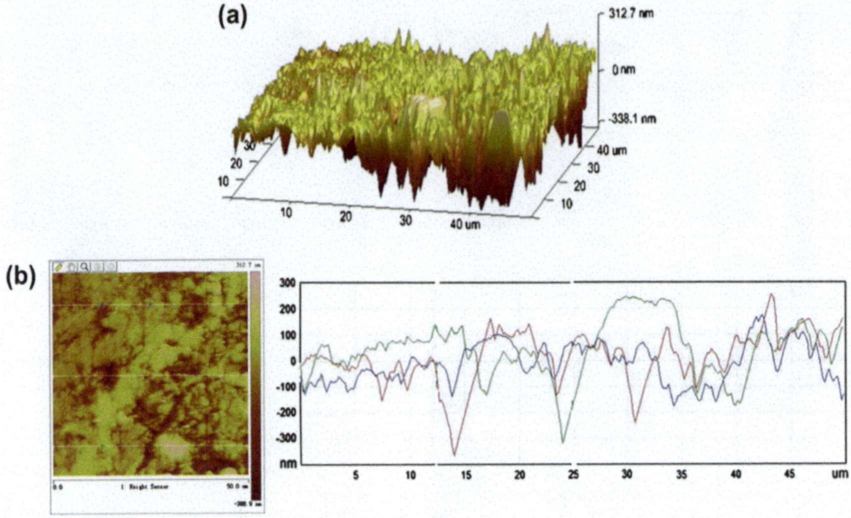

Fig. 3.4 Surface of cement paste sample after preparation detected by AFM [6]: **a** AFM 3D topography image of ITZ; **b** ITZ height distribution

3.3 Sample Surface Roughness Detection and Evaluation

After ground and polished, the sample is then usually examined under AFM. Based on the surface topography, the surface roughness of the tested sample can be calculated. As shown in Eq. 3.1, a root-meansquare (RMS) method is widely adopted for the roughness calculation [8]. Other methods like arithmetic average of the roughness profile in Eq. 3.2 have also been used [9]. Then, some criterions are applied to determine whether or not the surface situation of the samples is acceptable for nanoindentation. A RMS roughness criterion proposed by Miller et al. [2] has been widely used. This criterion indicates that the scanning size should be at least 200 times more than the average depth h of the first phase (the most left phase in PDF curves), thus making the measurement size on an order greater than the length scale of the largest material heterogeneity. In addition, the h should be more than 5 times of the RMS roughness to assure the self-similar properties of indentation analysis. When expanded applying this criterion to some other studies, the average depth has not been emphasized or limited to the specific first phase [10].

$$\mathrm{RMS} = \sqrt{\frac{1}{n \cdot m}\sum_{i=1}^{n}\sum_{j=1}^{m}\left(h_{ij} - h_{\mathrm{mean}}\right)^2} \tag{3.1}$$

$$R_a = \frac{1}{N}\sum_{j=1}^{N}\left|Z_j - Z_m\right| \tag{3.2}$$

where n, m represent pixel size of the image; h_{ij} is a height reading in pixel (i, j); h_{mean} is the mean value of all height obtained; N is the amount of data points; the Z_j is the height at a certain point j and Z_m is the mean value of height for the specific sample.

However, the existing criteria for evaluating the sample surface condition are considered to have some limitations. It is well known that mechanical polishing could cause work-hardening and residual stress for the surface preparation of metal samples [11]. These adverse effects are also analysed, for instance assessing of residual stress using XRD, to help understand the influence of surface preparation on the nanomechanical testing results. Similar influence is not significant for the nanoindentation test of brittle materials like hardened cement paste. However, there are other concerns. One is that the mechanical polishing procedures involve artefacts (appearing of detritus in pores or collapsing of pores et al.) for cement-based materials, which has been suspected to bring errors in surface roughness evaluation and nanomechanical testing but not to be considered in the above RMS roughness criterion. In order to assess the surface roughness of hardened cement paste without the influence of the normal sample preparation, Trtik et al. [1] studied the surface roughness of hardened cement paste using a virtual topographic experiment based on focused ion beam nanotomography (FIB-nt) datasets, and they found that their obtained surface roughness based on the ideal artefact-free surface was higher than those obtained using AFM for real prepared samples from published literature. Considering the differences in measurements, Trtik et al. [1] proposed that in addition to the upper limit, the lower limit of RMS surface roughness should also be stipulated. The need for a lower limit was due to the intrinsic porosity of porous materials, which could generate an inevitable roughness. When the roughness value was lower than the lower limit, it may indicate the production of artefacts during sample preparation. Besides, Trtik et al. [1] put forward that the shape of the tip in AFM investigation, the threshold setting in FIB-nt technique, and the size, resolution and the position of the images would affect the RMS value.

Researchers proposed that proper surface preparation is also required for quantitative modulus mapping, PeakForce QNM and nanoscratch testing [12–14]. In some extent, the surface requirements for some of them may be more stringent because of the shallow penetration depth. Currently, surface preparations are normally based on the above cutting, grinding, polishing and cleaning preparation procedures previously developed for nanoindentation. RMS roughness criterion and the empirical results of researchers have been used for roughness evaluation.

Li et al. [15] conducted nanomechanical investigation on hardened cement paste by nanoindentation, PeakForce QNM and modulus mapping test. Using the same as the preparation procedures in Xiao et al. [6], Li et al. [15] ground the sample using paper disks, then an autopolisher with ethanol-based diamond suspension was used. In the last step, diamond lapping films was employed for further polishing. The surface of cement paste sample after surface preparation is shown in Fig. 3.4b. Comparing with other research results, Li et al. [15] found that the roughness of their samples was sufficiently low, and they could be satisfactorily used for all the above-mentioned nanomechanical characterizations.

Fig. 3.5 Condensed cement paste sample production procedure [16]: **a** Rectangular cement paste sample; **b** A thin lamella extracted by high-current beam of FIB; **c** The sample surface polished by low-current beam of FIB

There are also advanced technologies adopted for sample preparation, for instance, the focussed ion beam (FIB) technology [12, 16], as shown in Fig. 3.5. Trtik et al. [12] adopted the focussed ion beam (FIB) technology in sample surface preparation for PeakForce QNM test. They found that the RMS surface roughness reached a high value of 38.5 nm in a 20 μm × 20 μm area. To a large extent, the relative high value of RMS surface roughness could be attributed to the FIB waterfall artefacts [12]. The RMS roughness study found that the RMS roughness in areas of 156 nm × 156 nm achieved low mean and median values of 5.3 nm and 3.3 nm, respectively. In addition to surface preparation, FIB, with an ability of precise micromachining, has also been used to implement other challenging tests, for instance, fabricating micro-beams samples for bending test loaded by nanoindenter or AFM in hardened cement paste [17]. During the use of FIB, attentions shall be paid for potential interference created from the surface preparation and micromachining of samples. Re-deposition of sputtered materials, Ga^+ ion damage and heat damage of microstructure may occur, which may affect test results, especially for surface nanomechanical testing [18]. The statistical sample roughness results for modulus mapping and PeakForce QNM test in cement-based materials studies are displayed in Tables 3.1 and 3.2, respectively. For nanoscratch test, only Xu et al. [19] and Mao et al. [20] provided the RMS roughness value, which were 23.5 nm in scanning size of 20 μm × 20 μm and obviously below 400 nm, respectively. This may be partially because the prescratch scan can provide surface roughness information.

3.4 Effects of Surface Preparation on Nanomechanical Test Results

In theory, flat surface is vital for nanoindentation test to avoid introducing surface roughness as another length scale in the similarity analysis [27]. Further studies also demonstrated that the nanoindentation test is sensitive for the roughness of the sample. Some studies revealed that the increase of surface roughness leads to

Table 3.1 Surface roughness for modulus mapping testing

Literature	RMS (nm)	Scanning size (μm)	Displacement amplitudes (nm)	Testing materials
Xu et al. [21]	59.4 ± 17.0	40 × 40	0.7–1.0	C–S–H gel in OPC paste (w/c = 0.28)
Xu et al. [21]	12.7 ± 7.3	15 × 15	0.5–0.6	Cement grain in OPC paste (w/c = 0.28)
Xu et al. [22]	23.5 (mean)	20 × 20	–	OPC paste; OPC paste with 1% CNS (w/b = 0.28)
Li et al. [13]	93.9 ± 20	50 × 50	7–12	Epoxy-impregnated OPC paste (w/c = 0.4)
Wei et al. [23, 24]	Below 40	35 × 35	0.5–1.0	OPC paste; Slag-blended paste (w/b = 0.3)
Gao et al. [25]	30–50 (mean)	2.2 × 2.2	0.5–4	OP, IP and C3S phases in OPC paste (w/c = 0.3)

Table 3.2 Surface roughness for PeakForce QNM testing

Literature	RMS (nm)	Scanning size (μm)	Contact depth (nm)	Test materials
Trtik et al. [12]	38.5	20 × 20	–	Epoxy-impregnated OPC paste (w/c = 0.5)
Trtik et al. [12]	5.3 (mean)	0.156 × 0.156	–	Epoxy-impregnated OPC paste (w/c = 0.5)
Li et al. [13]	93.9 ± 20	50 × 50	2–8	Epoxy-impregnated OPC paste (w/c = 0.4)
Zhu et al. [26]	87.4	50 × 50	–	OPC paste with 0.1% CNF (w/c = 0.51)
Zhu et al. [26]	96.6	50 × 50	–	OPC paste (w/c = 0.51)
Asgari et al. [16]	8.31–18.2	10 × 10	–	Epoxy-impregnated OPC paste with NS

increase the scatter degree in indentation results [28]. Decrease of nanomechanical properties with the increase of roughness was also found by some researchers [29]. However, there are different opinions about the influence of surface roughness on nanomechanical testing results by the new techniques. Trtik et al. [12] pointed out that the surface roughness would affect the nanomechanical measurements of multiphase materials. Because of intrinsic porosity, the unimpregnated hardened cement pastes dooms to have relatively high roughness, which leads to some challenges for use of PeakForce QNM for it. Howind et al. [30] proposed that due to the small and sharp tip in PeakForce QNM test, the surface roughness would not influence the test results significantly and the modulus results are mainly affected by the nanoscale defects. For modulus mapping test, Xu et al. [22] suggested that the rougher surface would cause higher feedback error signal in measurement. However, experimental investigation by Gao et al. [25] found that the surface roughness seemed not to significantly influence the measured storage modulus.

In addition to surface roughness, other factors like moisture content in samples, which could be affected by the sample preparation (such as using water for cutting, polishing and cleaning), would also influence the results of the subsequent nanomechanical analysis. Moisture was reported to have a significant impact on elastic modulus, viscoelastic performance, strength and interface strength of C–S–H [31–33]. Besides, the water used in sample preparation could affect the hydration of cement in the samples. Furthermore, comparison results revealed that diamond suspensions in water had advantage over diamond suspensions in oil for samples cleaning, and clear samples in water was a feasible way; but damage was found for sample cleaned more than 3 min [5]. Therefore, it is very important to control the water in sample preparation procedures to achieve expected testing condition and unbiased nanomechanical testing results. With increasing applications of various kinds of nanomechanical techniques into cement-based materials research, more efficient procedures for sample surface preparation are expected to be established. More rigorous criteria for different nanomechanical testing conditions and different testing techniques shall be developed. It shall also be noted that whether the surface assessed by AFM or virtual topographic experiment based on FIB-nt datasets, only several positions in dozens of microns are chosen. Therefore, the roughness measured with many positions is necessary to reflect the overall information of the heterogeneous, porous cement-based materials.

3.5 Summary

The new advanced nanomechanical characterization techniques including modulus mapping, PeakForce QNM and nanoscratch have provided new horizons for micro- and nano-scale characterization of cement-based materials. The current surface preparation procedure and roughness evaluation criteria of the samples for modulus mapping, PeakForce QNM and nanoscratch tests are primarily based on the previously established ones for nanoindentation. The influences of the sample surface roughness on the modulus mapping and PeakForce QNM tests are still controversial. As an important source of moisture in samples, water used in sample preparation (cutting, surface polishing and cleaning) should be considered according to the required test conditions to prevent its influence on nanomechancal test results. The inherent limitations processing facilities may also introduce micro-cracks or disrupt delicate microstructures in porous or brittle cementitious samples. In addition, the softer components such as C–S–H gel and N–A–S–H gel can smear across the surface due to overtime grinding and polishing, leading to inaccurate imaging or chemical analysis. Therefore, sample processing method should be carefully matched with the intended test, keeping in mind the specific properties of the cementitious materials and the sensitivity of the microscopic techniques. A combination of methods may often be needed to overcome individual limitations.

References

1. P. Trtik, J. Dual, B. Muench, L. Holzer, Limitation in obtainable surface roughness of hardened cement paste: 'virtual' topographic experiment based on focussed ion beam nanotomography datasets. J. Microsc. **232**(2), 200–206 (2008)
2. M. Miller, C. Bobko, M. Vandamme, F.-J. Ulm, Surface roughness criteria for cement paste nanoindentation. Cem. Concr. Res. **38**, 467–476 (2008)
3. A.R. Sakulich, V.C. Li, Nanoscale characterization of engineered cementitious composites (ECC). Cem. Concr. Res. **41**, 169–175 (2011)
4. C. Hu, B. Xu, H. Ma, B. Chen, Z. Li, Micromechanical investigation of magnesium oxychloride cement paste. Constr. Build. Mater. **105**, 496–502 (2016)

5. P. Mondal, S.P. Shah, L.D. Marks, Nanoscale characterization of cementitious materials. ACI Mater. J. **105**(2), 174–179 (2008)
6. J. Xiao, W. Li, Z. Sun, D.A. Lange, S.P. Shah, Properties of interfacial transition zones in recycled aggregate concrete tested by nanoindentation. Cement Concr. Compos. **37**, 276–292 (2013)
7. S. Zhao, W. Sun, Nano-mechanical behavior of a green ultra-high performance concrete. Constr. Build. Mater. **63**, 150–160 (2014)
8. J. Němeček, Creep effects in nanoindentation of hydrated phases of cement pastes. Mater Charact **60**, 1028–1034 (2009)
9. M.S.I. Zyganitidis, N. Kalfagiannis, S. Logothetidis, Nanomechanical characterization of cement-based pastes enriched with SiO_2 nanoparticles. Mater. Sci. Eng. B **176**, 1580–1584 (2011)
10. C. Hu, Z. Li, A review on the mechanical properties of cement-based materials measured by nanoindentation. Constr. Build. Mater. **90**, 80–90 (2015)
11. A.C. Fischer-Cripps, *Factors Affecting Nanoindentation Test Data* (Springer, New York, NY, 2011)
12. P. Trtik, J. Kaufmann, U. Volz, On the use of peak-force tapping atomic force microscopy for quantification of the local elastic modulus in hardened cement paste. Cem. Concr. Res. **42**, 215–221 (2012)
13. W. Li, S. Kawashima, J. Xiao, D.J. Corr, C. Shi, S.P. Shah, Comparative investigation on nanomechanical properties of hardened cement paste. Mater. Struct. **49**, 1591–1604 (2016)
14. S. Zhao, E.V. Dam, D. Lange, W. Sun, Abrasion resistance and nanoscratch behaviour of an ultra-high performance concrete. J. Mater. Civ. Eng. **29**(2), 04016212 (2017)
15. W. Li, J. Xiao, S. Kawashima, G.S. Shekhawat, S.P. Shah, Experimental investigation on quantitative nanomechanical properties of cement paste. ACI Mater. J. **112**(2), 229–238 (2015)
16. H. Asgari, A. Ramezanianpour, H.-J. Butt, Effect of water and nano-silica solution on the early stages cement hydration. Constr. Build. Mater. **129**, 11–24 (2016)
17. S. Chen, W. Duan, Z. Li, T. Sui, New approach for characterisation ofmechanical properties of cement paste at micrometre scale. Mater. Des. **87**, 992–995 (2015)
18. J. Nemecek, V. Kralík, V. Smilauer, L. Polívka, A. Jager, Tensile strength of hydrated cement paste phases assessed by microbending tests and nanoindentation. Cement Concr. Compos. **73**, 164–173 (2016)
19. J. Xu, D.J. Corr, S.P. Shah, Nanoscratch study of the modification effects of nano SiO_2 on C–S–H gel/cement grain interfaces. J. Mater. Civ. Eng. **29**(9), 04017093 (2017)
20. Y. Mao, W. Yao, J. Xu, Study on the unhydrated cement grain/C–S–H gel interface in cement paste by use of nano-scratch technique. Key Eng. Mater. **539**, 84–88 (2013)
21. J. Xu, D.J. Corr, S.P. Shah, Nanomechanical properties of C–S–H gel/cement grain interface by using nanoindentation and modulus mapping. J. Zhejiang Univ. -Sci. **16**(1), 38–46 (2015)
22. J. Xu, D.J. Corr, S.P. Shah, Nanomechanical investigation of the effects of nano SiO_2 on C–S–H gel/cement grain interfaces. Cement Concr. Compos. **61**, 7–17 (2015)
23. Y. Wei, S. Liang, X. Gao, Phase quantification in cementitious materials by dynamic modulus mapping. Mater Charact **127**, 348–356 (2017)
24. Y. Wei, X. Gao, S. Liang, A combined SPM/NI/EDS method to quantify properties of inner and outer C–S–H in OPC and slag-blended cement pastes. Cement Concr. Compos. **85**, 56–66 (2018)
25. X. Gao, Y. Wei, W. Huang, Critical aspects of scanning probe microscopy mapping when applied to cement pastes. Adv. Cem. Res. **30**(7), 293–304 (2018)
26. X. Zhu, Y. Gao, Z. Dai, D.J. Corr, S.P. Shah, Effect of interfacial transition zone on the Young's modulus of carbon nanofiber reinforced cement concrete. Cem. Concr. Res. **107**, 49–63 (2018)
27. G. Constantinides, F.-J. Ulm, The nanogranular nature of C–S–H. J. Mech. Phys. Solid. **55**(1), 64–90 (2007)
28. Y. Xia, M. Bigerelle, J. Marteau, P.E. Mazeran, S. Bouvier, A. Iost, Effect of surface roughness in the determination of the mechanical properties of material using nanoindentation test. Scanning **36**(1), 134–149 (2014)

29. L. Chen, A. Ahadi, J. Zhou, J.-E. Stahl, Modeling effect of surface roughness on nanoindentation tests. Procedia CIRP **8**, 334–339 (2013)
30. T. Howind, J. Hughes, W. Zhu, Mapping of mechanical properties of cement-based materials at micro/nano-scale. J. Innovative Eng. **2**(1) (2014)
31. C. Hu, Z. Li, Property investigation of individual phases in cementitious composites containing silica fume and fly ash. Cement Concr. Compos. **57**, 17–26 (2015)
32. R. Alizadeh, J.J. Beaudoin, L. Raki, Viscoelastic nature of calcium silicate hydrate. Cement Concr. Compos.Concr. Compos. **32**(5), 369–376 (2010)
33. C.A. Jones, *Contact Mechanics Based Mechanical Characterization of Portland Cement Paste* (Texas A&M University, 2011)

Chapter 4
Comparative Investigations on Nanomechanical Properties of Hardened Cement Paste

4.1 Introduction

Quantitative characterization of the nanomechanical properties of cement-based materials is essential, providing not only a better understanding of material behavior but also a basis for the design and preparation of novel cement composites with well-defined properties [1, 2]. Nanomechanical test methods can assess the properties and nanostructure of individual constituents in heterogeneous materials, providing valuable experimental data for multi-scale numerical modelling parameters. Nanoindentation is one of the most widely adopted techniques for this purpose of nanoscale characterization, and it determines elastic modulus and hardness based on the measured indenting force–displacement curve. Among various methods to calculate the elastic modulus and hardness from the curve, Oliver and Pharr's method has been widely adopted because of its reliable/concise hypothesis [3, 4]. The modulus mapping technique employs aspects of nanoDMA and in situ SPM imaging to realise characterization capabilities to oscillate the indenter tip with small forces while monitoring the resultant displacement and phase lag due to material responses [4, 5]. In situ SPM imaging allows the indenter tip to be rasterized across and produce a topographic image of the material surface. Using the modulus mapping technique, contact stiffness can be obtained continuously during scanning using a smaller contact force, which could be an ideal technique for multiphase and composite materials. The peak-force QNM is a new operating mode of AFM technique that can operate with a wide variety of standard AFM probes [6, 7]. In peak-force QNM, the probe is oscillated at a typical frequency of 2 kHz (far below the resonance of the cantilever) with typical peak-to-peak amplitudes in air of 300 nm. This brings the probe periodically into contact with the sample for a short period of time (less than 100 ls) and a periodically oscillating force is exerted onto the sample surface. Peak-force QNM has been recognized as a useful method for the assessment of local mechanical properties of materials, which can also provide phase contrast maps for qualitative characterization of materials with a very high spatial resolution [8].

W. Li et al., *Nanocharacterization of Cementitious Composites*,
https://doi.org/10.1007/978-981-96-2917-6_4

4.2 Detailed Information on Nanoindentation, Modulus Mapping and Peak-Force QNM Devices

In this section, the nanomechanical properties of cement pastes determined from three different methods, nanoindentation, modulus mapping and peak-force QNM, and the test results are compared. The merits of these methods are specifically discussed for the determination of the elastic modulus of cement paste from a 1-μm length scale to a tens of nanometres scale. These measurements can help clarify the multiple modal distributions of nanomechanical properties in cement paste.

4.2.1 *Nanoindentation*

A static nanoindentation experiment can be performed with a nanomechanical testing system (Hysitron) equipped with an in-place imaging mode (Fig. 4.1) [9, 10]. The nanoindentation instrument consists of a force–displacement transducer with electrostatic force actuation and displacement-sensing electronics. A two-dimensional Berkovich indenter with a high elastic modulus is used. The system is calibrated with standard quartz crystal before performing the actual indentation on the sample to ensure that the probe is not contaminated or damaged.

A typical representation of the indentation load versus indentation depth or displacement for an indentation experiment is shown in Fig. 4.2. The elastic modulus E of the sample is calculated as follows:

$$S = \frac{\mathrm{d}P}{\mathrm{d}h} = \frac{2}{\sqrt{\pi}} E_r \sqrt{A} \tag{4.1}$$

$$E_s = \left(1 - v_s^2\right)\left[\frac{1}{E_r} - \frac{1 - v_i^2}{E_i}\right]^{-1} \tag{4.2}$$

where P is the indentation load; h is the indentation depth; S is the initial unloading stiffness; A is the projected contact area; E_r is the reduced elastic modulus; E_s and v_s are elastic modulus and Poisson's ratio for the sample, respectively; and E_i and v_i are the corresponding parameters for the indenter, respectively. For the indenter used in this study, the elastic modulus $E_i = 1140$ GPa, and the Poisson's ratio $v_i = 0.07$.

4.2.2 *Modulus Mapping*

Quantitative modulus mapping in the form of SPM images is acquired using the direct force modulation (nanoDMA) operating mode of the nanomechanical testing

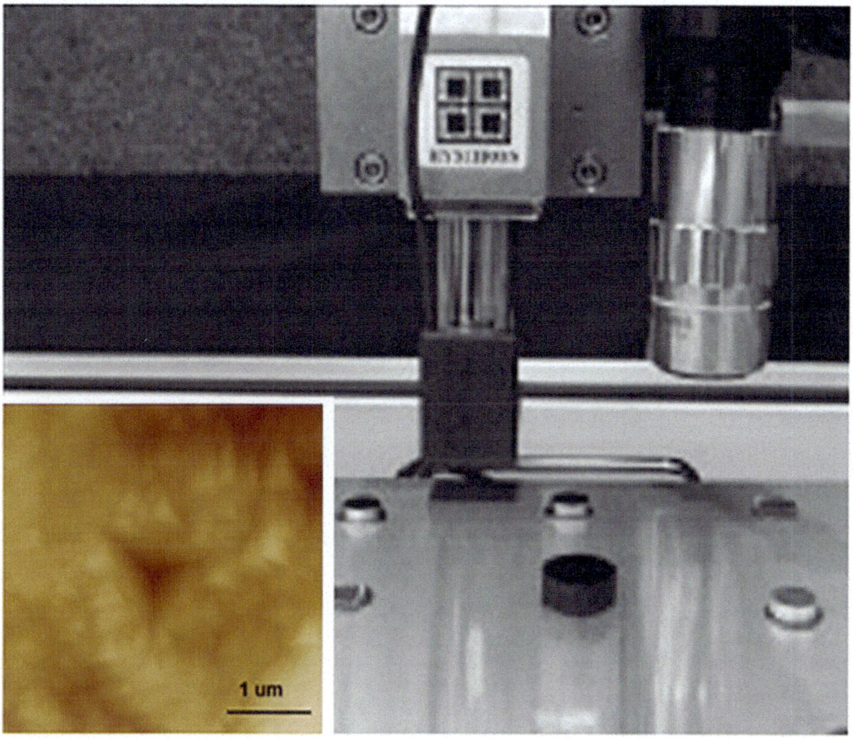

Fig. 4.1 Nanoindentation device

Fig. 4.2 Typical representation of indentation load versus indentation depth

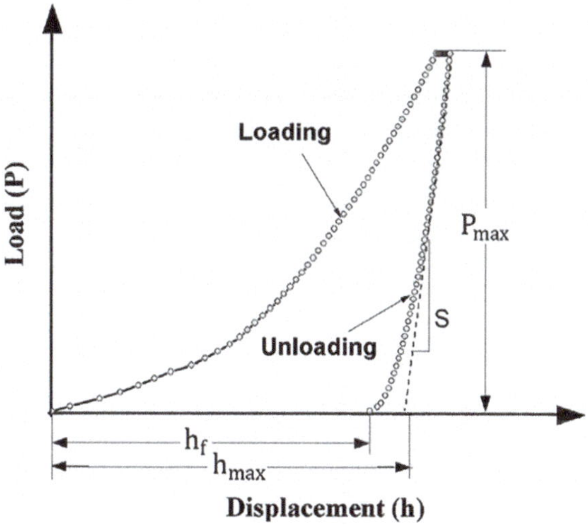

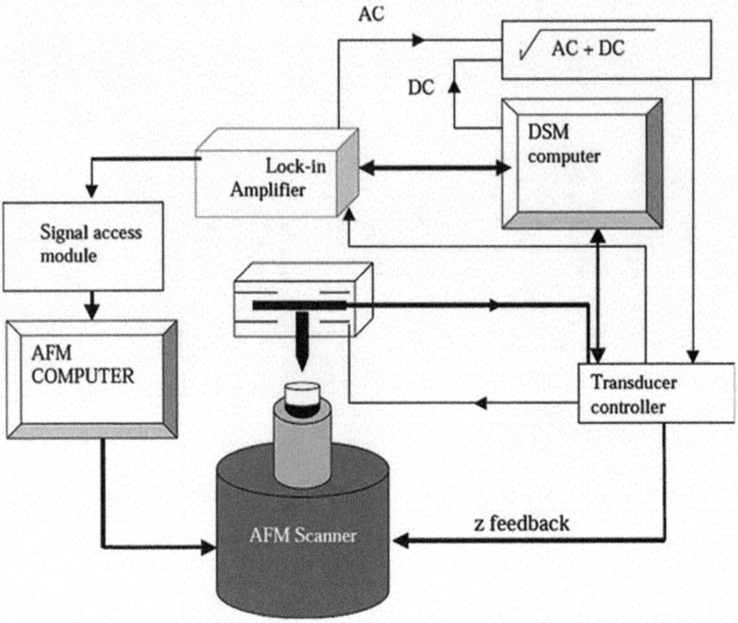

Fig. 4.3 Schematic of modulus mapping technique [9]

system [11, 12]. The schematic representation of the instrument configuration and dynamical mode are shown in Fig. 4.3. A cube-corner diamond tip is attached to a force–displacement transducer. As with nanoindentation, the system was calibrated with a standard quartz crystal. A lock-in amplifier is used to analyse the sample response, yielding displacement amplitude, and the phase shift between the contact force and the displacement. The contact stiffness and material damping are calculated from the amplitude and phase shift using a dynamic model.

To apply a load, a voltage is applied to generate an electrostatic force F_e. The mode of direct force modulation applies a parallel loading configuration. The relevant equation of motion is:

$$F_e(t) = F_0 \sin(\omega t) = m\left(\frac{dz^2}{dt^2}\right) + \beta\left(\frac{dz}{dt}\right) + kz \tag{4.3}$$

where ω is the angular frequency; F_0 is the amplitude of the applied sinusoidal force; m is the mass of the moving transducer; $\beta = \beta_i + \beta_s$ is the combined damping coefficient of the indenter and the sample; $k = k_i + k_c$ is the combined stiffness of the indenter and the sample; z is the resulting sinusoidal displacement; and t is time. The solution to this differential equation is

$$z = z_0 \sin(\omega t + \varphi) \tag{4.4}$$

where the displacement amplitude z_0 is given by the dynamical compliance function

$$\frac{z_0}{F_0} = \left[\left(k - m\omega^2\right)^2 + (\beta\omega)^2\right]^{-\frac{1}{2}} \tag{4.5}$$

where the phase φ of the displacement is dictated by

$$\tan(\varphi) = -\frac{\beta\omega}{k - m\omega^2} \tag{4.6}$$

Determination of m, β, and k_i is accomplished by nonlinear curve fitting of the dynamic compliance versus the frequency trace obtained in air. The storage K' and loss K'' components of the complex contact stiffness are extracted from the spatially correlated set of amplitude and phase images to generate quantitative stiffness maps

$$K' = k_c \tag{4.7}$$

$$K'' = \beta_c\omega \tag{4.8}$$

Rearrangement of the familiar equations of Hertzian theory yields the following relationship for the contact radius a

$$a = \sqrt{\frac{3FR}{2K'}} \tag{4.9}$$

where F corresponds to the nominal contact force, and R is the tip radius. The modulus then can be calculated as

$$E_s = \frac{1}{2}\frac{K'}{a} \tag{4.10}$$

In static nanoindentation, a minimum separation of 3 μm between indentations is required to avoid possible interactions between adjacent indentions. The modulus mapping technique in the form of scanning probe microscopy (SPM) images, however, can overcome these shortcomings. This is because the material under the tip is probed within the elastic region, and once the tip force is removed, the material can fully recover without local residual stress or permanent deformation.

4.2.3 PeakForce QNM

Recently, a high-resolution method, called PeakForce QNM, has been used to study the quantitative mechanical properties of cement-based materials at approximately 10 nm (0.4 μin.) resolution [7, 13, 14]. PeakForce QNM is a mode with improved force resolution combined with real-time calculation of elastic modulus at each surface contact, which can detect local elastic modulus [15, 16]. This technique (Bruker AXS) can provide compositional mapping of a complex composite material while providing high resolution. In PeakForce tapping, the probe and sample are intermittently brought together (similar to tapping mode) to contact the surface for a short period, which eliminates lateral forces. Unlike tapping mode, where the feedback loop keeps the cantilever vibration amplitude constant, peak-force tapping controls the maximum force on the tip. This protects the tip and sample from damage while allowing the tip sample contact area to be minimized.

For the PeakForce QNM, the schematic diagram of the collected force-separation curve is shown in Fig. 4.4. To elucidate the elastic properties of the PeakForce QNM, the Derjaguin-Muller-Toporov (DMT) model can be used for the estimation of reduced elastic modulus [17]. The forces during the tip-surface interaction can be described by

$$F_{int} = \frac{4}{3}E^* \sqrt{Rd^3} + F_{adh} \tag{4.11}$$

where F_{int} is the tip-sample force during the tip-surface interaction; E^* is the combined elastic modulus of the tip and the sample; R is the tip radius; d is the sample deformation; and F_{adh} is the adhesion force during the contact. The elastic modulus of the sample E_s can be derived from the following equation

$$\frac{1}{E^*} = \frac{1 - v_s^2}{E_s} + \frac{1 - v_{tip}^2}{E_{tip}} \tag{4.12}$$

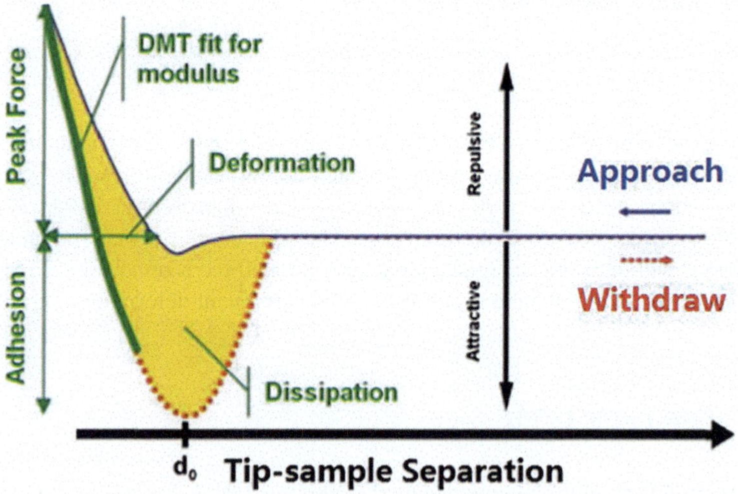

Fig. 4.4 PeakForce QNM schematic [15]

Table 4.1 Testing parameters of the three techniques

Probe type	Nanoindentation	Modulus mapping	PeakForce QNM
	Berkovich (two-dimensional, 142.3 degrees)	Berkovich (one-dimensional, cube-corner)	AFM cantilever
Tip radius	600 nm	–	5 nm
Tip material	Diamond	Diamond	Diamond
Indent depth	200–800 nm	7–12 nm	2–8 nm
Radius of curvature	–	557.14 nm	–
Applied force	1200 μN	20 μN	50 nN
Deflection sensitivity	–	–	70 nm/V
Maximum resolution	800 nm	60 nm	5 nm

where v_s is Poisson's ratio of the sample; v_{tip} is Poisson's ratio of the tip; E_s is the elastic modulus of the sample; and E_{tip} is the elastic modulus of the tip. Calibration was also performed by the relative method [18] using a standard quartz crystal. The testing parameters of the three techniques are listed in Table 4.1.

4.3 Cement Paste Sample Preparation for Micromechanical Property Tests

To compare the results obtained from Nanoindentation, Modulus Mapping and Peak-Force QNM test methods, these tests shall be conducted at the same locations of the same sample. As the sample surface could potentially be damaged by the larger force in static nanoindentation, peak-force QNM and modulus mapping tests shall be performed before nanoindentation test. Samples should be prepared to ensure a flat surface, repeatable test results, and minimal disturbance [19, 20].

This chapter studies the nanomechanical properties of a paste made with Type I Portland cement (see Table 4.2), w/c of 0.40, and water cured under 20 °C and 95% relative humidity condition for 90 days. To prepare a sample, in this study, a 5 mm (thickness) × 10 mm × 10 mm block of a hardened cement paste was firstly produced and submerged into epoxy. After the epoxy impregnation, a small cylindrical sample, 20 mm in diameter, containing the paste block, was cored. It was then affixed with cyanoacrylate glue to a circular steel disc 15 mm in diameter. The exposed flat surface of the cement paste sample was initially ground on paper disks with a gradation of grits. Then an auto-polisher was used with diamond suspensions in ethanol with gradations. Each polishing step lasted approximately 30 min. Finally, diamond-lapping films with various gradations were used to refine the surface to a mirror finish [19–21].

The completed polished surface is shown in Figs. 4.5a, b. The surface topography of a square region of the polished sample as acquired by high-resolution tapping

Table 4.2 Physiochemical compositions of Portland cement

Materials	Type I Portland cement
SiO_2	20.2
Al_2O_3	4.7
Fe_2O_3	3.3
SO_3	3.3
CaO	62.9
MgO	2.7
Na_2O	–
K_2O	–
LOI	1.1
Total	98.2
Density (g/cm^3)	3.10

mode AFM. Figure 4.5b shows the surface topography of a porous square region of the polished sample acquired by high-resolution tapping mode AFM. As expected, the root-mean-square surface roughness of the sample based on the topography map is 93.9 $\pm$ 20 nm for a scanning size of 50 μm $\times$ 50 μm. The RMS value can be used to evaluate the surface roughness because there is no standard criterion of roughness of cement pastes for nanomechanical testing due to the inherent roughness of hardened cement pastes, particularly which would be applicable to the three methods presented [22]. Miller et al. [19] used nanoindentation to characterize cement hydration products and they suggested that the RMS roughness of the sample measured over an area with one side 200 times the average depth of indentation should be less than one-fifth of the average indentation depth. This suggested sample surface roughness is comparable to those used by other investigators [15, 16, 23]. The three nanomechanical methods are applied to measure the nanomechanical properties on the same location of the sample surface, respectively.

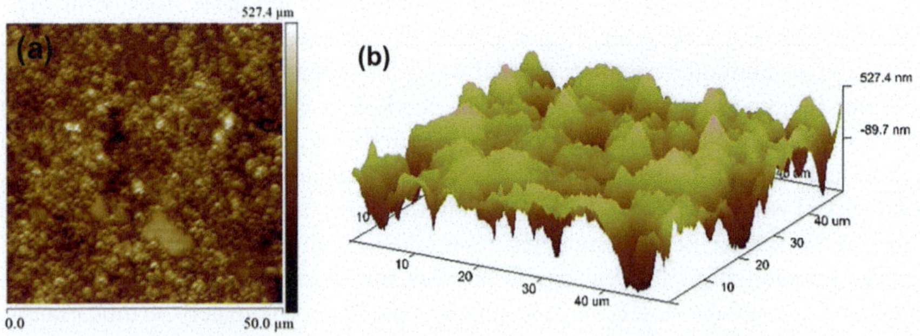

Fig. 4.5 Cement paste sample after polishing: **a** an AFM image; **b** surface topography (50 μm × 50 μm)

As mentioned earlier, the modulus mapping and peak-force QNM can be performed with two devices (Hysitron and Bruker AXS, respectively). For this reason, when switching peak-force QNM testing to modulus mapping from one method to the other, it is necessary to fully confirm that the measurements are each performed at the same position on the same sample. To achieve this, extra high-resolution images were captured, and image analysis was applied to measure the distance and adjust the test position and location distance to make sure all measurements were conducted at the same region of the tested sample. As for the tested area, the deviation should be acceptable less than 0.5 μm.

4.4 Micromechanical Properties of Cement Paste Samples Obtained by Different Microscopic Characterisation Techniques

4.4.1 Nanoindentation

The nanoindentation is performed on the cement paste sample by single loading with a maximum load of 1200 μN, which is similar to those presented in [3, 12, 24]. During the first 5 s, the indenter was driven into the sample at a 240 μN/s loading rate. After reaching the maximum load of 1200 μN/s, the indenter was held in place for 2 s to reduce the creep effect before unloading at a rate of 240 μN/s. Indentations were made on separate 15 × 15 μm and 100 × 100 μm regions to compare profiles and cover more phases for statistical analysis. The completed polished surface is shown in Fig. 4.6, which indicates the surface topography of a square region of the tested sample. The distance between indents was set at 3 μm, properly avoiding potential overlapping of the plastic deformation zone onto neighbouring indents. To achieve a more precise profile, two different regions were indented to cover more phases for statistical analysis.

As identified by the past works [2, 12, 20], the areas with relatively high indentation modulus values (higher than 50.0 GPa) are ascribed to the existence of unhydrated, or partially hydrated cement particles, and the regions that have indentation modulus values ranging from 10.0 to 50.0 GPa are hydration products. Moreover, the lower values of indentation modulus (less than 10 GPa) are predominantly due to the porous regions in cement paste. Based on the nanoindentation measurements results, different representative indent load-depth cures which located in different modulus distribution ranges were shown correspondingly. Figure 4.7 shows typical loading and unloading curves during indentations on different phases of cement paste, including porosity, hydrates (predominantly calcium silicate hydrate and others such as slight ettringite), calcium hydroxide (CH), and unhydrated cement grains. Each load/displacement diagram was plotted and inspected for signs of problems due to surface roughness, such as abnormal or discontinuous shapes. The grid indentation method was performed on a large number of indents on the sample, and statistical

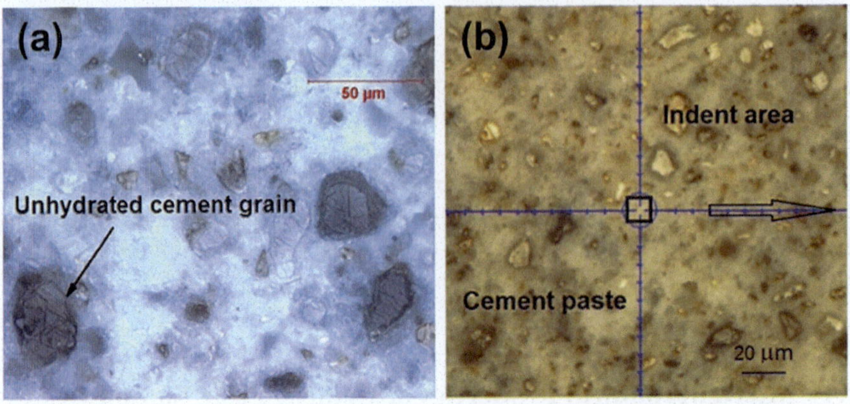

Fig. 4.6 Cement paste sample after polishing: **a** optical image; **b** indent location

analysis of the data obtained from all the indentations provides information about the overall nanomechanical properties of the sample. Figures 4.8 and 4.9 are contour maps and statistical analysis of indentation modulus for the areas of 100 μm × 100 μm and 15 μm × 15 μm, respectively. It can be seen from the figures that the general quantitative correlation between the mechanical properties and phase features can be obverted clearly. It shows that the peak value of the modulus probability plot for the area of 100 μm × 100 μm falls between 10.0 and 15.0 GPa, while the peak value of the modulus probability plot for the area of 15 μm × 15 μm is between 15 and 20 GPa. From the probability distribution, it is found that the peak value of the modulus probability plot in the area of 15 μm × 15 μm is visually higher than that in the area of 100 μm × 100 μm. This indicates that more volume fraction of hydration products, such as the C–S–H phase, exist in the smaller indent area. The distribution of phases is clearly influenced by the size of the scanned area: for the larger area, the amount of cement clinker and porosity phase increase, which results in a greater variation in the indentation modulus distribution. Although many studies have been conducted on the nanomechanical properties of phases in cement paste, little has been reported to characterize and provide quantitative measurements considering different sizes of the investigated area. Moreover, the modulus distributions that correspond to the porosity lower than 10.0 GPa and cement clinker above 50.0 GPa are significantly increased for the larger indent area, resulting in a greater indentation modulus distribution range and variation. The phases in cement paste can be divided into three types based on elastic modulus tested by nanoindentation, as shown in Table 4.3. The areas with a relatively high indentation modulus value, higher than 50 GPa, would be unhydrated or partially hydrated cement particles. This value is in agreement with the values of the elastic modulus measured experimentally for partially hydrated clinker residues by other researchers [21, 25]. The regions with indentation modulus values ranging from 10 to 50 GPa are hydration products [6, 10, 17, 23, 26]. The lower values of indentation modulus, less than 10 GPa, are likely due to porous regions in the cement paste [19, 27]. The illustration of the cumulative probability of a cement paste in the areas of 100 μm × 100 μm and 15 μm × 15 μm are shown in Fig. 4.10. It indicates that modulus results ranging from 10 to 50 GPa are representative of hydration phases, and the main effect on the volume fractions of the different phases presented by nanoindentation in the hardened cement paste in the measurement area [26, 28]. Although many studies have been conducted on the nanomechanical properties of cement paste, little has been reported to characterize and provide quantitative measurements considering the different sizes of the investigated areas. From the measurements, the basic distribution range and volume fraction of the phases in cement paste are obviously influenced by the tested areas, and the nanoindentation measurements should further consider the different sizes of investigated areas.

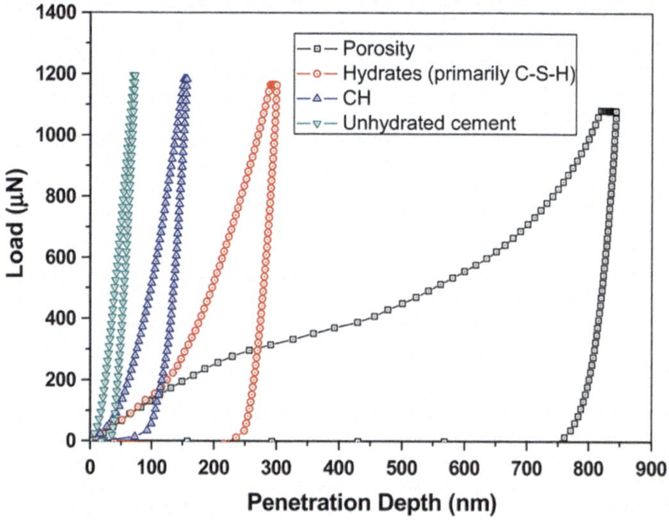

Fig. 4.7 Representative load-depth curves of different phases in cement paste

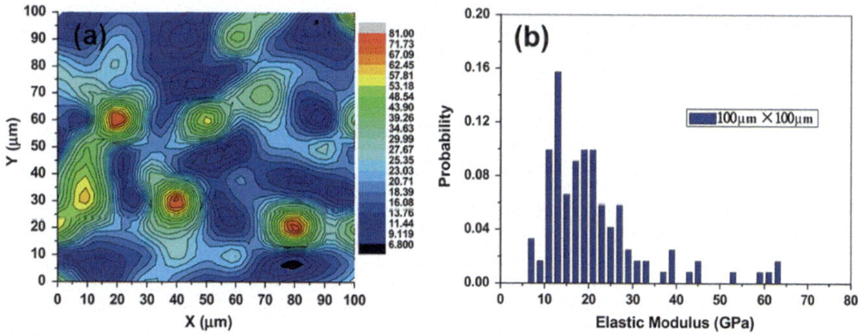

Fig. 4.8 Elastic modulus distribution obtained by nanoindentation (area of 100 μm × 100 μm): **a** indentation modulus contour map; **b** probability distribution (bin size = 2.0 GPa)

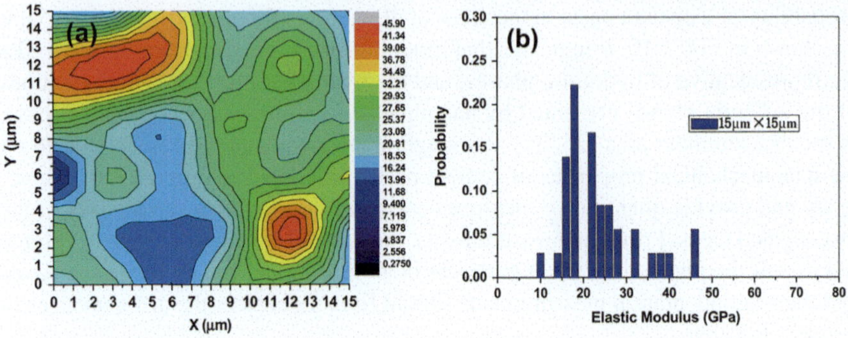

Fig. 4.9 Elastic modulus distribution obtained by nanoindentation (area of 15 μm × 15 μm): **a** indentation modulus contour map; **b** probability distribution (bin size = 2.0 GPa)

Table 4.3 Phases identified by nanoindentation tests on cement paste

Phases	Elastic modulus
Cement clinker	>50 GPa
Hydration products	10–50 GPa
Pores and cracks	<10 GPa

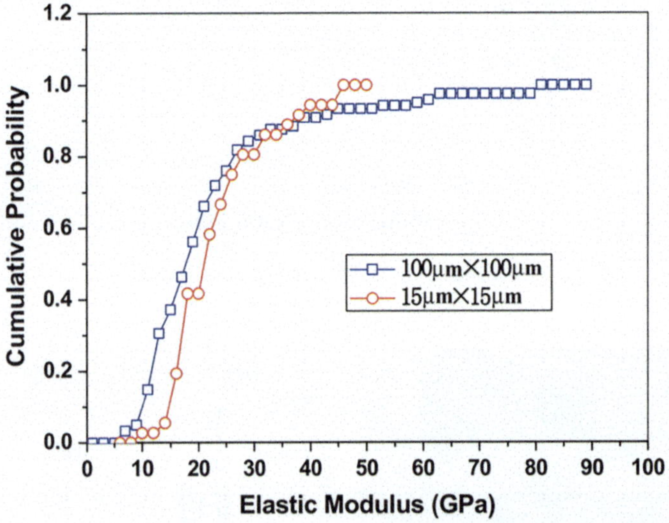

Fig. 4.10 Cumulative probability distribution of elastic modulus obtained by nanoindentation

4.4.2 Modulus Mapping

The modulus mapping was further performed to study the nanomechanical properties with higher resolution compared to static nanoindentation [12, 16]. When switching the modulus mapping testing mode to nanoindentation mode to study the nanomechanical properties of cement paste, the position of the tested sample remains unchanged. Eventually, it is efficient to conduct a reliable comparison between the three different method measurements on the same region of the same sample. As the sample surface could potentially be damaged by the larger force required by static nanoindentation, modulus mapping was always performed first on the sample.

Figure 4.11 presents the elastic modulus contour maps and statistical analysis of the area of 15 μm × 15 μm. The range of measured elastic modulus is found to be significantly greater than that obtained by nanoindentation. The range of elastic moduli, from 0.2 to 124.5 GPa, is found to be significantly greater than that obtained by static nanoindentation. This can be attributed to the higher resolution and smaller contact area of modulus mapping compared with nanoindentation. The results of modulus mapping for the scanned area 5 μm × 5 μm are shown in Fig. 4.12. It is found that the quantitative modulus variations in modulus mapping are larger than those in nanoindentation. It reveals that the modulus mapping technique is more sensitive than nanoindentation with a lower contact force. In general, obvious local elastic modulus variations were observed at the sample surface, which comprised different phases and interfaces between hydration products, such as inner and outer hydration products. Previous workers also have identified variations in elastic modulus as being due to different phases in the cement [18, 20]. On the other hand, clear abrupt changes in the elastic modulus could be due to the interfaces between the porous areas and different hydration products. The peak of the modulus probability plot for the area of 15 μm × 15 μm falls between 15 and 20 GPa, while the peak of the modulus probability plot for the area of 5 μm × 5 μm is between 20 and 25 GPa. More importantly, it seems that the peak value of elastic modulus in the probability plot in the area of 5 μm × 5 μm is somewhat higher than that in the area of 15 μm × 15 μm. Moreover, it is needed to note that it is likely for the probability distributions of elastic modulus measured using modulus mapping to provide an obvious indication for the different types of hydration products. However, the probability distribution of elastic modulus cannot exactly indicate how many different types of C–S–H exist in cement paste, such as high-density and low-density C–S–H [4, 21]. The difference in the cumulative probability distribution of phases between the different measured areas of 15 μm × 15 μm and the area of 5 μm × 5 μm is typically consistent, as shown in Fig. 4.13. Assuming that the C–S–H phase can be identified with two characteristic forms, low-density C–S–H and high-density C–S–H [2, 4], the area of 5 μm × 5 μm is found to have a smaller volume fraction of low-density C–S–H, compared to the area of 15 μm × 15 μm.

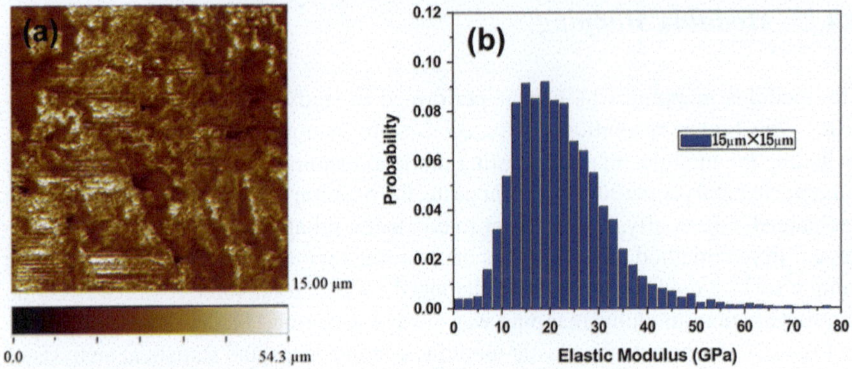

Fig. 4.11 Elastic modulus obtained by modulus mapping (area of 15 μm × 15 μm): **a** elastic modulus contour map; **b** probability distribution (bin size = 2.0 GPa)

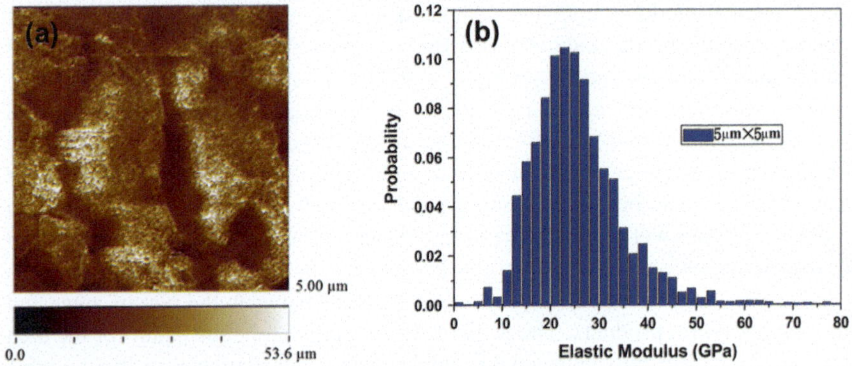

Fig. 4.12 Elastic modulus obtained by modulus mapping (area of 5 μm × 5 μm): **a** elastic modulus contour map; **b** probability distribution (bin size = 2.0 GPa)

4.4.3 PeakForce QNM

In addition to nanoindentation and modulus mapping, PeakForce QNM technique was used to evaluate nanomechanical properties at a very high resolution (5–10 nm). Modulus mapping and PeakForce QNM were performed on two separate devices. When switching peak-force QNM testing to modulus mapping from one method to the other, it is necessary to fully confirm that the measurements are each performed at the same position on the same sample. To achieve this, extra high-resolution images with distinguished features were captured, and image analysis was applied to measure the distance and adjust the test position based on a precise distance and direction from the features to make sure all measurements were conducted at the same region of the tested sample.

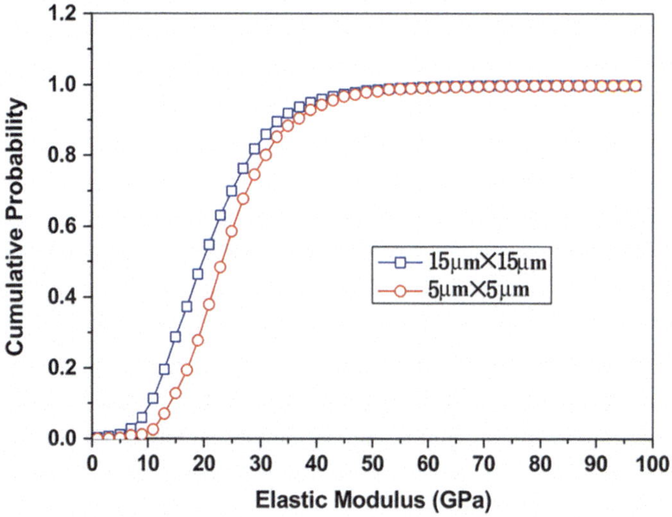

Fig. 4.13 Cumulative probability distribution of elastic modulus obtained by modulus mapping

Figure 4.14 shows the 512 × 512 pixel topography images of two scanning areas, 5 × 5 μm and 1 × 1 μm, of a cement paste sample obtained by PeakForce QNM. The microstructural and nanostructural contrast can be observed clearly in the surface topography. It should be noted that the intrinsic porosity in cement paste poses some constraints on the use of PeakForce QNM, for which some noise could be expected. The images indicate that the height distribution of local areas varies widely for different measuring areas. The values of all local roughness, however, are sufficiently low for PeakForce QNM testing [13, 29]. The topography images also demonstrate that the cement paste is a granular material, where the sub-micron scale grains or basic nanoscale units pack together densely [7, 13, 15–17].

Based on the elastic modulus contour image of cement paste offered in Fig. 4.15, visual nanostructural characteristics of the area of 5 μm × 5 μm can be obtained. There is a significant difference observed in modulus distributions between the multiphases in cement paste. The porosity and the interfaces between nanoscale grains with a lower elastic modulus appear darker, while the calcium hydroxides, unhydrated

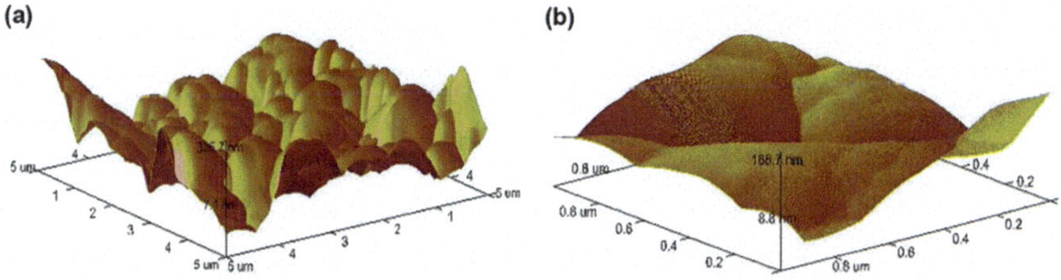

Fig. 4.14 Surface topography obtained by PeakForce QNM: **a** 5 μm × 5 μm; **b** 1 μm × 1 μm

cement particles and other phases with higher elastic modulus exhibit brighter. Based on previous studies, the elastic modulus of the measuring area identified as the pore space is typically below 10.0 GPa, while the elastic modulus of the measuring areas identified as hydration products were found to vary approximately between 10.0 and 50.0 GPa. In order to analyse the nanostructure and nanomechanical properties in more detail, the results of a smaller area of 1 μm × 1 μm were presented in Fig. 4.16. The elastic modulus contour maps also demonstrate that the cement paste is a granular material, where the sub-micron scale grains or basic nanoscale units pack together densely, which has also been reported in other researches [15, 16]. Based on the basic composition of the selected cement, the hydration products should be predominantly composed of C–S–H grains. Hence, the discussion was focused on the aforementioned C–S–H grains and their shape characteristics. Compared to the different scanned areas with PeakForce QNM, the elastic modulus distribution range of the area of 5 μm × 5 μm is slightly larger than that of the area of 1 μm × 1 μm because of the different nanoscale grains experiencing different intrinsic elastic modulus, respectively. The probability distributions of elastic modulus for cement paste seem to provide an indication for the occurrence of different types of C–S–H but cannot exactly distinguish present there are how many types of C–S–H which are representative of low-density and high-density C–S–H, as reported in previous research [21, 29]. When comparing the analogous probability distribution results of the two different scanning areas, the peak value of the modulus probability plot for the area of 5 μm × 5 μm appears to be consistent with that of 1 μm × 1 μm, which is found to be around 20–25 GPa. It indicates that the measurement result of the area of 5 μm × 5 μm has a similar cumulative probability distribution with the area of 1 μm × 1 μm, as shown in Fig. 4.17. Furthermore, it can be found that the phase volume fractions in cement paste typically keep constant with each other for different scanning areas using PeakForce QNM testing.

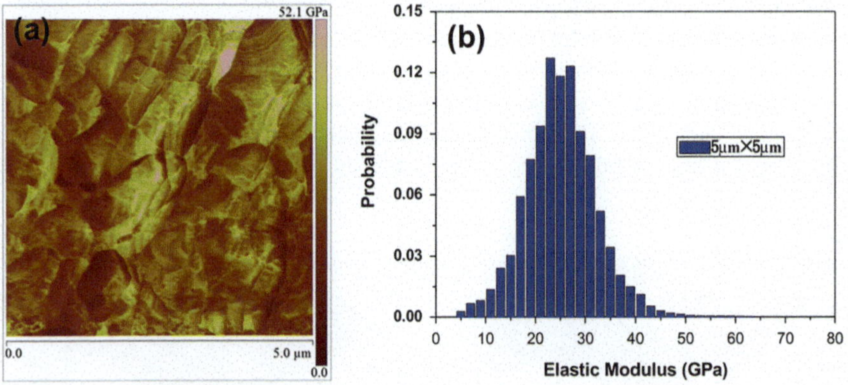

Fig. 4.15 Elastic modulus distribution obtained by peak-force QNM (area of 5 μm × 5 μm): **a** elastic modulus contour map; **b** probability distribution (bin size = 2.0 GPa)

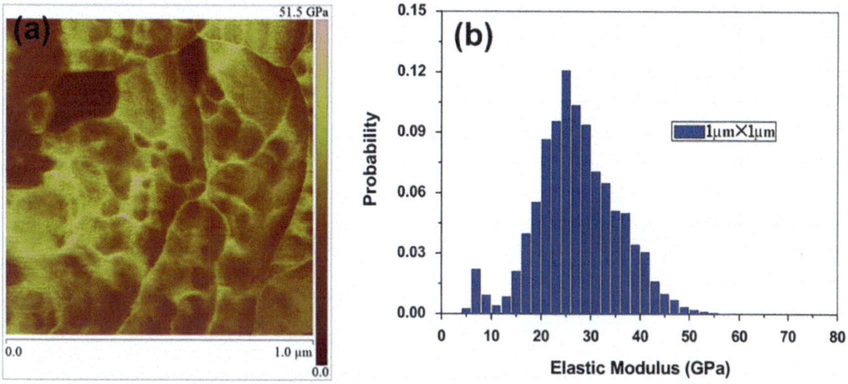

Fig. 4.16 Elastic modulus distribution obtained by peak-force QNM (area of 1 µm × 1 µm): **a** elastic modulus contour map; **b** probability distribution (bin size = 2.0 GPa)

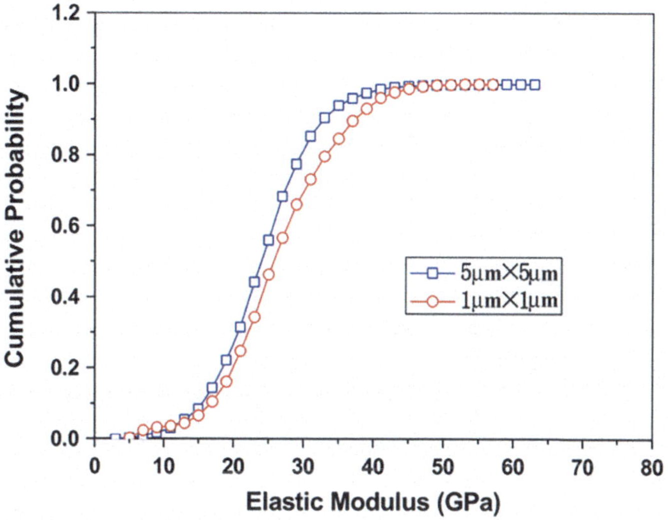

Fig. 4.17 Cumulative probability distribution of elastic modulus obtained by peak-force QNM

As shown in Fig. 4.18, it is suggested that the PeakForce QNM technique can reveal the different types of grains in hardened cement paste. This observation also highlights the advantage of evaluating the interfaces between grains by detecting the elastic modulus distribution. According to the quantitative modulus value of the presented results, there is a clear jump in the elastic modulus distribution across the interfaces between grains. The elastic modulus of the grains themselves, however, typically exhibits a constant value. Moreover, abrupt changes in the elastic modulus across the interfaces were clearly observed. The information stemming from the PeakForce QNM technique can reasonably identify the grains with sizes from 20 to

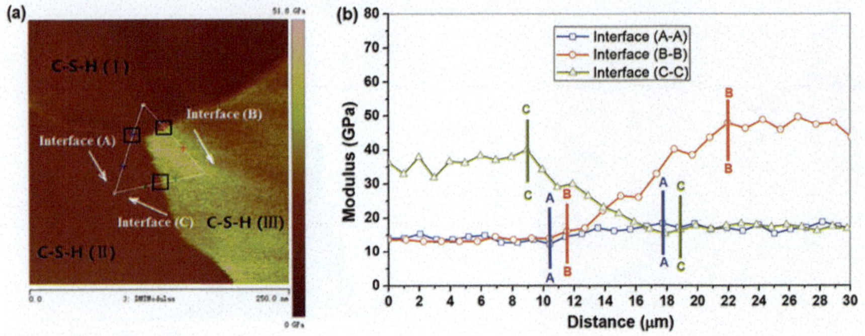

Fig. 4.18 DMT modulus obtained by PeakForce QNM (0.25 μm × 0.25 μm): **a** interface position; **b** modulus distribution

400 nm in the nanostructure. To provide quantitative measurements of the interface thickness between grains, radial lines (interface A–A, B–B, and C–C) across the grain boundaries were applied to indicate the elastic modulus variation corresponding to the interface thickness. The interface starts from the point where the elastic modulus begins to change from that of one grain and extends to another point where the local elastic modulus is equal to that of another grain. The distance between these two points corresponds to the interface thickness. The thickness of the interface obtained by averaging these lines, and the most probable thickness of the interface between grains, was found to be 15 ± 5 nm.

4.5 Comparison Analysis of Nanoindentation, Modulus Mapping and Peak-Force QNM

4.5.1 Nanoindentation and Modulus Mapping

Compared to nanoindentation, modulus mapping allows for the evaluation of local mechanical properties over a smaller area with higher spatial resolution at the nanoscale [5, 24]. In nanoindentation, although the nanomechanical properties of different phases in cement paste can be estimated by statistical analysis, the separation between indentations is required to be large enough to avoid the effects of residual stress and interaction between the neighbouring indents on the results. For Berkovich diamond tips, the lateral dimension of the indentation is greater than the depth of indentation. Therefore, to avoid error in elastic modulus due to possible interaction between adjacent indentions, a separation of 5–10 times the lateral indentation dimensions is employed as a general guide, which means a minimum separation of 3 μm for nanoindentation testing. As a result, determination of the local quantitative variation at the nanometer level is practically impossible for nanoindentation;

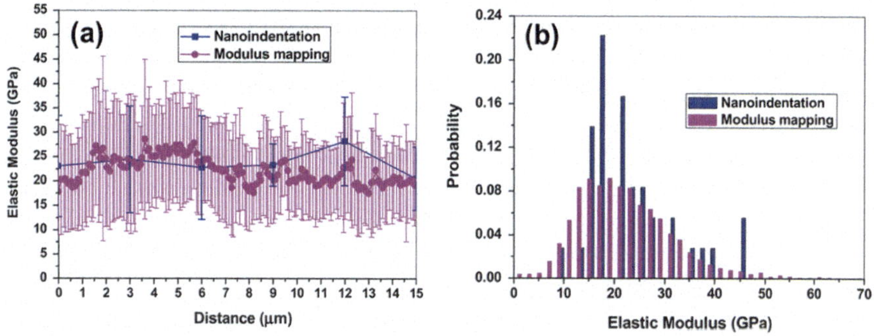

Fig. 4.19 Elastic modulus distribution of nanoindentation and modulus mapping (area of 15 μm × 15 μm): **a** elastic modulus distribution; **b** probability distribution (bin size = 2.0 GPa)

however, the modulus mapping technique in the form of SPM images can overcome these shortcomings [20]. This is because the sample under the tip is probed at a shallow indent depth (elastic deformation), and once the tip force is removed, the probed materials can fully recover without local residual stress or permanent deformation. Moreover, modulus mapping can provide measurements on independent phases with greater precision.

From Fig. 4.19a, the elastic modulus distribution obtained by modulus mapping is generally consistent with that by nanoindentation at the same area of 15 μm × 15 μm. Specifically, the average modulus by modulus mapping was around 22.2 ± 9.8 GPa, and the average nanoindentation modulus was around 23.8 ± 8.7 GPa. Generally, the average elastic modulus measured by nanoindentation is consistent with those measured by modulus mapping. Nonetheless, visually, the slightly higher average modulus value in nanoindentation could be a result of the local confinement effect and interaction between multiple phases in the indent area because the indent depth of nanoindentation is much higher than the probed depth in modulus mapping. Furthermore, Fig. 4.19b shows that the peak value of the modulus probability plot extracted from nanoindentation is somewhat higher than the corresponding peak value measured with modulus mapping. More importantly, the large indent area can reduce the influence of porosity, and unhydrated cement residuals in the nanoindentation measurement compared with modulus mapping measurement could explain the generally concentrated contribution in the measured elastic modulus by modulus mapping. However, the smaller indent area can achieve a denser grid nanoindentation tests, and detect nanoparticles in cement paste.

4.5.2 Modulus Mapping and Peak-Force QNM

Compared to the contact radius (60 nm) of modulus mapping testing, the probe radius of peak-force QNM mode is significantly sharper (5 nm), which makes the spatial resolution significantly higher [7, 13]. For this reason, the peak-force QNM appears more useful when investigating local nanomechanical properties of a small area in more details. For peak-force QNM, a typical scan takes 10 min, which is significantly faster than the modulus mapping. After a scan is completed, 512 lines of data are available with 512 data points in each line. On the other hand, compared to peak-force QNM mode, the modulus mapping will be more convenient for a larger measured area.

As indicated in Fig. 4.20a, except for the variation characterization, the modulus distribution of the same measuring area of 5 μm × 5 μm obtained by peakforce QNM is typically consistent with those by modulus mapping. The average elastic modulus (25.6 ± 9.2 GPa) measured by modulus mapping is basically consistent with the average elastic modulus (25.2 ± 6.6 GPa) obtained by peak-force QNM. At the same time, the modulus probability characterization measured from peak-force QNM is similar to that measured with modulus mapping, which can be clearly seen in Fig. 4.20b. Furthermore, the peak value of the modulus probability plot measured by peakforce QNM is typically higher than the value obtained by modulus mapping. Consequently, the plausible explanation for this slight difference could be related to the different related nanomechancial theories and contact forces during testing for these two nanomechanical methods. Overall, it seems that the mean value and distribution characteristics of the measured values obtained by peak-force QNM are in good agreement with the measured results by the modulus mapping. Based on the probe radius of the peak-force QNM is quite sharper than that of modulus mapping (as shown in Table 4.1), the measured values are less likely to involve influence from the interaction between multiple phases and are thus likely more representative of the intrinsic properties of the C–S–H particle. In particular, the homogenized elastic modulus of the composite obtained from peak-force QNM measurements should present a relatively smaller error when compared to the modulus mapping.

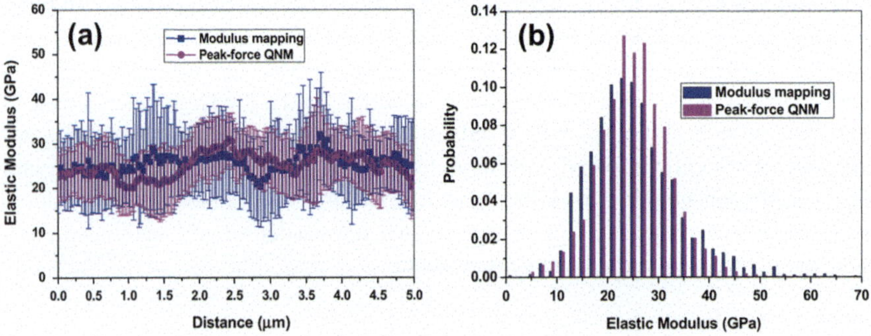

Fig. 4.20 Elastic modulus distributions of modulus mapping and peak-force QNM (area of 5 μm × 5 μm): **a** elastic modulus distribution; **b** probability distribution (bin size = 2.0 GPa)

4.5.3 Peak-Force QNM and Nanoindentation

Based on the existing literature [12, 15, 29], the C–S–H phase is the primary phase of the hydration products in cement paste, and the properties of cement paste are governed through the nucleation, packing, and stiffness of the nanoscale C–S–H particles. As the grains form at random sizes, improving the diversity in the size of nanoscale units can eventually lead to a denser and disorderly packing of the particles, which corresponds to a stronger and higher-performance cement paste. The topography features of cement paste are of different sizes to the nanoscale grains hypothesized in the colloidal spheres model of granular particles, as shown in Fig. 4.21. It indicates that cement paste is a granular material, where the particles or basic nanoscale units pack together most orderly. When comparing the nanostructure characterization of cement paste from peak-force QNM measurement to the analogous results Refs. [13, 15, 17], it demonstrates that the observed spatial variety of the grain packing fraction and grain shape can be associated with the diversity model of granular particles (predominately C–S–H). On the other hand, the topographical image obtained from peak-force QNM has not only great intrinsic quantitative information on the size and shape of the gains but also insight into nanostructural morphology for cement paste.

The grain packing density considering a colloidal sphere structure in cement paste at the nanoscale seems to provide a significant framework for reconciling the differences between the peak-force QNM measurements and those obtained by nanoindentation. Therefore, it is hypothesized that the significant difference between the elastic modulus measured from peak-force QNM and those from nanoindentation is primarily due to the differences in grains packing at two different length scales (from one micrometer to tens of nanometers scale). With regard to the peak-force QNM testing, which probes a much shallower depth than nanoindentation one, the peakforce QNM is likely more sensitive to surface grinding and polishing artifacts which seem to influence the packing density of the sample surface. During the sample preparation, the grinding and polishing may cause the nanoscale grains compaction

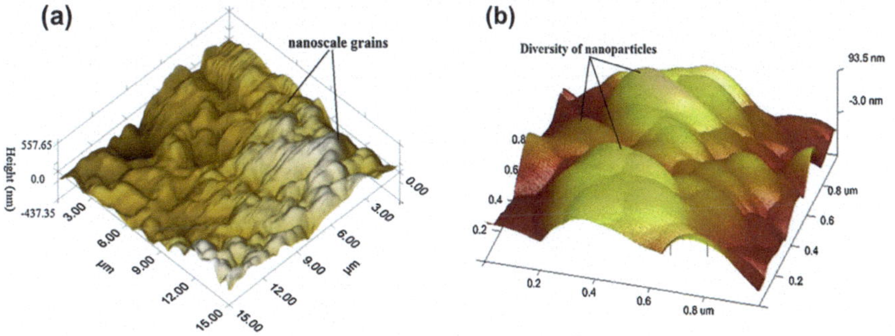

Fig. 4.21 Nanoscale grains characteristic of hardened cement paste (area of 15 μm × 15 μm and 1 μm × 1 μm): **a** topography by nanoindentation mapping; **b** topograghy by peak-force QNM

on the sample surface. Such influences explain why the distribution range of elastic modulus measured by peak-force QNM in this study is found to be greater than those obtained from nanoindentation testing. It is hypothesized that the difference between the measured modulus obtained from nanoindentation and from peak-force QNM is also due primarily to differences in particle packing between the two length scales. One possibility should be noted that the increased grain packing is confined to the surface layer of the sample due to the artifact of the polishing process. As a result, whether the granular particles in cement paste are of colloidal sphere structure will still be a debate, as the granular particles may be formed from the layered model C–S–H, which can be stacked together during the sample grinding and polishing [10, 22].

4.6 Conclusions

The investigations on nanostructure and nanomechanical properties of cement paste were investigated using three correlative nanomechanical methods. Based on the analysis and comparisons on these different measurements, the following conclusions can be drawn:

(1) Compared with nanoindentation, modulus mapping and peak-force QNM allow for evaluating local mechanical properties at smaller measuring areas. The range of elastic modulus distribution measured by modulus mapping is relatively greater than that obtained from nanoindentation.
(2) The average value of elastic modulus measured by nanoindentation is generally consistent with those obtained using modulus mapping and PeakForce QNM. The different modulus probability between them may be due to a result of the local confinement effect and interaction between multiple phases.
(3) The PeakForce QNM can provide local mechanical properties in more details with a higher spatial resolution than modulus mapping. The peak value of modulus probability distribution plot measured by PeakForce QNM is typically higher than the value obtained by modulus mapping, which could be related to the different related nanomechancial theories and contact forces during testing.
(4) At nanocalse, cement paste contains the sub-micron scale or nanoscale grain units packed together. Compared with the nanoindentation, the PeakForce QNM measurements can provide more insight into the characteristics of nanostructures and nanoscale granular particles in cement paste.

References

1. R. Alizadeh, J.J. Beaudoin, L.R. Raki, Mechanical properties of calcium silicate hydrates. Mater. Struct. **44** (2011).
2. G. Constantinides, F.J. Ulm, The effect of two types of C–S–H on the elasticity of cement-based materials: results from nanoindentation and micromechanical modelling. Cem. Concr. Res. **34** (2004)
3. G. Constantinides, F. Ulm, K. Vliet, On the use of nanoindentation for cementitious materials. Mater. Struct. **36** (2003)
4. G. Constantinides, F.J. Ulm, The nanogranular nature of C–S–H. J. Mech. Phys. Solids **55** (2007)
5. G. Balooch, G.W. Marshall, S.H. Marshall, O.L. Warren, S.A.S. Asif, M. Balooch, Evaluation of a new modulus mapping technique to investigate microstructural features of human teeth. J. Biomech. **37** (2004)
6. H.M. Jennings, J.J. Thomas, J.S. Gevrenov, G. Constantinides, F.J. Ulm, A multi-technique investigation of the nanoporosity of cement paste. Cem. Concr. Res. **37** (2007)
7. C.A. Jones, Z.C. Grasley, J.A. Ohlhausen, Measurement of elastic properties of calcium silicate hydrate with atomic force microscopy. Cem. Concr. Compos. **34** (2012)
8. K. Sobolev, SP. Shah, Nanotechnology of Concrete: Recent Developments and Future Perspectives (American Concrete Institute, Detroit SP-254, 2008)
9. W.G. Li, J.Z. Xiao, S. Kawashma, G.S. Shekhawat, S.P. Shah, Experimental investigation on quantitative nanomechanical properties of cement paste. ACI Mater. J. **111** (2014)
10. E. Masoero, E. Del Gado, R.J.M. Pellenq, F.J. Ulm, S. Yip, Nanostructure and nanomechanics of cement: polydisperse colloidal packing. Phys. Rev. Lett. **109** (2012)
11. F. Sanchez, K. Sobolev, Nanotechnology in concrete—a review. Constr. Build. Mater. **24** (2010)
12. P.S. Uskokovic, C.Y. Tang, C.P. Tsui, N. Ignjatovic, D.P. Uskokovic, Micromechanical properties of a hydroxyapatite/poly-L-lactide biocomposite using nanoindentation and modulus mapping. J. Eur. Ceram. Soc. **27** (2007)
13. J.H. Kim, O. Balogun, S.P. Shah, Atomic force acoustic microscopy to measure nanoscale mechanical properties of cement pastes. J. Transp. Res. Board **2141** (2010)
14. W.G. Li, J.Z. Xiao, Z.H. Sun, S. Kawashima, S.P. Shah, Interfacial transition zones in recycled aggregate concrete with different mixing approaches. Constr. Build. Mater. **35** (2012)
15. P. Trtik, J. Kaufmann, U. Volz, On the use of peak-force tapping atomic force microscopy for quantification of the local elastic modulus in hardened cement paste. Cem. Concr. Res. **42** (2012)
16. T.J. Young, M.A. Monclus, T.L. Burnett, W.R. Broughton, S.L. Ogin, P.A. Smith, The use of the PeakForce™ quantitative nanomechanical mapping AFM-based method for high-resolution Young's modulus measurement of polymers. Meas. Sci. Technol. **22** (2011)
17. A. Peled, J. Castro, J. Weiss, Atomic force microscopy examinations of mortar made by using water-filled lightweight aggregate. J. Transp. Res. Board **2141** (2010)
18. O. Sahin, N. Erina, High-resolution and large dynamic range nanomechanical mapping in tapping-mode atomic force microscopy. Nanotechnology **19** (2008)
19. M. Miller, C. Bobko, M. Vandamme, F.J. Ulm, Surface roughness criteria for cement paste nanoindentation. Cem. Concr. Res. **38** (2008)
20. N.X. Randall, M. Vandamme, F.J. Ulm, Nanoindentation analysis as a two dimensional tool for mapping the mechanical properties of complex surfaces. J. Mater. Res. **24** (2009)
21. M. Vandamme, F.J. Ulm, P. Fonollosa, Nanogranular packing of C–S-H at substochiometric conditions. Cem. Concr. Res. **40** (2010)
22. K.L. Scrivener, R.J. Kirkpatrick, Innovation in use and research on cementitious material. Cem. Concr. Res. **38** (2008)
23. J.Z. Xiao, W.G. Li, Z.H. Sun, D.A. Lange, S.P. Shah, Properties of interfacial transition zones in recycled aggregate concrete tested by nanoindentation. Cem. Concr. Compos. **37** (2013)

24. L. Sorelli, G. Constantinides, F.J. Ulm, F. Toutlemonde, The nano-mechanical signature of ultra high performance concrete by statistical nanoindentation techniques. Cem. Concr. Res. **38** (2008)
25. X.H. Wang, S. Jacobsen, S.F. Lee, J.Y. He, Z.L. Zhang, Effect of silica fume, steel fiber and ITZ on the strength and fracture behavior of mortar. Mater. Struct. **43** (2010)
26. B. Šavija, M. Luković, S.A.S. Hosseini, J. Pacheco, E. Schlangen, Corrosion induced cover cracking studied by X-ray computed tomography, nanoindentation, and energy dispersive X-ray spectrometry (EDS). Mater. Struct. (2014)
27. P. Mondal, S.P. Shah, L.D. Marks, Nanoscale characterization of cementitious materials. ACI Mater. J. **105** (2008)
28. W.R.L. Silva, J. Němeček, P. Štemberk, Methodology for nanoindentation-assisted prediction of macroscale elastic properties of high performance cementitious composites. Cem. Concr. Compos. **45** (2014)
29. M.R. Vanlandingham, S.H. McKnight, G.R. Palmese, J.R. Elings, X. Huang, T.A. Bogetti, R.F. Eduljee, J. Gillespie, Nanoscale indentation of polymer systems using the atomic force microscope. J. Adhes. **64** (1997)

Chapter 5
Applying Grid Nanoindentation and Maximum Likelihood Estimation for N–A–S–H Gel in Geopolymer Paste

5.1 Introduction

Nano/micromechanical properties are one of the essential aspects for the understanding of materials. The advances in micro and nanoscale mechanical characterization techniques and theory can enable various measurement and open new doors for material analysis [1–4]. In the field of cement-based materials, they have been typically used to test micro- and nanomechanical properties of reacted phases [5–7]. Among micro- and nano-characterization techniques, nanoindentation has been most widely used and usually combined with deconvolution techniques [8–11], which is known as statistical nanoindentation technique (SNT) [12]. One of the most important deconvolution techniques, the maximum likelihood estimation (MLE) method (Expectation–Maximization algorithm) [13], to the best of knowledge, applied in cement paste for the first time by Ulm et al. [14]. Based on the MLE method, one study found that the results obtained were not a pure phase but a mixture of phases, particularly calcium silicate hydrate (C–S–H) and (calcium hydroxide) CH [15]. Another important deconvolution technique, the least-square estimation (LSE) method, which is the mainstream approach for the deconvolution of nanoindentation data, however, the validity of this kind of method is sometimes questioned for the analysis of cement-based materials.

Alkali-activated materials are well known for their benefits of reusing industrial by-products and reduced carbon dioxide emission, which are promising supplement or even substitute of Portland cement paste and have attracted extensive relevant studies [16–20]. Cement and geopolymer pastes are compared in Table 5.1 from chemistry and microstructure. Fly ash based geopolymer is one of the most representative alkali-activated materials [21, 22]. The micromechanical properties of alkali-activated fly ash (AAFA) were investigated by previous studies [23–26]. However, inconsistent results were presented by different studies. Typically, the modulus of sodium aluminosilicate hydrate (N–A–S–H) gels vary from 4.44 to 16.78 GPa (28 days of curing) in [27], 11 GPa to 25 GPa (28 days of curing) in [28], and

W. Li et al., *Nanocharacterization of Cementitious Composites*,
https://doi.org/10.1007/978-981-96-2917-6_5

Table 5.1 Comparison between cement and geopolymer paste in chemistry and microstructure

Objects	Cement paste	Geopolymer paste
Raw materials	Portland cement (calcium silicates, tricalcium aluminate and tetra-calcium alumino-ferrite	Aluminosilicate sources (fly ash, slag, or metakaolin); alkali activator (usually sodium or potassium silicates and hydroxides)
Binding chemistry	Calcium silicate hydrate (C–S–H), calcium hydroxide ($Ca(OH)_2$)	Amorphous aluminosilicate gel (N–A–S–H)
pH environment	High-pH environment due to hydration products ($Ca(OH)_2$)	High-pH environment due to the use of alkali activators
Main gel phases	Partially crystalline with some amorphous regions	Predominantly amorphous
Porosity	Capillary pores and gel pores within the C–S–H structure, influenced by the water-to-cement ratio	A denser matrix with less capillary porosity, comes from the gel structure and residual unreacted particles

achieve the similar value of about 17–18 GPa (6 months of curing) in [29], presenting significantly different results. It should be noticed that different testing and analysis methods and parameters were adopted in the above studies. For the statistical analysis of AAFA, the LSE method is still normally adopted with different bin sizes such as 1.0 GPa [30, 31] and 2.0 GPa [32] and different phase numbers of 3 [33] or 4 [34, 35] were used. This again raised questions regarding the influence of the different choice of parameters in the deconvolution process on the final deconvolution results and also the accuracy and reliability of the LSE method itself.

Based on the AAFA sample, via investigating the effect of bin size, number of phases and grid number on the results, the validity of the LSE method and the MLE method are examined in this chapter. The MLE method is found to have great advantages over the LSE method. Besides, the gel phase with significant inclusions removed is obtained by MLE in this study by increasing the number of phases for deconvolution. The advantages of the MLE method can be displayed based on a premise that the "compromise approach" is adopted.

5.2 Geopolymer Paste Sample Preparation for Nanoindentation Tests

Fly ash used was low calcium fly ash with the chemical composition shown in Table 5.2. Alkali solution is sodium hydroxide or sodium silicate with different silica modulus of 1 or 1.5. Sodium hydroxide pellets were dissolved in water and cooled down. Then, it was mixed with sodium silicate to adjust silica modulus. The original modulus of the commercial sodium silicate is 2.07 with 14.7% of Na_2O and 29.4% of SiO_2. Cooled sodium hydroxide solution was incorporated into sodium silicate solution to adjust the silica modulus of the solution to the designed value of 1 or 1.5. Different samples with Na_2O to fly ash weight ratio of 6%, 8%, and 10% were prepared. For these samples, constant silica modulus of 1.0 in alkali solution, as well as constant water to solid ratio of 0.338 were adopted to keep other factors similar for each kind of sample. The mixture ratios for samples are shown in Table 5.3. Fly ash was mixed with corresponding alkali solution for 5 min and cast into cubic plastic moulds with the size of 50 mm × 50 mm × 50 mm. The samples were vibrated for 3 min and sealed by plastic film. Then, they were put into oven with heat treatment of 70 °C for 24 h. After that, samples were taken out of the oven and put into a standard curing cabinet for further curing (temperature of 20 °C and RH of 95%). The standard cured sample was always stored in the standard curing cabinet once finished casting and vibration of the sample.

Microstructure observation was conducted on small pieces of crushed samples while BSE imaging and nanoindentation test were operated on polished samples. Cutting machine was used to take out the core part of the cubic samples, which was then cast in epoxy resin and subjected to grind and polish. Abrasive papers with grits of 320, 600 and 1200 were used to grind the samples with each grade lasted for 10 min. 0.3 μm alumina and 0.05 μm cerium oxide slurry were used for further polishing samples with each grade lasted for 40 min. Residual particles present on the surface of samples after each polish procedure were removed by ultrasonic cleaning of samples in isopropanol for 3 min. After finishing the sample preparation, samples

Table 5.2 Chemical composition of fly ash for geopolymer

Oxide	Al_2O_3	CaO	SiO_2	Fe_2O_3	K_2O	MgO	Na_2O	MnO	P_2O_5	TiO_2	LOI
Weight (%)	25.21	1.73	64.55	2.85	1.47	0.41	0.48	0.07	0.19	0.91	1.54

Table 5.3 Mixture design and curing condition of geopolymer

Samples	SiO_2/Na_2O (silica modulus)	Na_2O/fly ash (%)	w/s	Curing condition
AAFA-M1-S	1	8	0.338	Standard curing
AAFA-M0-H	0 (NaOH)	8	0.338	Heat curing at 70 °C for 24 h
AAFA-M1-H-10%	1	10	0.338	Heat curing at 70 °C for 24 h
AAFA-M1-H-8%	1	8	0.338	Heat curing at 70 °C for 24 h
AAFA-M1-H-6%	1	6	0.338	Heat curing at 70 °C for 24 h
AAFA-M1.5-H	1.5	8	0.338	Heat curing at 70 °C for 24 h

Note w/s is water to solid ratio. 'M0', 'M1' or 'M1.5' behind AAFA means the silica modulus of alkali solution to make AAFA is 0, 1 or 1.5. 'H' or 's' behind AAFA denotes the corresponding curing condition is heat curing or always standard curing were dried by vacuum oven under 50 °C for 72 h, and then stored in a vacuum desiccator.

5.3 Nanoindentation Test Setup on Geopolymer Paste

After 28 days, Agilent G200 Nano Indenter with a Berkovich tip in The Sydney Centre in Gemechanics and Mining Materials (SciGEM) within The University of Sydney was used to conduct nanoindentation to investigate the micromechanical properties of AAFA. Nine 10 × 10 grid nanoindentation were performed with a grid spacing of 15 μm. The nanoindentation depth can be determined by satisfying both the scale separability condition and roughly one-tenth rule of thumb as the below Eq. 5.1 [36], where *d* is the largest heterogeneity of geopolymer phases and *D* is the characteristic size of microstructure. Besides, enough depth is also necessary to avoid interference from surface roughness [37, 38]. For the investigation of N–A–S–H gel, it was reported that the N–A–S–H gel consists of particles with the diameter of about 5 nm [39]. As for characteristic microstructure size, more than about 4 μm can be found even just 7% Na_2O (Ms = 1) is used [40]. After several attempts, trapezoid loading with the peak force of 2 mN was adopted as shown in Fig. 5.1, with an average penetration depth of 228.21 nm. The loading procedure adopted in this test would be suitable for achieving balance to satisfy the scale separability condition and avoid the effect of surface roughness and multiple phases' responses. Similar loading procedures have also been adopted by other studies for Portland cement paste and AAFA [41–44], making it easy for comparison.

In this study, the constant Poisson's ratio of 0.2 was set for testing. According to nanoindentation load-penetration curves and some basic parameters, nanoindentation modulus *E* and hardness *H* for material at each testing point can be calculated by Eqs. 5.2–5.4, based on the initial unloading stage and peak load of the load-depth

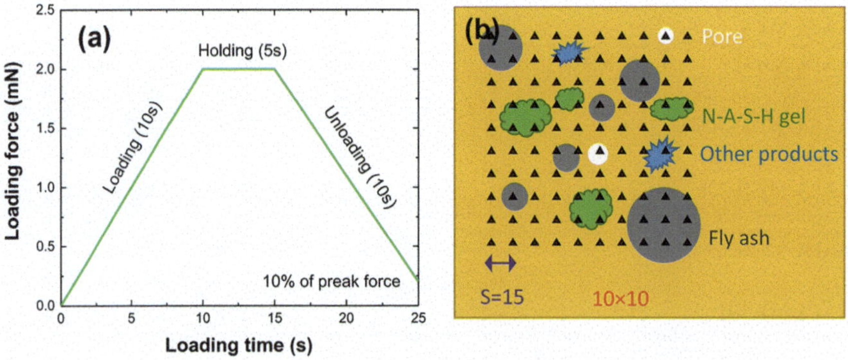

Fig. 5.1 Nanoindentation testing on geopolymer: **a** test procedure of individual nanoindentation; **b** schematic of grid nanoindentation

curve, respectively [45, 46]. All testing points with abnormal load-penetration curves were deleted before the subsequent deconvolution analysis. The following equations summarise the fundaments of nanoindentation tests used in this study:

$$d \ll h_{\max} < D/10 \tag{5.1}$$

$$H = \frac{P_{\max}}{A} \tag{5.2}$$

$$S = \left.\frac{dP}{dh}\right|_{h=h_{\max}} = \frac{2}{\sqrt{\pi}} E_r \sqrt{A} \tag{5.3}$$

$$\frac{1}{E_r} = \frac{1-\nu^2}{E} + \frac{1-\nu_i^2}{E_i} \tag{5.4}$$

where P and h are the indentation load and indentation depth, respectively; S is the initial unloading stiffness; E_r is the reduced elastic modulus, which contains the elastic response of both indenter tip and tested material; E_i and ν_i are the elastic modulus and Poisson's ratio of the indenter, respectively; E and ν are the corresponding parameters for sample; the A is the projected contact area.

5.4 Deconvolution for Nanoindentation Statistical Analysis

Maximum likelihood estimation [47] was adopted for the result analysis. The micromechanical properties distribution of AAFA was assumed as a Gaussian mixture model (GMM) as in Eq. 5.5. Each phase was treated as two-dimensional Gaussian distribution with probability density function shown as in Eq. 5.7, where $x = (M, H)^T$ is a column vector.7

$$p(x) = \sum_{k=1}^{K} \pi_k \mathcal{N}\left(x|\mu_k, \sum_k\right) \tag{5.5}$$

$$\sum_{k=1}^{K} \pi_k = 1 \tag{5.6}$$

$$\mathcal{N}\left(\left(x|\mu, \sum\right) = \frac{1}{\sqrt{\det(2\pi \sum)}} \exp\left(-\frac{1}{2}(x-\mu)^T\right) \sum^{-1}(x-\mu)\right) \tag{5.7}$$

where π_k, μ_k and Σ_k are the corresponding weighting coefficient, mean value and covariance of the kth component, respectively.

The log-likelihood function is given by Eq. 5.8 and the mean value (μ), weighting coefficient (π) and covariance matrix (Σ) that make the likelihood function achieve the maximum value for the data set of observations are the estimated parameter values.

$$\ln L = \sum_{n=1}^{N} \ln\left(\sum_{k=1}^{K} \pi_k \mathcal{N}\left(x_n|\mu_k, \sum_k\right)\right) \tag{5.8}$$

EM algorithm was used to achieve this purpose. The E step is shown in Eq. 5.9 and M step shown in Eqs. 5.10–5.12, where $x_n = (M_n, H_n)$ is a row vector with the modulus and hardness data from the nth nanoindentation point. In the E step, $\gamma(z_{nk})$ is a posterior probability that can be regarded as the responsibility that component k takes for 'explaining' the observation x_n [48], or understood as the probability of the given nth nanoindentation point belongs to the kth component in this case. In the M step, the parameters were re-estimated by the current responsibilities. Iteration calculation of E and M steps was made until the convergence of the log-likelihood. K-means algorithm was combined used to find the suitable initial values for parameters to decrease the time for iteration calculation. At least 1000 times of repeated calculation with random initial input values were made to find global maximum log-likelihood function value. Then the final estimated parameter values and GMM model were determined.

$$\gamma(z_{nk}) = \frac{\pi_k \mathcal{N}(x_n|\mu_k, \sum_k)}{\sum_{j=1}^{K} \mu_j \mathcal{N}\left(x_n|\mu_j, \sum_j\right)} \tag{5.9}$$

$$\mu_k = \frac{1}{N_k}\sum_{n=1}^{N} \gamma(z_{nk})x_n \tag{5.10}$$

$$\sum_k = \frac{1}{N_k}\sum_{n=1}^{N} \gamma(z_{nk})(x_n - \mu_k)(x_n - \mu_k)^T \tag{5.11}$$

$$\pi_k = \frac{N_k}{N} \tag{5.12}$$

$$N_k = \sum_{n=1}^{N} \gamma(z_{nk}) \tag{5.13}$$

$$BIC = klnn = 2\ln L \tag{5.14}$$

where π_k, μ_k and $\sum_k$ are the corresponding weighting coefficient, mean value and covariance of the kth component, respectively. N is the total number of observed data, corresponding to nanoindentaion testing data here.

In theory, GMM model can fit any type of probability distribution and higher order of model tends to fit given distribution better. In order to penalize the overfitting errors, Bayesian Information Criterion (BIC) was adopted [49, 50], which is known for more heavily penalizing the model complexity than the Akaike Information Criterion especially when the amount of data N is huge. The model with minimum BIC value is the target model that has a suitable number of components. After the above calculation, the raw testing data was clustered, which was based on the estimated parameters and model by MLE. For a given nanoindentation data point, which belongs to the component where achieves the maximum posterior probability. Confidence ellipses under different confidence levels of 95%, 80% and 75% for each estimated component model were also plotted.

5.5 Nanoindentation Deconvolution Analysis and Results of Alkali-Activated Fly Ash Geopolymer Paste

5.5.1 Compromise Approach for the Mismatch Between Real Components and Micromechanical Peaks

Normally, the number of phases set for the deconvolution of statistical nanoindentation data is the real number of components in the sample [51–53]. However, this kind of match is not always reasonable. For this study, different kinds of fly ash particles composed of different elements would have different nano/micromechanical properties. The micromechanical distribution of fly ash may vary in a large range and overlap other phases such as crystal and mixed phase. The involved range of gel is known as 3–4 times of the nanoindentation depth [54]. Crystals and unreacted particles at the submicron and nano-scale would be mixed with N–A–S–H gel to be identified as mixed phases. For crystal and unreacted particles at a larger scale, the interaction of the different phases (e.g. gel and unreacted FA) at the region around interface would also lead to the identification of mixed phases. Another factor that cannot be ignored is the limited test number in any real test, which would be not able to fully present the real information of micromechanical distribution. Because of very few data acquired, some real phases with a very small proportion may be mixed into other phases instead of identifying as an individual peak in the deconvolution process. In the meantime, due to discrete data, some phases with large proportion and broad distribution are possibly breakdown into several sub-phases in deconvolution. Both the mixed phases and sub-phases are spurious phases and can appear simultaneously. When increasing the number of components for deconvolution, the difference is the sub-phases would increase while the mixed phases would decrease.

The nature of the sample, the test factors and the deconvolution parameters (number of components for MLE) would be responsible for the generation of mixed phases and sub-phases. It is virtually impossible to reconcile these two kinds of spurious phases to obtain the accurate nano/micromechanical properties of each real component in geopolymer by statistical nanoindentation. The strategy of this study is using a "compromise approach", focusing on the most crucial phase N–A–S–H gel merely. The nano/micromechanical properties of some phases like crystals are not able to be obtained by statistical nanoindentation since there are intrinsic limitations from the size and proportion of the phase (nature of the sample), and the involved volume (test factors) as well. However, as a gel phase, N–A–S–H has a large proportion, suitable characterization size. In addition, the nano/micromechanical properties (elastic modulus and hardness) of it would be significantly smaller than the unreacted fly ash particles, crystal, and significant mixed phases. Those factors ensure that the micromechanical peak of the N–A–S–H gel can be separated from mixed phases by increasing the number of components in the deconvolution. In this process, the individual phase N–A–S–H is sometimes obtained by sacrificing the accuracy of other phases as spurious phases (sub-phases) may be introduced.

5.5.2 Deconvolution Results for AAFA-M1-S

For AAFA cured under standard curing condition, the deconvolution results of the phase ($k \leq 9$) or the two phases ($k > 9$) which show the minimum nano/micromechanical properties are listed in Table 5.4. Some key deconvolution results are revealed in Fig. 5.2. These results indicate that when k (the number of phases) is 2 or 3 or 4, the corresponding phase has high variance values. For instance, for the model with 4 components, the elastic modulus of the "blue phase" ranges from very small values to almost 50 GPa, while the hardness of it ranges from around 0 GPa to more than 2 GPa. Besides, high average elastic modulus and hardness values are observed. When the number of components reaches 5, a new phase, phase 2 in Fig. 5.2d with small nano/micromechanical properties and variances is separated from the phase 1 in the model with 4 components. For this new phase, the elastic modulus and hardness standard deviation of it are 3.61 and 0.25, respectively, which are similar to the value reported for the C–S–H gel in cement pastes [55]. In addition, the elastic modulus of 11.20 GPa and hardness of 0.41 GPa of this phase are also close to lots of results reported for low-density C–S–H [56], although a bit low. Therefore, the big phase, phase 1, in the model with 4 components should be a phase mixed by gel and other inclusions, while the phase 2 separated from it in the model with 5 components should be the gel phase. The reinforcing effect of the inclusions enhances the nano/micromechanical properties of the mixed phase. The presence of multiple phases induces a large range of variation for both elastic modulus and hardness.

As revealed in Table 5.4, there is a slight decrease in the elastic modulus and hardness of the gel phase when the number of components in the model increases from 5 to 9. A more significant change occurs when the number of components reaches 10, where the phase 3 in the model with 9 components is decomposed into two small phases, phase A and phase B shown in Fig. 5.2h. The properties of the phase A and phase B remain stable even the number of components increases to 12. Increasing the number of components further makes the calculation difficult to converge with lots of random initial input values and hard to find the real global

Table 5.4 Deconvolution results for AAFA-M1-S (clustered blue points and red points)

k	M (GPa)	H (GPa)	F (%)	BIC	C		
					C11	C12 = C21	C22
2	32.96	2.95	65.30	16,050.70	296.21	32.10	5.26
3	22.32	1.13	29.73	15,568.81	116.77	5.87	0.42
4	23.56	1.18	30.09	15,466.12	126.07	6.66	0.48
5	11.20	0.41	7.27	15,403.81	13.03	0.62	0.06
6	11.18	0.41	7.28	15,363.51	13.01	0.61	0.06
7	10.76	0.36	7.16	15,326.70	11.85	0.49	0.05
8	10.76	0.36	7.16	15,298.94	11.87	0.49	0.05
9	10.76	0.36	7.23	15,278.04	11.92	0.49	0.05
10-phase A	11.08	0.45	5.15	15,255.38	14.24	0.62	0.04
10-phase B	9.87	0.13	1.86	15,255.38	5.66	0.12	0.003
11-phase A	11.37	0.46	5.57	15,243.42	14.93	0.66	0.04
11-phase B	9.89	0.13	1.86	15,243.42	5.70	0.12	0.003
12-phase A	11.10	0.45	5.10	15,233.15	14.19	0.63	0.04
12-phase B	9.86	0.13	1.86	15,233.15	5.63	0.12	0.003
Ave_5-7	11.05	0.40	7.24	–	12.63	0.57	0.06

Note k is the number of phases. The numbers in the column of "M" and "H" refer to mean value of elastic modulus and hardness, respectively. f is the proportion of the phase. C is covariance matrix, where C11 is the variance of elastic modulus, C22 is the variance of hardness and C12 is the covariance of modulus and hardness

optimal solution. Therefore, the calculation is stopped after the model reaches 12 components even if it does not reach the optimal Bayesian value (minimum value).

5.5.3 *Deconvolution Results for AAFA-M0-H*

Table 5.5 and Fig. 5.3 present the deconvolution results for the NaOH activated fly ash sample. For this sample, a phase with nano/micromechanical properties conforming to the empirical knowledge (low-density (LD) C–S–H and N–A–S–H identified in AAFA-M1-S) of the gel phase presents when k is 3. Then, the properties of the phase decrease slightly with the increase in the number of components until 6 components are given to the model. For the models with 3-6 components, the properties of the phase listed in Table 5.5 are similar to some results for standard cured sample AAFA-M1-S in Table 5.4. However, this phase is decomposed into two smaller phases when k is 7, which can be observed from Fig. 5.3b, d, f. These two phases persist even if there are 12 components in the model as shown in Table 5.5.

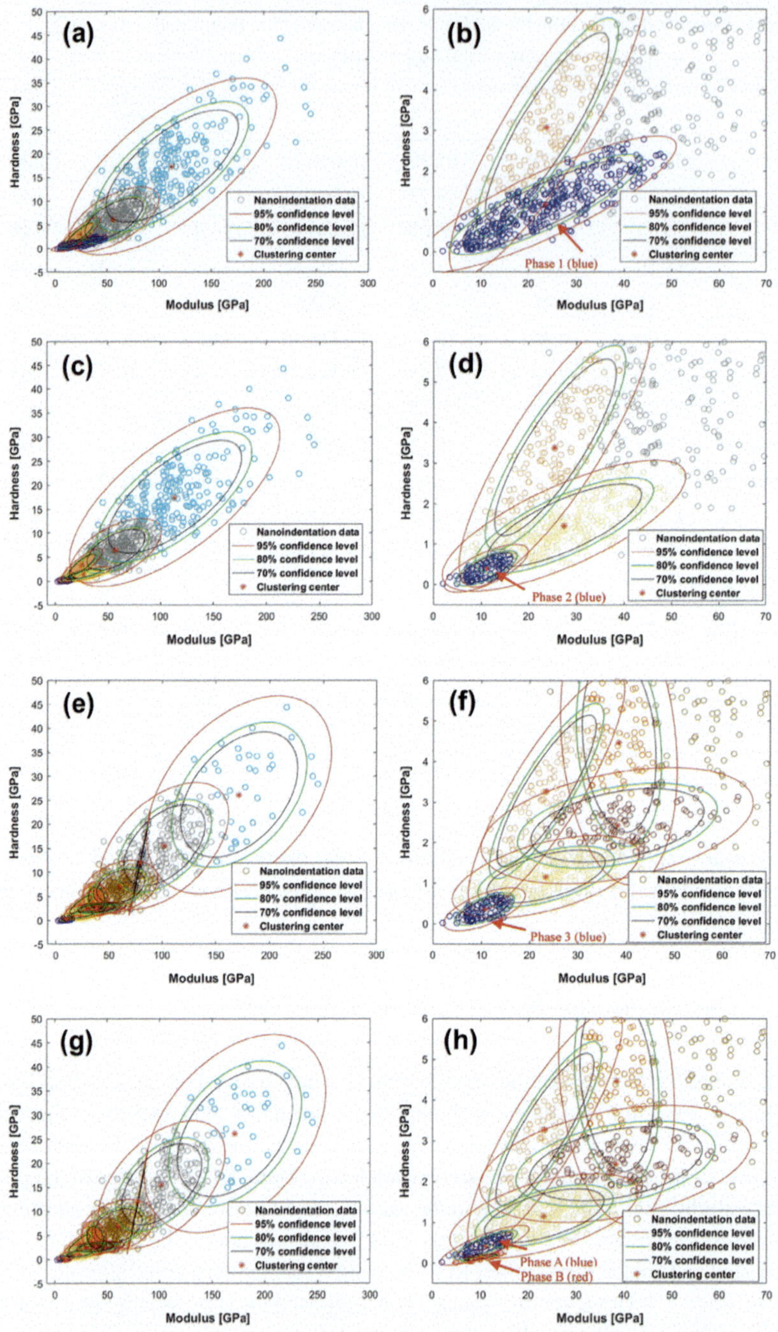

Fig. 5.2 Deconvolution of nanoindentation data (elastic modulus and hardness) for AAFA-M1-S: **a** k = 4; **b** local details for k = 4; **c** k = 5; **d** local details for k = 5; **e** k = 9; **f** local details for k = 9; **g** k = 10; **h** local details for k = 10

Table 5.5 Deconvolution results for AAFA-M0-H (clustered blue points and brown points)

k	M (GPa)	H (GPa)	*f* (%)	BIC	C		
					C11	C12 = C21	C22
2	20.74	1.33	65.81	10,247.74	91.06	6.07	0.64
3	11.49	0.48	17.77	10,054.60	19.84	1.04	0.07
4	10.22	0.39	14.11	9961.75	14.01	0.67	0.05
5	9.78	0.36	12.10	9930.88	12.40	0.57	0.04
6	9.78	0.36	12.13	9911.21	12.41	0.57	0.04
7-phase A	7.84	0.25	8.23	9893.07	6.26	0.24	0.02
7-phase B	13.28	0.60	8.24	9893.07	5.58	0.01	0.03
8-phase A	7.81	0.25	8.13	9882.40	6.32	0.24	0.02
8-phase B	13.40	0.61	9.71	9882.40	7.11	0.08	0.04
9-phase A	7.76	0.25	7.98	9880.84	6.14	0.23	0.02
9-phase B	13.29	0.60	8.26	9880.84	5.77	0.05	0.03
10-phase A	7.83	0.25	8.23	9874.01	6.25	0.24	0.02
10-phase B	13.24	0.60	8.19	9874.01	5.59	0.01	0.03
11-phase A	7.21	0.20	5.78	9865.82	6.17	0.19	0.01
11-phase B	12.80	0.56	10.10	9865.82	7.46	0.11	0.03
12-phase A	7.78	0.25	8.10	9857.43	6.14	0.23	0.02
12-phase B	13.23	0.60	8.60	9857.43	6.05	0.03	0.03
Ave_3-5	10.50	0.41	14.66	–	15.42	0.76	0.05

Note k is the number of phases. The numbers in the column of "M" and "H" refer to mean value of elastic modulus and hardness, respectively. *f* is the proportion of the phase. C11 is the variance of elastic modulus, C22 is the variance of hardness and C12 is the covariance of modulus and hardness

5.5.4 Deconvolution Results for AAFA-M1-H-10%

The deconvolution results for AAFA-M1-H-10% were summarized in Table 5.6. In Fig. 5.4, some critical deconvolution processes were presented to reveal the variation of phases and corresponding micromechanical properties with the increase of components. It is clear that there is a good consistency between clustered data and the GMM model determined by the EM algorithm. The tilt of the axis of the confidence ellipses means that the covariance of modulus and hardness for phases are not zero, namely the corresponding correlation coefficients are not zero and there is a linear relationship between these two properties.

As a gel phase, N–A–S–H would have lower elastic modulus and hardness than unreacted fly ash particles. This feature acted as a criterion in the initial stage to judge whether a new phase presented in the deconvolution process is the possible N–A–S–H phase. When the number of components in the GMM model was set as 2, there is a component with the modulus of 20.80 GPa and hardness of 1.26 GPa (clustered blue points). The properties of this phase change just slightly when the number of

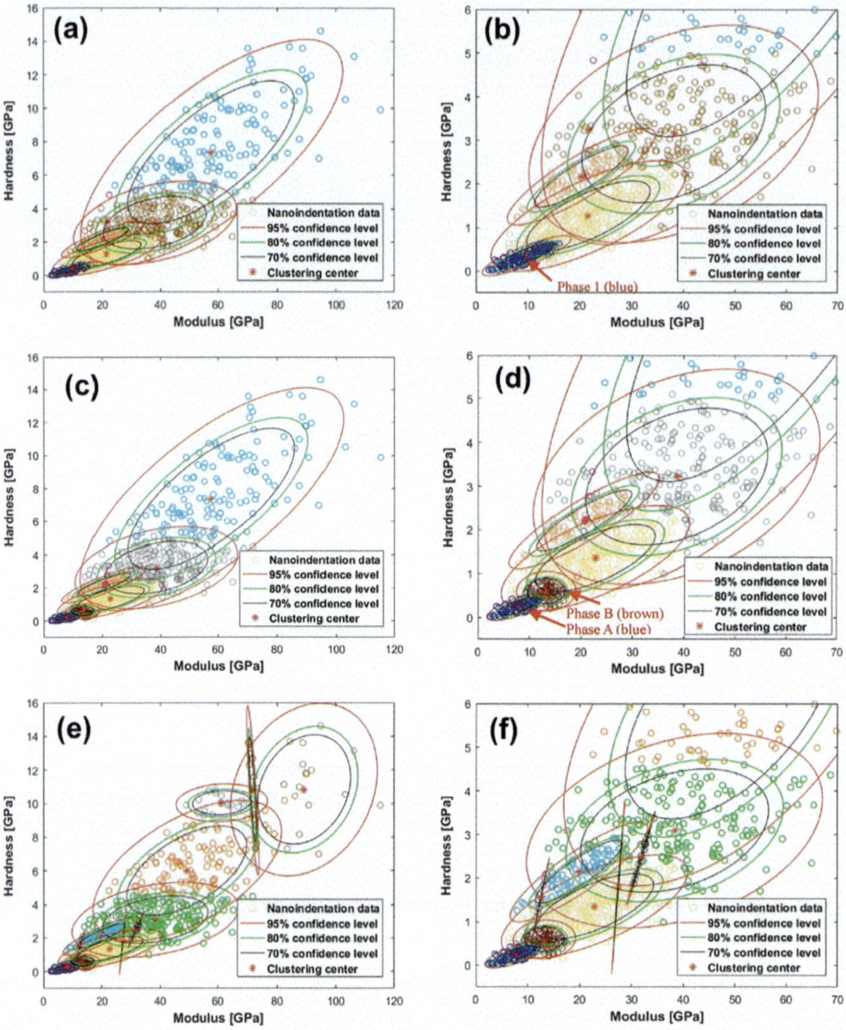

Fig. 5.3 Deconvolution of nanoindentation data (elastic modulus and hardness) for AAFA-M0-H: **a** k = 6 (purple dots is a phase); **b** local details for k = 6 (purple dots is a phase); **c** k = 7; **d** local details for k = 7; **e** k = 12; **f** local details for k = 12

components increases from 2 to 4. However, a new phase (clustered blue points) with a lower modulus of 15.63 GPa and hardness of 0.77 GPa emerges when the number of components in the model reaches 5. More details can be found by the magnified area in Fig. 5.4d. The new phase even presents when the number of components for the GMM model increased to 12 as shown in Fig. 5.4f and Table 5.6. For this study, nine 10 × 10 grids were set for AAFA-M1-H-10%. The total number of test points here is more than those usually used in the nanoindentation test for Portland cement

Table 5.6 Deconvolution results for AAFA-M1-H-10% (clustered blue points for k = 2, 4, 5 and 12)

k	M (GPa)	H (GPa)	*f* (%)	BIC	C	
2	20.80	1.26	44.64	9033.55	53.86	3.33
					3.33	0.27
3	20.26	1.20	40.29	8825.91	49.13	3.01
					3.01	0.23
4	19.83	1.16	38.04	8762.12	46.91	2.85
					2.85	0.22
5	15.63	0.77	10.30	8733.89	12.13	0.72
					0.72	0.06
6	15.73	0.77	10.41	8715.35	12.29	0.73
					0.73	0.06
7	15.46	0.73	8.67	8704.53	11.62	0.67
					0.67	0.05
8	15.53	0.74	8.77	8690.73	11.77	0.68
					0.68	0.05
9	15.39	0.71	8.95	8683.30	12.49	0.72
					0.72	0.05
10	15.43	0.74	9.06	8679.30	11.72	0.68
					0.68	0.05
11	15.32	0.70	7.38	8671.01	12.71	0.71
					0.71	0.05
12	15.21	0.70	7.57	8662.68	11.44	0.66
					0.66	0.05
Ave-grid 5–12	15.46	0.73	8.89	–	12.02	0.70
					0.70	0.05

paste, AAFA and alkali-activated slag paste. However, it should be noted that the BIC still has not reached the optimum value even 12 phases were assumed in the reacted AAFA system (for k = 13, BIC is 8647.16). Further calculation with more components becomes more and more difficult because it takes a longer time and hard to be calculated successfully for many random initial input values. It leads to more difficult to find the real global optimal solution. What more important is when increasing the number of components, the insufficient test data for some phases leads to greater error for estimated results. Therefore, the calculation was stopped at 12 components. For the constant phase existing from 5 to 12 components, the average modulus and hardness are 15.46 GPa and 0.73 GPa, and the corresponding standard deviation of them are 3.47 and 0.22, respectively, which are all close to the properties of low-density (LD) C–S–H reported [56, 57]. It is therefore empirically regarded as the possible N–A–S–H gel phase, and subjected to further discuss in the next

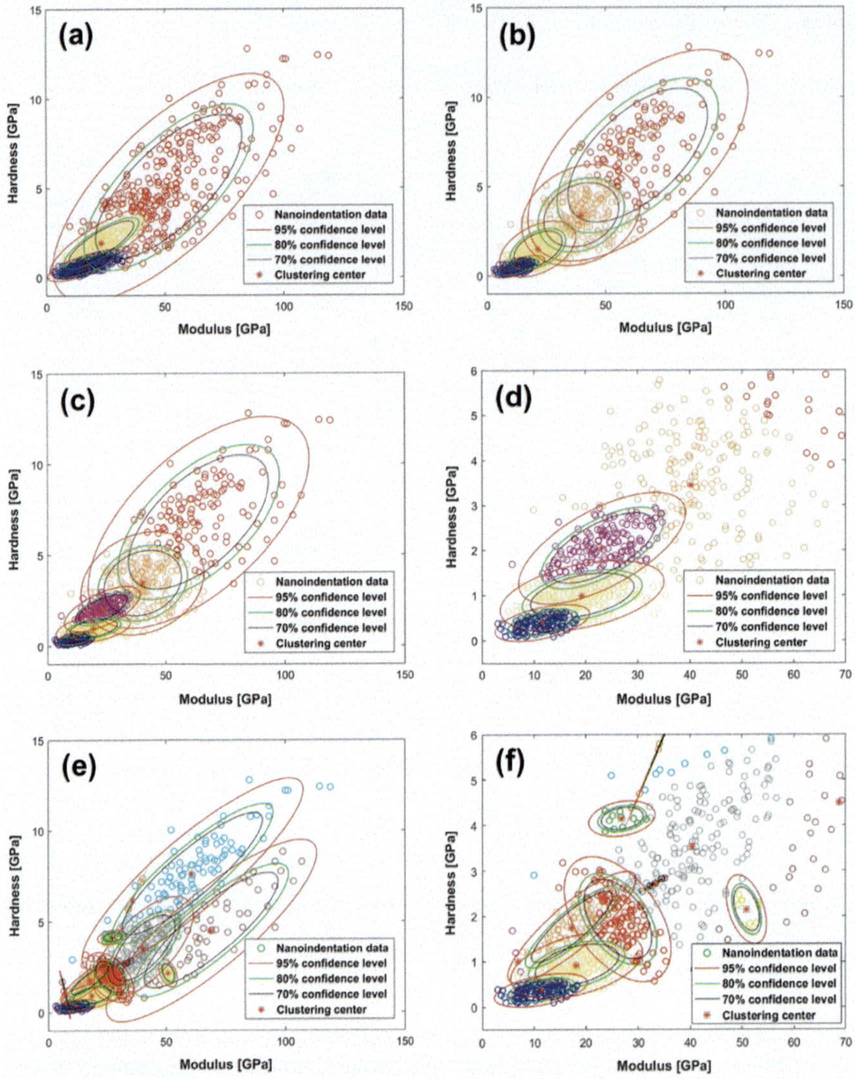

Fig. 5.4 Deconvolution of nanoindentation data for AAFA-M1-H-10%: **a** k = 2; **b** k = 4; **c** k = 5; **d** local details for k = 5 (only partial ellipses); **e** k = 12; **f** local details for k = 12 (only partial ellipses)

section for whether it is a pure phase, why it should be an individual phase instead of spurious phase, and the possible reasons of the small proportion of this phase.

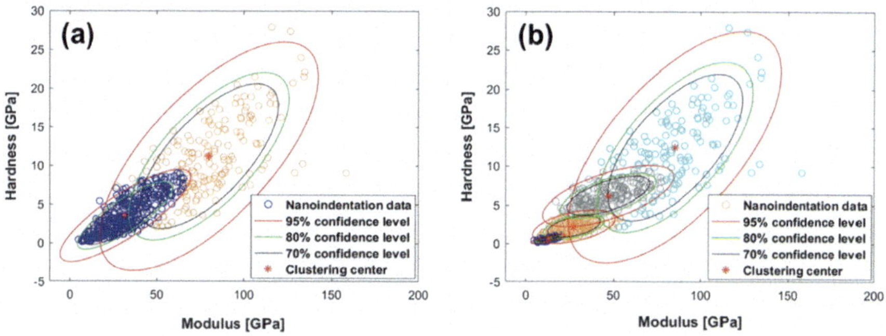

Fig. 5.5 Deconvolution of nanoindentation data for AAFA-M1-H-6%: **a** k = 2; **b** = 4

5.5.5 Deconvolution Results for AAFA-M1-H-6%

The deconvolution results of AAFA-M1-H-6% are shown in Fig. 5.5 and summarized in Table 5.7. For this sample, the deconvolution process is similar to AAFA-M1-H-10% and the possible N–A–S–H phase appears when k is 3 earlier than that of AAFA-M1-H-10%. It becomes a stable phase when k is 4 and shows almost the same value even k reaches 12. This stable phase has average modulus, hardness and fraction of (12.70, 0.62, and 10.67%). The average standard deviation for modulus and hardness are 4.06 and 0.24, respectively, which are similar to that of AAFA-M1-H-10%.

5.5.6 Deconvolution Results for AAFA-M1-H-8%

For the deconvolution of AAFA-M1-H-8%, the results are different from those of AAFA-M1-H-10% and AAFA-M1-H-6%. As shown in Table 5.8 and Fig. 5.6a–d, with the increase of the number of components, a phase presents when k is 3 and shows as a stable phase in the GMM model with 8 components to the model with 12 components. However, based on the tested results for AAFA-M1-H-6% and AAFA-M1-H-10%, this phase is not accepted as a possible N–A–S–H gel phase. The modulus and hardness of this phase are more than normal properties found for the possible gel phase in this study. Besides, the high variance of 61.96 and 0.27 for modulus and hardness also indicate that this phase may not be a pure N–A–S–H gel phase. Typically, when comparing with the deconvolution results for AAFA-M1-H-10%, there is a phase with very similar properties. This similar phase can be seen in Table 5.6 and Fig. 5.4a, b (clustered blue points) when k is 2, 3 and 4. Actually, as can be seen from Fig. 5.4b–f, the possible N–A–S–H phase is separated from this phase. In order to acquire the corresponding N–A–S–H phase as AAFA-M1-H-10%, further deconvolution for AAFA-M1-H-8% is necessary. However, further increase of components in this model is not efficient to deal with this problem as illustrated

Table 5.7 Deconvolution results for AAFA-M1-H-6% (clustered blue points for k = 2 and 4)

k	M (GPa)	H (GPa)	*f* (%)	BIC	C	
2	31.74	3.55	65.78	7315.92	234.90	31.32
					31.32	5.98
3	13.42	0.68	11.66	7158.23	21.23	1.17
					1.17	0.09
4	12.75	0.63	10.83	7086.09	16.98	0.87
					0.87	0.06
5	12.74	0.63	10.83	7052.31	16.95	0.87
					0.87	0.06
6	12.71	0.62	10.76	7035.45	16.77	0.86
					0.86	0.06
7	12.73	0.62	10.81	7014.24	16.85	0.86
					0.86	0.06
8	12.55	0.62	10.14	7001.98	15.88	0.82
					0.82	0.06
9	12.71	0.62	10.20	6993.97	15.95	0.84
					0.84	0.06
10	12.72	0.63	10.54	6984.93	15.59	0.84
					0.84	0.06
11	12.54	0.61	10.60	6979.63	15.93	0.80
					0.80	0.06
12	12.88	0.63	10.97	6938.78	17.43	0.92
					0.92	0.07
Ave-grid 4–12	12.70	0.62	10.67	–	16.48	0.85
					0.85	0.06

previously, and attempts for the GMM models with 13, 14 and 15 components also failed to separate the possible N–A–S–H gel phase.

In theory, upper-level data would be better to be used as the original data of the second deconvolution. Namely, choosing the clustered data of the phase which corresponds to the lowest modulus and hardness in the GMM model with $1 \leq k < m$. m is the number of components where the stable phase with the average modulus of 22.90 GPa and hardness of 1.31 GPa is observed for the first time. This would reduce the risk of information loss of meaningful data but didn't achieve efficient results in this case. In order to simplify the deconvolution process, the nanoindentation data that have been clustered to the phase with the average modulus of 22.90 GPa and hardness of 1.31 GPa were used as the data directly and subjected to the second deconvolution. The clustered data obtained at 12 components are adopted for this deconvolution process.

Table 5.8 Deconvolution results for AAFA-M1-H-8% (clustered blue points for k = 8 and 12)

k	M (GPa)	H (GPa)	*f* (%)	BIC	C	
2	31.66	2.94	72.15	11,191.64	178.21	17.28
					17.28	3.03
3	23.39	1.66	45.36	11,004.61	65.51	3.66
					3.66	0.54
4	24.73	1.45	37.08	10,930.86	85.58	4.82
					4.82	0.39
5	24.62	1.44	36.82	10,893.79	84.00	4.72
					4.72	0.38
6	23.95	1.39	36.83	10,870.81	71.05	3.80
					3.80	0.33
7	24.09	1.40	37.34	10,864.59	73.57	3.95
					3.95	0.34
8	22.95	1.31	30.71	10,850.75	61.36	3.32
					3.32	0.27
9	23.20	1.33	33.42	10,831.49	67.43	3.53
					3.53	0.29
10	22.73	1.30	30.57	10,823.63	60.10	3.22
					3.22	0.27
11	22.96	1.31	30.84	10,819.10	60.60	3.16
					3.16	0.26
12	22.69	1.29	30.25	10,811.48	60.32	3.27
					3.27	0.27
Ave-grid 8–12	22.90	1.31	31.16	–	61.96	3.30
					3.30	0.27

The results from the second deconvolution display in Fig. 5.6e–g and Table 5.9. Figure 5.6e shows the clustered data that belong to the stable phase when k is 12 in the first deconvolution and subjected to the second deconvolution with one component. As the results listed in Table 5.8 (k = 12) and Table 5.9 (k = 12-1), there is a good agreement between the previous Gaussian model for this stable phase and the second time's Gaussian model estimated based on the clustered data. This means it is reasonable to use this data to represent the original stable phase. Similar to the first deconvolution, there is also a stable phase in the second deconvolution that starts at the model with 3 components and lasts until model with 7 components. When comparing with the results of AAFA-M1-H-10% and AAFA-M1-H-6%, this phase was regarded as the possible N–A–S–H phase as they show similar modulus, hardness and covariance matrix value. When k is higher than 7, although smaller BIC value still can be obtained, there will be no more stable phase, and phases with smaller

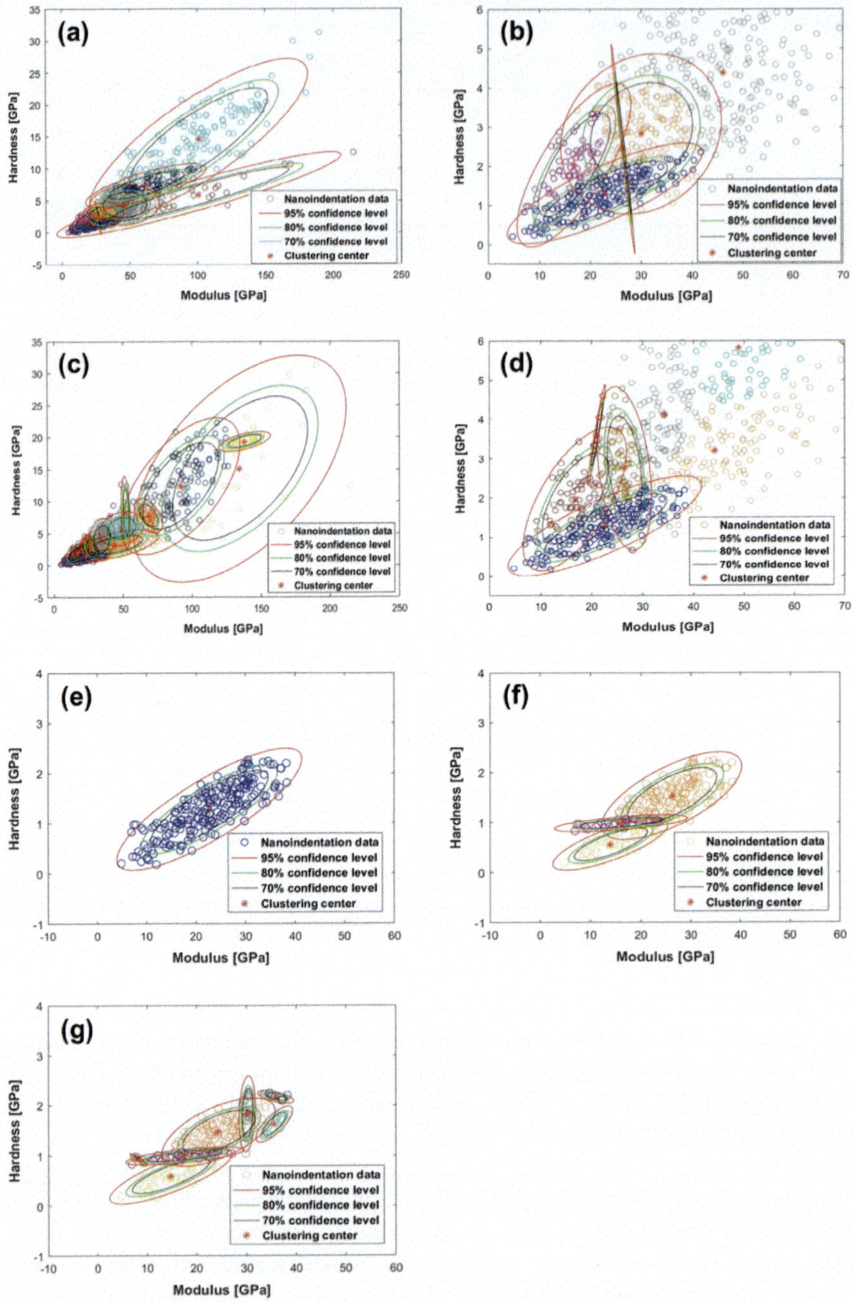

Fig. 5.6 Deconvolution of nanoindentation data for AAFA-M1-H-8%: **a** k = 8; **b** local magnification for k = 8; **c** k = 12; **d** local magnification for k = 12; **e** 2th deconvolution with k = 1 (12-1); **f** 2th deconvolution with k = 3 (12-3); **g** 2th deconvolution with k = 7 (12-7)

Table 5.9 Deconvolution for clustered data belong to stable phase when k is 12 (AAFA-M1-H-8%, clustered yellow points for k = 12-3 and 12-7)

k	M (GPa)	H (GPa)	*f* (%)	BIC	C	
12-1	22.71	1.29	100.00	1473.28	58.54	3.14
			(30.25)		3.14	0.25
12-2	22.20	1.24	93.54	1460.87	58.45	2.96
			(28.30)		2.96	0.22
12-3	14.03	0.56	19.90	1446.38	16.84	0.75
			(6.02)		0.75	0.04
12-4	14.19	0.58	21.06	1435.05	16.90	0.79
			(6.37)		0.79	0.05
12-5	14.67	0.58	20.17	1425.61	19.83	0.84
			(6.10)		0.84	0.04
12-6	15.26	0.63	19.99	1419.11	19.36	0.81
			(6.05)		0.81	0.04
12-7	14.79	0.60	22.74	1416.02	19.68	0.91
			(6.88)		0.91	0.05
Ave-grid 3–7	14.59	0.59	20.77 (6.28)		18.53	0.82
					0.82	0.05

Note The fraction without brackets is the proportion of the component to all components in the second deconvolution, and the fraction within brackets is the proportion of the component to all components in original AAFA reaction system

modulus, hardness and very low percentage occur. As the limited number of data here, more components for deconvolution were regarded as an excessive division.

5.5.7 Deconvolution Results for AAFA-M1.5-H

The deconvolution results for the AAFA-M1.5-H sample are shown in Fig. 5.7 and Table 5.10. For this sample, the variation of the properties of phases starts in a small range when the number of components reaches 3. The more stable phase presents from the model with 5 components. Afterwards, the properties of this phase do not change evidently even if the number of components reaches 12.

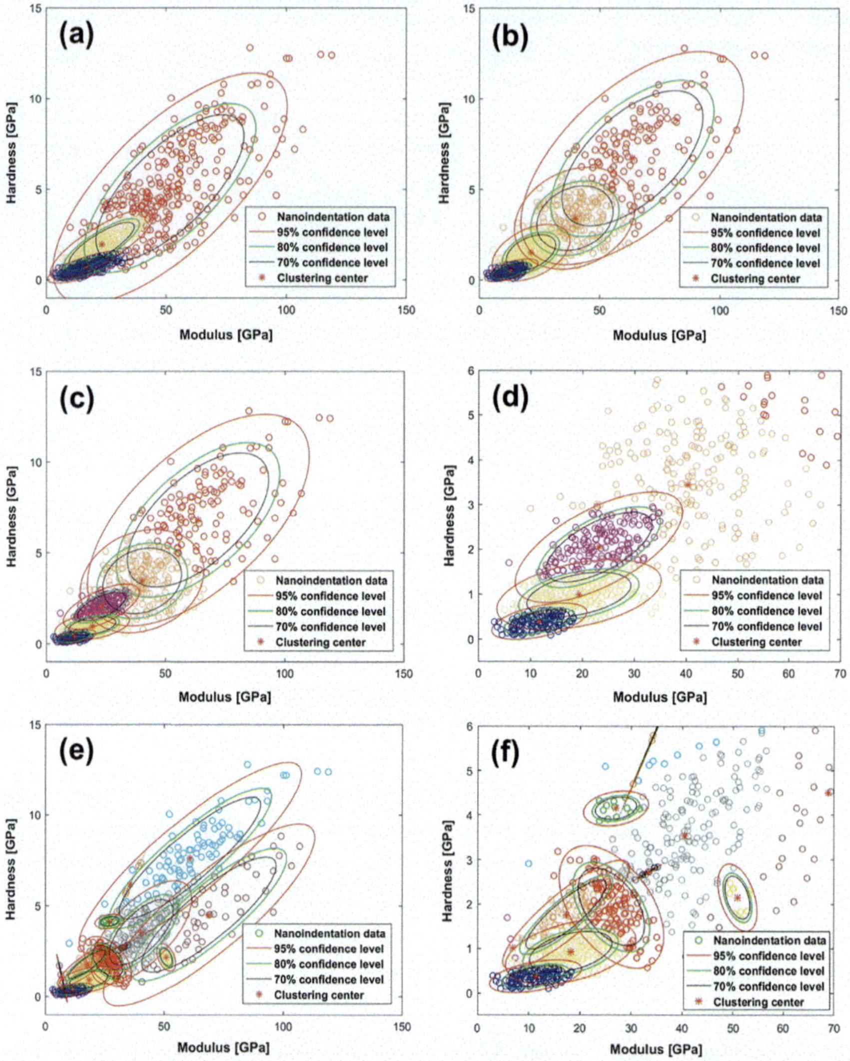

Fig. 5.7 Deconvolution of nanoindentation data (elastic modulus and hardness) for AAFA-M1.5-H: **a** k = 3; **b** k = 4; **c** k = 5; **d** local details for k = 5; **e** k = 12; **f** Local details for k = 12

Table 5.10 Deconvolution result for AAFA-M1.5-H (clustered blue points)

k	M (GPa)	H (GPa)	*f* (%)	BIC	C		
					C11	C12 = C21	C22
2	18.68	1.19	60.21	9330.89	54.53	3.96	0.55
3	16.57	0.71	33.69	9185.27	43.51	1.79	0.13
4	13.21	0.54	22.40	9135.45	20.70	0.76	0.07
5	11.71	0.40	15.21	9088.36	14.02	0.31	0.03
6	11.70	0.40	15.10	9066.53	13.98	0.31	0.03
7	11.72	0.40	15.40	9048.99	14.02	0.31	0.03
8	11.36	0.39	14.63	9033.79	12.81	0.27	0.03
9	11.72	0.40	15.41	9026.87	14.00	0.31	0.03
10	11.37	0.38	13.81	9011.51	13.50	0.26	0.02
11	11.58	0.39	13.95	8996.77	12.14	0.26	0.02
12	11.37	0.38	13.62	8995.79	13.53	0.26	0.02
Ave_4-6	12.21	0.45	17.57	–	16.23	0.46	0.04

Note The numbers in the column of "M" and "H" refer to mean value of elastic modulus and hardness, respectively. f is the proportion of the phase. C is covariance matrix, where C11 is the variance of elastic modulus, C22 is the variance of hardness and C12 is the covariance of modulus and hardness

5.6 Nanoindentation Data Analysis and Discussion for Alkali-Activated Fly Ash Geopolymer Paste

5.6.1 Determination of Gel Phases in Geopolymer

For the deconvolution results given above, one crucial step is to determine the gel phase correctly. Theoretically, the optimal Gaussian mixture model can be determined by the Bayesian Information Criterion (BIC). In this study, the BIC does not reach the optimum value, which means more components are needed. Although the number of test points in this study is even more than three times of some other studies, the data is still sparse at some locations, which would lead to peaks for subphases. The presence of phases with a very small proportion in models indicates that some points are misidentified as an individual phase. In fact, BIC evaluates the whole model. It is not necessary to achieve the optimal as this study focuses on the gel phase only. The empirical knowledge of the gel phase is used for determining the N–A–S–H gel.

As typically analyzed above for AAFA-M1-S, the phase in the model with 4 components and has high elastic modulus and hardness of 23.56 GPa and 1.18 GPa should be the mixed phase, while the phase in 5 components model with elastic modulus and hardness of 11.20 GPa and 0.41 GPa, respectively, is consistent with the features of gel phase. It is easy to find that there are similar mixed phase and gel phase in the other three kinds of samples. When further increasing the number of components, the nano/micromechanical properties and proportion of the gel phase in AAFA-M1-S and especially AAFA-M0-H are decreased. In the deconvolution process, few components for a model may make the phase that has minimum nano/micromechanical properties a mixed phase as mentioned. Actually, too many components may result in the excessive separation of the gel phase, breaking the single gel phase into subphases. The gel phase in 5–9 components models of AAFA-M1-S just has a small proportion of slightly more than 7.27%, the breakdown of the single phase into two phases with very small proportions (even 1.86%) should be more ascribed to the spurious phase generated when too much phase number assigned to the model. For the results slightly changed from the 5 components model to 9 components model, it is hard to ascertain if it is a refinement of the result by removing more inclusion or an excessive separation of the gel phase. In this study, if the possible gel phase lasts more than three models, then the final properties of the gel phase are determined by the average value of the first three results to reduce the impact of uncertainty (average all gel phases when less than 3). Considering the fact that geopolymer at least contains gel, mixed phase, and unreacted fly ash, the possible gel phase should be selected from the model by at least 3 components. For AAFA-M1-S, it is the average result from 5 components model to 7 components model as listed in the last row of Table 5.4. Correspondingly, it is the average result of 3 components model to 5 components model for AAFA-M0-H, 12-3 components model to 12-5 components model for AAFA-M1-H-8% and 4 components model to 6 components model for AAFA-M1.5-H. The phase in the 3 components model of AAFA-M1.5-H is not accepted as the gel phase as it has large variance and a large range of variation of both elastic modulus and hardness.

5.6.2 *Evaluation of Deconvolution Error*

In order to compare the nano/micromechanical properties of N–A–S–H in different geopolymers, one of the key issues is to evaluate the errors generated from the analytical technique itself. The first error should be the average method we adopted to represent the nano/micromechanical properties of gels. Illustrated by the case of AAFA-M1-S, the possible elastic modulus of gel for it is 11.20, 11.18 or 10.76 GPa. For the average value of 11.05 GPa adopted, the corresponding deviation is 0.15, 0.13 or 0.29 GPa, as shown in Fig. 5.8. For all these samples, the maximum possible deviations are 0.99 GPa for elastic modulus, 0.09 GPa for hardness and 4.83% for proportion. The average value of all those deviations (absolute value) for elastic modulus, hardness, and proportion are 0.44 GPa, 0.04 GPa, and 1.37%, respectively.

Another possible error can be ascribed to the test data, although the number of points (Nine grids) we tested is more than conventional research. The number of grids was increased to 18 for AAFA-M0-H to check if there are significant different results for the properties obtained from 9 and 18 grids. Since this study aims at the gel phase only, the additional grids were intentionally put on the areas that very rich in gel to

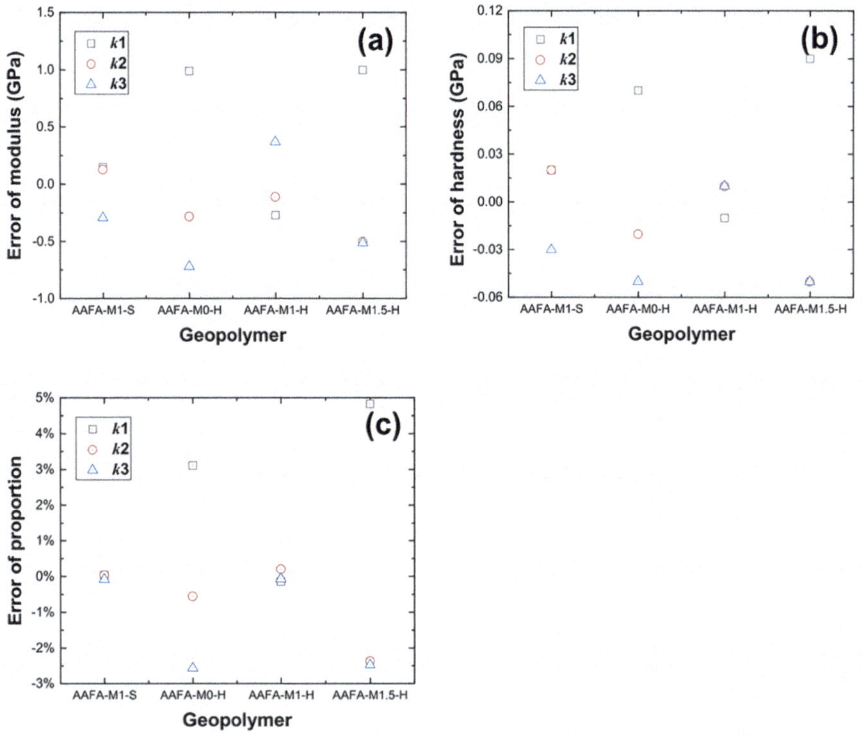

Fig. 5.8 Errors result from the average method to determine the nano/micromechanical properties of gel: **a** error of modulus; **b** error of hardness; **c** error of proportion

provide more data for the analysis of gel. Some typical results for AAFA-M0-H based on 18 grids are compared with the results from 9 grids and shown in Fig. 5.9. The micromechanical results from models are (11.08, 0.57) for 3 components and (9.43, 0.42) for 4 components, respectively. There are close to the corresponding values obtained from the deconvolution of 9 grids as given in Table 5.5. When increasing the number of components, there will be no more similar gel phase. Namely, the average properties for N–A–S–H gel based on 18 grids are (10.26, 0.50).

One special phenomenon for AAFA-M0-H is the presence of two phases when the number of components is more than 7, while both of these two phases have a considerable proportion. For the sake of comparison, the 7 components model based on 18 grids is provided in Fig. 5.9g, h. Similar to that in Fig. 5.3, there are also two phases. The blue phase and the brown phase have the properties of (6.99, 0.25) and (11.27, 0.56), respectively. It indicates that the two phases are not generated due to the limited test data. It is easy to track that these two phases are mainly decomposed by the gel phase in models that have few components, although some points are removed or included during this process. The average properties of these two phases in the 7 component model based on both 9 grids and 18 grids are obtained by the cluster of data for each phase. As shown in Fig. 5.9, the nanoindentation data that belong to

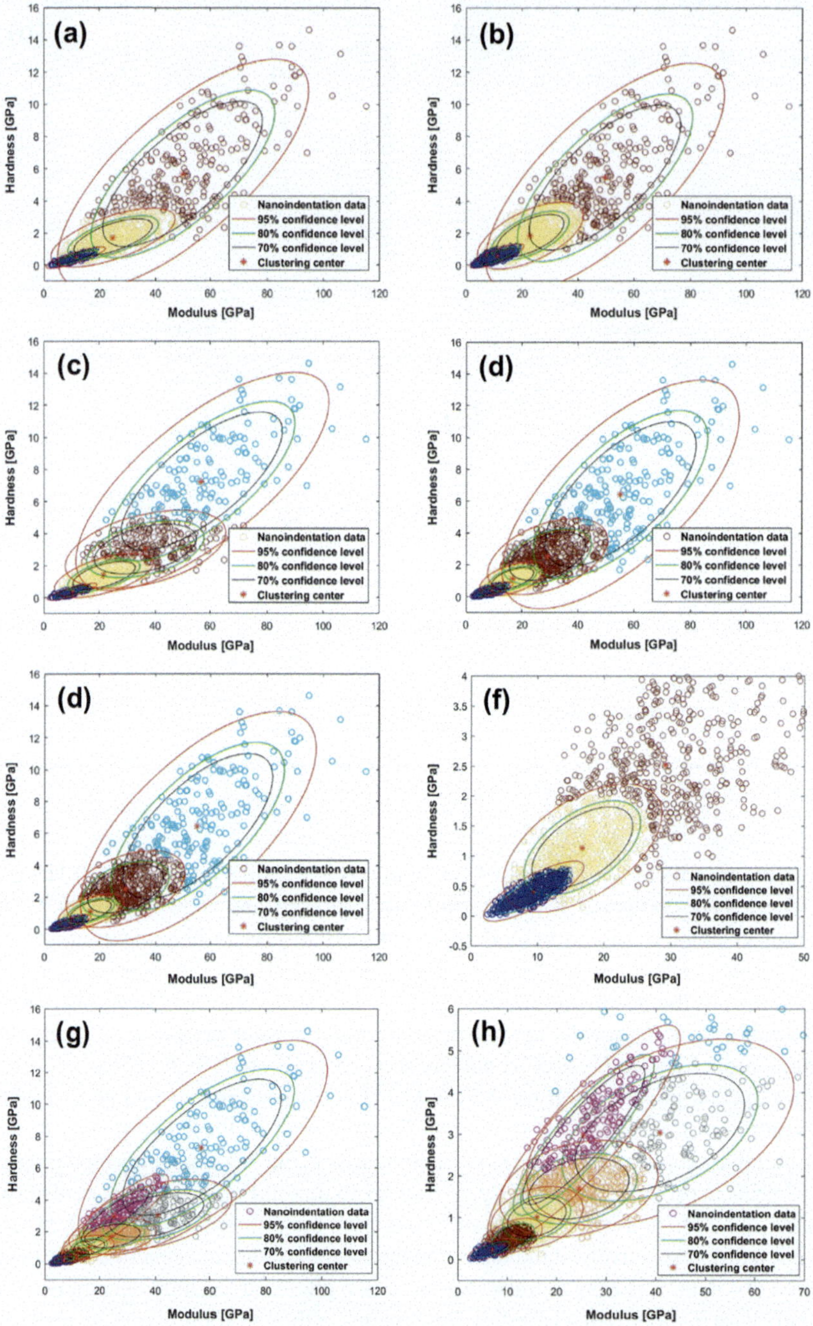

Fig. 5.9 Errors result from the limited test data of statistical nanoindentation: **a** k = 3 (9 grids); **b** k = 3 (18 grids); **c** k = 4 (9 grids); **d** k = 4 (18 grids); **e** local details for k = 4 (9 grids); **f** local details for k = 4 (18 grids); **g** k = 7 (18 grids); **h** local details for k = 7 (18 grids)

the phases with nano/micromechanical properties of (6.99, 0.25) and (11.27, 0.56) are clustered into blue points and brown points according to the maximum posterior probability. The blue points and brown points are collected and plotted in Fig. 5.10a, b, with their average value denoted by the clustering center, which are (6.75, 0.23) and (11.32, 0.55), respectively. Namely, the clustered data can well represent the estimated phase in the deconvolution of AAFA-M0-H based on 18 grids. Then, the blue points and brown points are combined together. The average properties of all those points are (9.60, 0.43) as shown in Fig. 5.10c. Similarly, the average properties of all those points in AAFA-M0-H based on 9 grids are (10.66, 0.42). The average properties of blue and brown phase are similar in models with different grid numbers. Moreover, the average properties of these two phases are very close to the properties of the gel phase. It is possible that the blue phase is the gel with higher porosity than the brown phase.

The errors result from limited test data are summarized in Fig. 5.10d. For the elastic modulus and hardness of N–A–S–H we study, the deviation of results is just − 0.24 and 0.09 GPa, respectively. Taking into account the above errors caused by

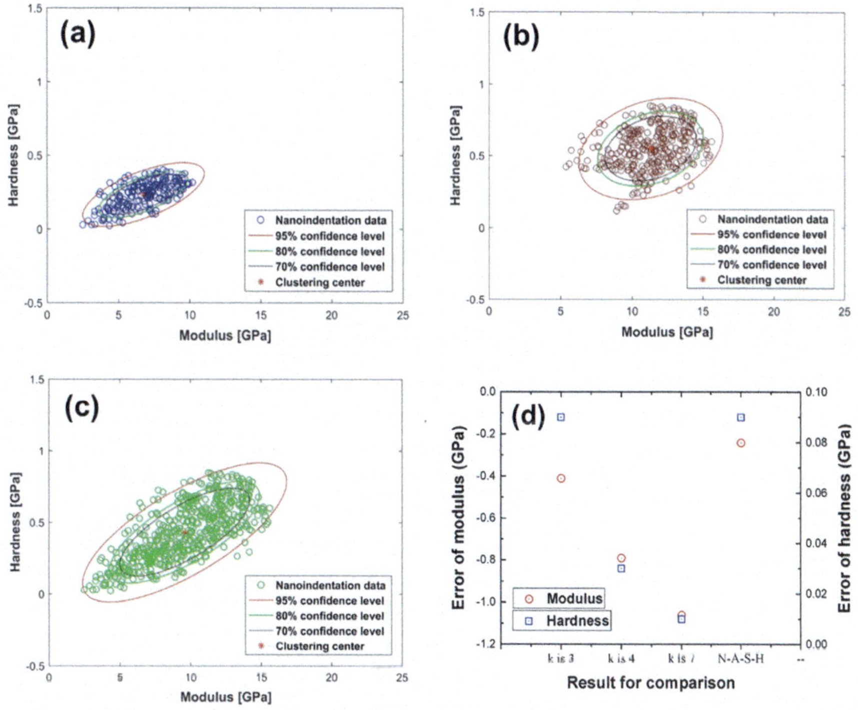

Fig. 5.10 Average properties of the decomposed phases in the 7 component model of AAFA-M0-H (18 grids), and results of error evaluation: **a** clustered data for estimated phase with M of 6.99 and H of 0.25; **b** clustered data for estimated phase with M of 11.27 and H of 0.56; **c** average properties of two decomposed phases; **d** deviation of results between deconvolution of 9 grids and 18 grids data

the average method to determine the properties of gel phase, the approximate errors for elastic modulus and hardness are 0.68 and 0.13 GPa. Meanwhile, the maximum deviations are obtained as 1.23 GPa for elastic modulus and 0.18 GPa for hardness. One thing to note is the results are based on the sum of the absolute values that would enlarge the evaluation of the errors. Namely, the positive and negative errors do not cancel each other out.

5.6.3 Pure N–A–S–H Phase Properties in Alkali-Activated Fly Ash Geopolymer Paste

When conducting the nanoindentation test for C–S–H or N–A–S–H in cement paste material, there are several critical issues that determine whether the micromechanical properties of gel detected are based on pure individual phase [58]. Typically, intermix pores or other small phases in the gel phase are suspected to influence the accurate measurement of modulus and hardness of the gel phase. Choosing the suitable indentation depth and deleting the abnormal indentation test points help partially avoid these problems but may not be all.

In order to verify these issues, Davydov et al. [9] conducted comparative research and pointed that part of the CH is mixed with C–S–H because the percentage of CH detected by nanoindentation is lower than that by Back-scattered Electron Detector (BSE) and XRD. Besides, even test points precisely conducted on the same target phase C–S–H_{HD}, big scatter properties results were obtained and the peak corresponding to CH can even sometimes be observed directly on the frequency densify figure. The number of component k was set as 4 to keep it the same as the k usually used in the nanoindentation test [9]. The total proportion of low-density (LD) C–S–H and high-density (HD) C–S–H reached a high level of 85% with only 10% CH. This proportion is not reasonable and means the existence of mixed phase as put forward by Davydov et al. [9]. These facts also hint that when the nanoindentation data is sufficient, increasing the number of components may be able to separate purer gel phase since there is a difference between the micromechanical properties of gel and gel with significant inclusions. At the same time, it should be noted that the modulus and hardness of LD C–S–H and HD C–S–H reached a high value of (30.10, 1.25) and (36.23, 1.55), respectively in that study [9]. For the other studies [11, 22] that also identified and verified a mixed phase, that phase also has high modulus and hardness and is called ultra-high density (UHD) phase. The UHD phase is an intimate nanocomposite where nanoscale CH reinforces C–S–H by partially filling the latter's gel pores. The modulus and hardness of this phase achieved a high value of 47.2 GPa and 1.6 GPa, respectively. Therefore, mixed phase is characterized by high mechanical properties (e.g. H reflects strength) as the reinforcing effect by inclusions. As typically shown in Fig. 5.4, the clustered sandy brown points span a very large range of modulus and hardness value. Some of the points reach high modulus and hardness of almost 40 GPa and 2 GPa, respectively. Besides, the similar phase does not

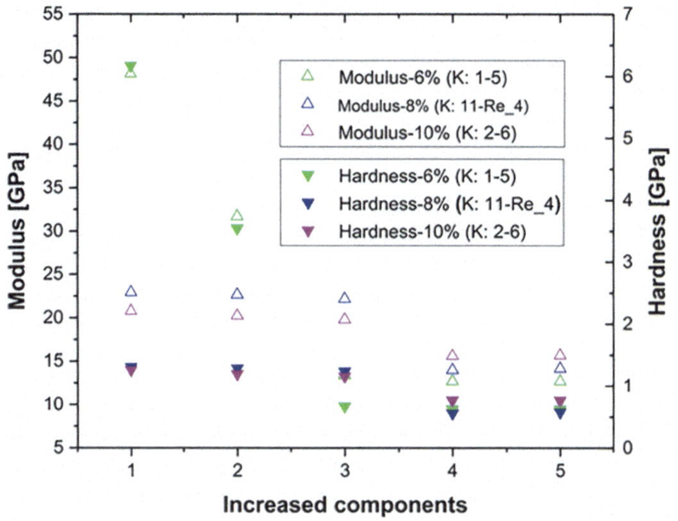

Fig. 5.11 Separate N–A–S–H from mixed phase in AAFA

exist in AAFA-M1-H-6%. These features indicate that the phase shows high average modulus and hardness of (22.90, 1.31) in AAFA-M1-H-8% as shown in Table 5.8 and exists in AAFA-M1-H -10% when k is 2, 3 or 4 as shown in Table 5.6 is one of the mixed phases instead of a single gel phase. The final stable phases separated from mixed phase with small modulus, hardness and proportion in the range of 12.7 to 15.46 GPa, 0.59 to 0.73 GPa and 6.28% to 10.67% in this study were therefore regarded as the purer gel phases. The separation process can be seen in Figs. 5.4, 5.5 and 5.6, and the variation of modulus and hardness in key separation steps is given in Fig. 5.11.

In order to verify if the possible pure phase obtained can be further divided into purer N–A–S–H phase, deconvolution analysis was conducted on the clustered data that belongs to the 'possible N–A–S–H phase', as typically for AAFA-M1-H-10% with 12 components shown in Fig. 5.12 and Table 5.11. This process is conducted for a maximum of 5 components and this number is considered as more than necessary. For 2 and 3 components, the 'possible N–A–S–H phase' ($k = 1$) is divided into a major phase and one or two phases with a small proportion. The major phases still show similar modulus and hardness as 'possible N–A–S–H phase'. When k reached 4 and 5, different results present and there are two phases with considerable proportion, the clustered blue points and the clustered sandy brown points.

For this study here, even if up to nine 10×10 nanoindentation grids were tested, due to the small proportion of the 'possible N–A–S–H phase', nanoindentation testing points for this deconvolution are still relatively sparse, especially when points away from the centre of the Normal model. The sparse data would undoubtedly influence the clustered results and result in unreliable new phases. Besides, in the deconvolution process, some new phases are generated based on only 3 tested points (e.g. $k = 2$) or

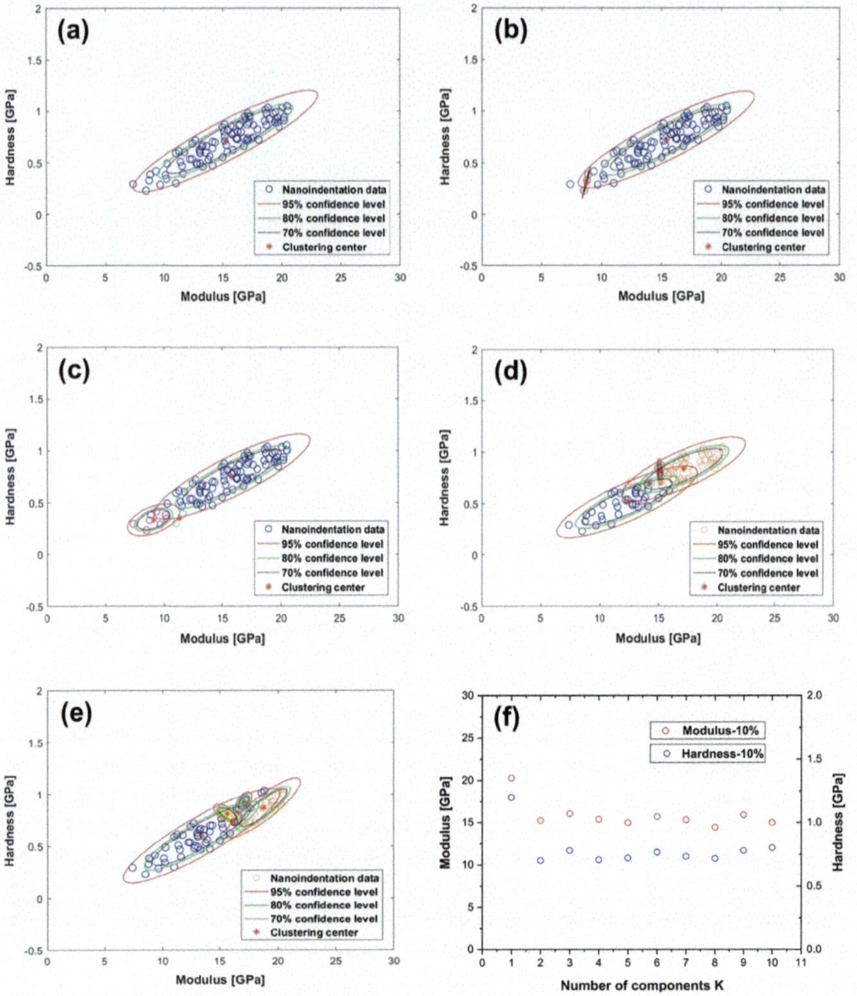

Fig. 5.12 Deconvolution of clustered data belongs to possible N–A–S–H (AAFA-M1-H-10%, k = 12) and mixed phases (AAFA-M1-H-10%, k = 3): **a** k = 1 (12-1); **b** k = 2 (12-2); **c** k = 3 (12-3); **d** k = 4 (12-4); **e** k = 5 (12-5); **f** deconvolution of mixed phase (3-1 to 3-10)

2 tested points (e.g. k = 3, 4, 5). That would not be able to be identified as a phase in a test that has near 800 valid tested points due to the negligible proportion. These phases consist of 2 or 3 points also cannot be regarded as testing errors as they are at varied locations shown in Fig. 5.12 in the deconvolution process. Although there is a phase with modulus and hardness similar to the average results in AAFA-M1-H-6% when k is 4 or 5, this phase does not show as a stable phase, and the fraction even increased with k. Besides, this phase shows much smaller proportion and variance

Table 5.11 Deconvolution for clustered data belong to possible N–A–S–H phase (AAFA-M1-H-10%, k = 12)

k	M (GPa)	H (GPa)	*f* (%)	BIC	C	
12-1	15.21	0.71	100.00	288.04	10.19	0.60
					0.60	0.04
12-2	15.41	0.72	96.88	280.26	8.76	0.47
					0.47	0.03
12-3	15.80	0.75	90.38	218.08	7.40	0.39
					0.39	0.03
12-4-1	17.16	0.84	56.75	209.82	4.66	0.22
					0.22	0.02
12-4-2	12.33	0.52	37.65	209.82	6.01	0.29
					0.29	0.02
12-5-1	13.38	0.60	58.26	178.59	7.65	0.46
					0.46	0.03
12-5-2	18.74	0.87	24.59	178.59	1.70	0.13
					0.13	0.01

than that in AAFA-M1-H-6%. When conducting deconvolution for 'possible N–A–S–H phase' in AAFA-M1-H-6%, the phase with the average modulus of 12.70 GPa and hardness of 0.62 GPa no more exists when reaching similar small proportion and variance, and also no stable phase appears. Therefore, the further deconvolution for the 'possible N–A–S–H phase' is regarded as an excessive division of this phase, at least for the number of available data here.

The second round of verification is conducted by deconvolution of clustered data that belongs to mixed phase for AAFA-M1-H-10% with 3 components. The clustered data contain all possible data for N–A–S–H but do not have lots of data that belong to unreacted fly ash, which are upper-level data for N–A–S–H gel and treated as a mixed phase. This deconvolution process can more strictly separate the N–A–S–H than deconvolution of all tested data, but less strict than the deconvolution of 'possible N–A–S–H phase' directly. The results in Fig. 5.12f show that the 'possible N–A–S–H phase' presents when k is 2 and is always there even the number of components set as 10, although some data within the ellipses of the 'possible N–A–S–H phase' are divided into individual phases when k is 8, 9 and 10 due to excessive division. This process still proved that the N–A–S–H phase is still a stable phase at a relatively stricter extent and verified the feasibility of the way of separating N–A–S–H phase from the mixed phase for AAFA-8%.

In summary, there is a clear separation of the 'possible N–A–S–H phase' from the mixed phase. This phase behaves as a relatively stable phase in the deconvolution process and is not suitable to be separated anymore based on the available data. Therefore, it is accepted as the pure N–A–S–H phase in this study. However, this is the pure phase that can be obtained by the deconvolution technique, not necessarily the real pure phase defined by the chemical composition. For the N–A–S–H phases obtained, they show similar modulus, hardness and standard deviation as previous research for the LD C–S–H gel phase in Portland cement paste.

5.6.4 Number of Phases, Bin Size and Feasibility of PDF by MLE

For the deconvolution of nanoindentation data, the number of phases is usually determined by empirical estimation or features of frequency distribution histogram. In this study, Bayesian Information Criterion is used for choosing the number of phases. It should be noted that the Bayesian Information Criterion aims at the whole model, thus it is not always ideal for the investigation of the micromechanical properties of pure N–A–S–H, and it can just be used for the preliminary estimation of the number of phases.

For alkali-activated fly ash, the raw material fly ash is a highly heterogeneous material composed of amorphous phases SiO_2 glass, Si-Al glass, Si-Al [Na, K, Ca] glass, etc. and also many crystalline phases such as mullite, quartz, and iron oxides hematite, with size ranges from 0.5 to 300 μm [30–32]. It makes the activated system more complex than the Portland cement paste. Besides, there are a considerable number of pores on the surface of fly ash particles and new pores can be generated in the dissolution process, which further increases the divergence when testing micromechanical properties. This can be observed directly from Figs. 5.4, 5.5 and 5.6 for the unreacted fly ash phases, the biggest ellipse means the biggest deviation of data. All these factors make the investigation of nanomechanical properties of pure N–A–S–H more difficult due to the presence of more kinds of partly-activated and unreacted particles and mixed phases.

For most of the cases, the number of components set for current research of modulus of Portland cement paste and AAFA is 3 or 4 and bin size is 1 or sometimes 2.0 GPa, based on the LSE method for deconvolution. In this study, when adopting the MLE method, the number of phases for AAFA-M1-H-8% and AAFA-M1-H-10% are more than the typically used number. Besides, The LSE method obtains the mechanical properties of different phases by fitting the probability density functions (PDF), or sometimes the cumulative distribution function (CDF). The former is the most commonly used one. It can display the distribution of modulus and hardness of different phases intuitively based on frequency distribution histogram, but is also questioned for the artificial choice of the bin size. In order to make some comparisons, the PDF obtained in this study by MLE is plotted with frequency distribution histograms as shown in Figs. 5.13 and 5.14.

PDF of GMM models obtained based on a different number of phases is given in Figs. 5.13 and 5.14. As listed in Table 5.6, the N–A–S–H gel appears when k is 5, which can also be observed by modulus frequency density histogram in Fig. 5.13a–c for the necessity of treating this phase as an individual phase. When k is 3 or 4, the

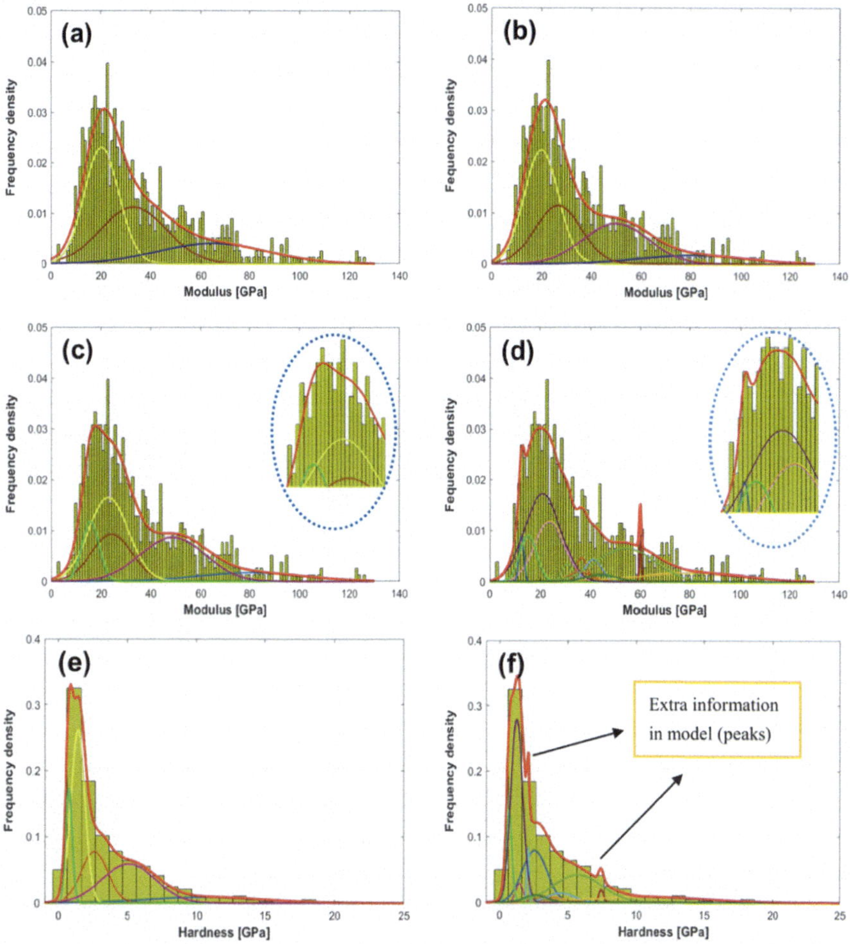

Fig. 5.13 Comparison on PDF obtained by MEL with frequency histogram: effect of number of components: **a** modulus k = 3 (Bin = 1 GPa); **b** modulus k = 4 (Bin = 1 GPa); **c** modulus k = 5 (Bin = 1 GPa); **d** modulus k = 12 (Bin = 1 GPa); **e** hardness k = 5 (Bin = 1 GPa); **f** hardness k = 12 (Bin = 1 GPa)

model failed to include an evident phase (gel phase) displayed by 'green' colour in Fig. 5.13c, which leads to worse matching between the two models and histograms in the adjacent area of this phase than that for 5 components. When increased the number of components in the GMM model to 12, the corresponding PDF in Fig. 5.13d can reflect information in the modulus histogram better than the GMM model with 5 components. For hardness frequency density histogram, as revealed in Fig. 5.13e, f, the GMM model with 12 components is even less match histogram than the GMM model with 5 components. For this phenomenon, before questioning the estimated

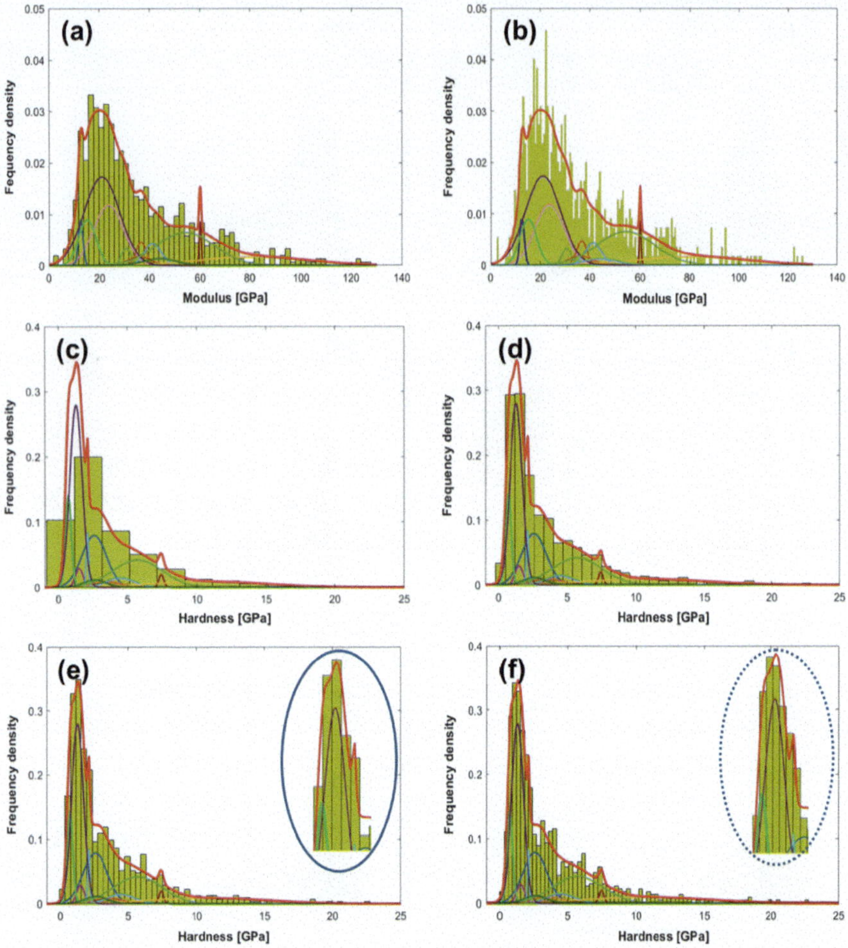

Fig. 5.14 Compare of PDF obtained by MLE with frequency histogram: effect of different bin size: **a** modulus k = 12 (Bin = 2 GPa); **b** modulus k = 12 (Bin = 0.7 GPa); **c** hardness k = 12 (Bin = 2 GPa); **d** hardness k = 12 (Bin = 0.7 GPa); **e** hardness k = 12 (Bin = 0.4 GPa); **f** hardness k = 12 (Bin = 0.3 GPa)

model, the more important concerns should be the different bin sizes needed for modulus histogram and hardness histogram.

The effect of bin size was studied. Firstly, the normally used bin size of 1.0 or 2.0 GPa in deconvolution of modulus data was investigated as shown in Figs. 5.13d and 5.14a. When the number of bin size only increased to 2.0 GPa, there is no distinct difference for the frequency distribution histogram. In order to reveal more information from the histogram, bin size was decreased to 0.7 GPa. Due to the large range of modulus value and corresponding low frequency density and especially the limited number of test points, the decrease of bin size leads to more discrete data instead of detailed frequency density information in Fig. 5.14b. Due to the significantly smaller distribution range and higher frequency density, the change of bin size in hardness frequency histogram leads to a different phenomenon. For large bin size of 2.0 GPa, as shown in Fig. 5.14c, it is difficult to identify hardness distribution information of individual phases from the histogram, and the peaks of the histogram clearly deviate from the peaks in the histogram with a bin size of 1 as shown in Fig. 5.13f. When decreased the bin size from 1 to 0.7 GPa as shown in Fig. 5.14d, the histogram is also markedly changed, which means the bin size adopted would influence the LSE results for hardness. Further decrease of bin size to 0.4 and 0.3 GPa is accompanied by a better match of the PDFs estimated by the MLE method with histograms. Some extra information (peaks) didn't show in the histogram in Fig. 5.13f has appeared. Namely, for the model that can match modulus histogram with the bin size of 1, it matches the hardness histogram well only when the bin size decreased to at least 0.4. For a given test data set, it is very difficult to specify the bin size reasonably for the LSE analysis. The needed bin size is at least determined by factors such as the nature of samples, the types of the histogram and the number of the test data. The inappropriate bin size may either lose the real information or lead to spurious peaks and irregular histogram.

When using the LSE method, enough test points are needed to allow reasonably small bin size to show the details of distribution and avoid the spurious peaks and irregular histogram. This point can be observed by the histograms from 5 grids in Fig. 5.15, where 5 grids-1st means 5 grids randomly chosen from 9 and 5 grids-2nd consists of the remaining 4 grids and 1 random grid from the 5 grids-1st. It is obvious that when the bin size is 1, the shape of the histogram of 5 grids-1st and especially the 5 grids-2nd is more discrete and irregular than the corresponding histogram of 9 grids in Fig. 5.13c, increasing the difficulty of fitting by LSE. Besides, as shown in Fig. 5.16a, b and Table 5.12, the deconvolution results of the 5 grids-1st based on 5 components are quite close to the results from 9 grids. It means that the PDF determined by the MLE method from 5 grids-1st still can reflect the micromechanical properties distribution of AAFA and is reliable. The mismatch of PDF and histogram in Fig. 5.15a implies that when the test data are not enough, specifying of small bin size forcibly for a histogram and then fitting it by the LSE method would cause deviation in results. In Fig. 5.15b, in the result of 5 grids-2nd, the corresponding deconvolution result with 5 components doesn't generate the gel phase. However, the gel phase occurs when the number of components specified for deconvolution is raised to 6 as can be observed in Fig. 5.16c, d and Table 5.12. Decrease of grid number from 9 to 5 significant increases the difficulty for LSE method as it is dependent on the histogram, while the MLE method helps to reveal the properties of gel phase and other key phases from all the three kinds of data sources and the results from 9 grids, 5 grids-1st, and 5 grids-2nd are similar to each other. It clearly displays the reliability and stability of the MLE method. In addition, the deconvolution process again indicates that the number of components needed to generate the gel phase in the model should not be a constant value of 3 or 4 and would change with both sample and test factors.

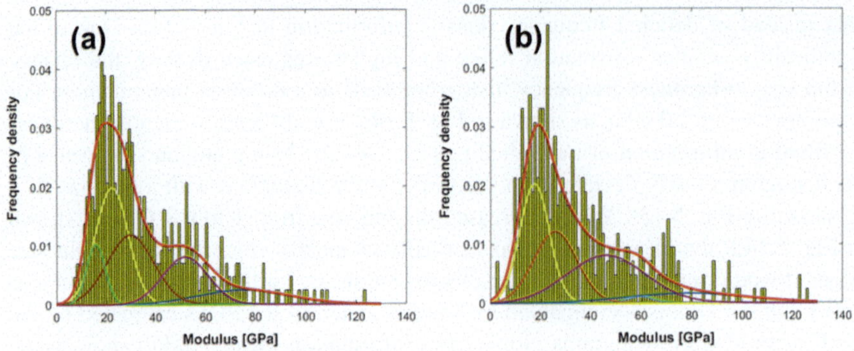

Fig. 5.15 Compare of Modulus PDF obtained by MLE with frequency histogram: 5 grids data: **a** 5 grids-1st (k = 5, Bin = 1 GPa); **b** 5 grids-2nd (k = 5, Bin = 1 GPa)

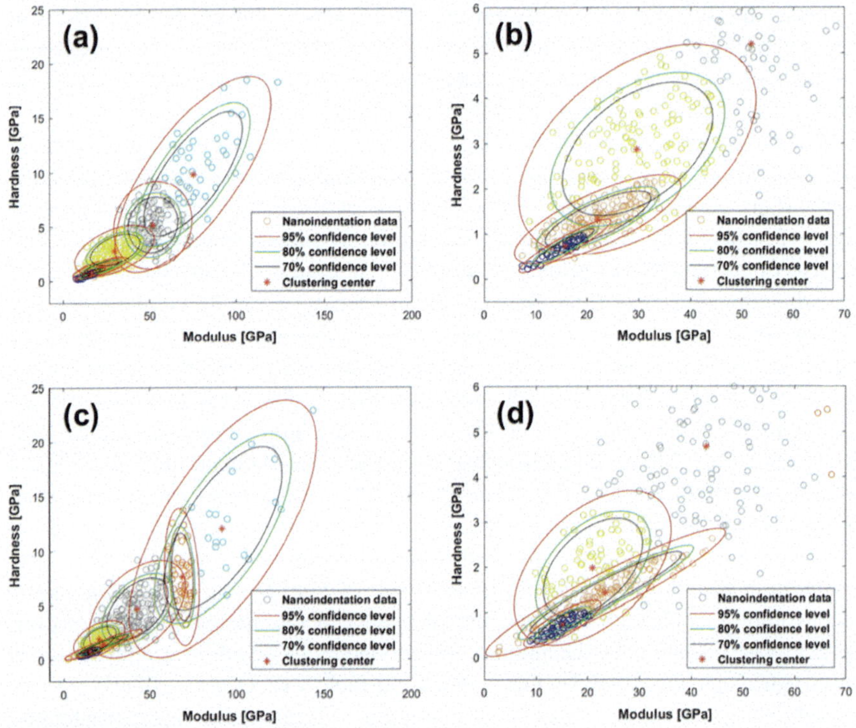

Fig. 5.16 Deconvolution of different 5 grids nanoindentation data in AAFA-M1-H-10%: **a** k = 5 (5 grids-1st); **b** local details for k = 5 (5 grids-1st); **c** k = 6 (5 grids-2nd); **d** local details for k = 6 (5 grids-2nd)

Table 5.12 Deconvolution results for AAFA-M1-H-10% based 5 grids (clustered blue points)

Data source	k	M (GPa)	H (GPa)	*f* (%)	C	
9 grids	5	15.63	0.77	10.30	12.13	0.72
					0.72	0.06
5 grids-1st	5	15.72	0.75	9.38	13.44	0.83
					0.83	0.06
5 grids-2nd	5	17.39	0.99	29.78	34.66	2.15
					2.15	0.16
5 grids-2nd	6	15.18	0.75	11.29	9.43	0.54
					0.54	0.05

In fact, even the suitable bin size based on enough tested points is specified, it is still difficult to identify the N–A–S–H phase from a histogram. It may be ignored or be combined into a big peak becoming properties of a mixed phase when using the LSE method. Typically, as shown in Figs. 5.13c, d and 5.14b, the 'green' Gaussian distribution is the N–A–S–H phase, but it does not correspond to any recognizable peaks in histogram or PDF of GMM model. At the same time, the 'blue' Gaussian distribution in Figs. 5.13d and 5.14b under a prominent peak is easy to be considered as the gel phase. In this study, a two-dimensional GMM model is adopted, which can reveal both modulus and hardness of each phase at the same time. The 'blue' Gaussian distribution phase corresponds to clustered purple points in Fig. 5.4f and has modulus and hardness of 12.62 GPa and 1.48 GPa, respectively. This phase shows high hardness, small proportion (2.06%) and negative correlation coefficient between modulus and hardness. Therefore, it is not accepted as a gel phase. As for hardness histogram, even the bin size decreased to 0.4 or 0.3 GPa, in Fig. 5.14e, f, it is still easy to include the N–A–S–H ('green' Gaussian distribution) into the big peak in the dotted ellipses rather than a single phase especially when a small number of phases is set. The results indicate that there are challenges for using the LSE method to investigate the N–A–S–H phase.

For this study, it is highly suspected that the BIC cannot reach optimal value even for a model with 12 components is caused by disturbance from limited and discrete data. However, what undoubtedly is that at least 5 components (for AAFA-M1-H-10%) are needed to obtain the pure N–A–S–H phase and more components are needed when describing the micromechanical distribution of the whole model. These are caused by the highly heterogeneous components of fly ash and the complex reacted system, and also the limited data. It should be mentioned that for the estimation method itself, better fitting of the model from the mathematical perspective should be at the cost of introducing spurious phases even if the BIC is used. Part of the analysis based on 12 components presented above aims to illustrate that the model estimated by MLE and BIC can reflect the overall micromechanical distribution of AAFA but does not mean that 12 components conform to the real situation in AAFA. The most reasonable model that can both reflect the real nano/micromechanical distribution and also conforms to the real number of components in AAFA may have the number of components > 5 and < 12, but that is not the focus of this study.

5.6.5 *Small Proportion of N–A–S–H Gel in Alkali-Activated Fly Ash Paste*

The deconvolution results for AAFA with different alkali concentrations are listed in Table 5.13 and shown in Fig. 5.17. All the fraction, average modulus and hardness vary in a small range. For the fraction of N–A–S–H gel, it is in the range of 6.28% to 10.67%. Even for AAFA-M1-H-6%, which achieved the maximum value, the proportion of N–A–S–H gel of it is still far less than the normally reported range for C–S–H gel. Easy to mix or interact of the gel phase with other phases is considered as the main reason that largely decreases the proportion of pure N–A–S–H gel detected. The "other phases" include partly-activated or un-activated small particles in the highly heterogeneous AAFA system. Besides, mixing of C–S–H gel with nanoscale CH crystal is the most common form of the mixed phase detected in the research of Portland cement paste [20, 47]. CH crystal does not exist in the AAFA and nanoscale crystal is not considered in any current nanoindentation investigation of AAFA. However, the AAFA has crystal phases and some micron-sized crystals can be observed directly by SEM as in Fig. 5.18.

Some of those crystals would also exit in smaller sizes like nano and sub-micron scale. In addition, based on an analysis of extensive existing results, Provis et al. [35] proposed that the geopolymer binders are formed by agglomerating of nanocrystalline zeolites and amorphous gel phase, and the nanocrystalline zeolites phase cannot be detected by XRD due to the small size. The nanoindentation results in this study agree well with the phenomenon of coexistence of crystal and amorphous gel phase in binders. As given in Tables 5.6 and 5.8, the upper-level mixed phases before generating the gel phase shows average modulus and hardness of (22.90, 1.31) in AAFA-M1-H-8% and (19.83, 1.16) in AAFA-M1-H-10% may be the mixture of crystal and gel phase, as the mechanical properties of it is similar and slightly lower than the phase mixed by low-density (LD) C–S–H with nanoscale CH crystal [9].

Table 5.13 Properties of N–A–S–H in AAFA with different alkali concentration

Sample	k	M (GPa)	H (GPa)	*f* (%)	C	
AAFA-M1-H-6%	≥ 4	12.70	0.62	10.67	16.48	0.85
					0.85	0.06
AAFA-M1-H-8%	> 12	14.59	0.59	6.28	18.53	0.82
					0.82	0.05
AAFA-M1-H-10%	≥ 5	15.46	0.73	8.89	12.02	0.70
					0.70	0.05

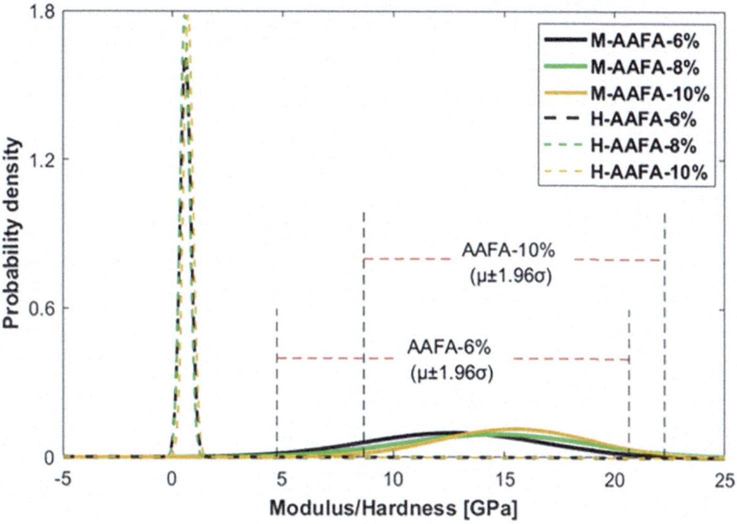

Fig. 5.17 Modulus and hardness probability density distribution of N–A–S–H in AAFA with different alkali concentration

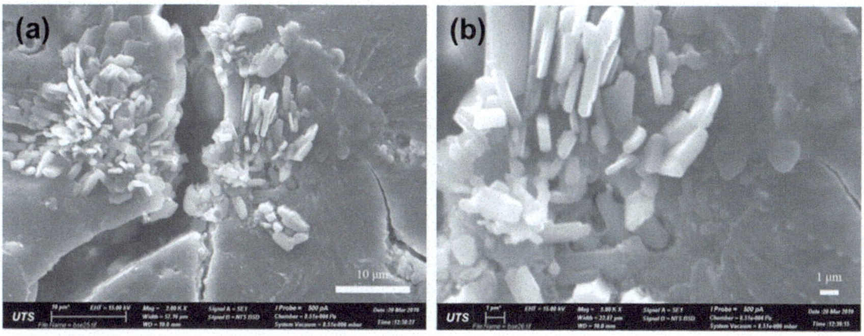

Fig. 5.18 Morphology of micron-sized crystals in AAFA: **a** magnification of 2000×; **b** magnification of 5000×

Due to the low concentration of AAFA-M1-H-6%, it may be difficult to form the supersaturation solution and the mixed phases mainly exist in this sample are the mixture of gel and partly-activated or un-activated small particles, thus no similar mixed phase with low mechanical properties ($1 < H < 2$) as in AAFA M1-H-8% and AAFA-M1-H-10% can be found. Therefore, for AAFA-M1-H-8% and AAFA-M1-H-10%, the nutrition consumed for forming of crystals and especially the interaction of gel with crystals in nanoindentation test are possibly important factors leading to the low proportion of pure gel detected, which are even lower than that in AAFA-M1-H-6% where the fly ash may be harder to be activated due to the low alkali concertation.

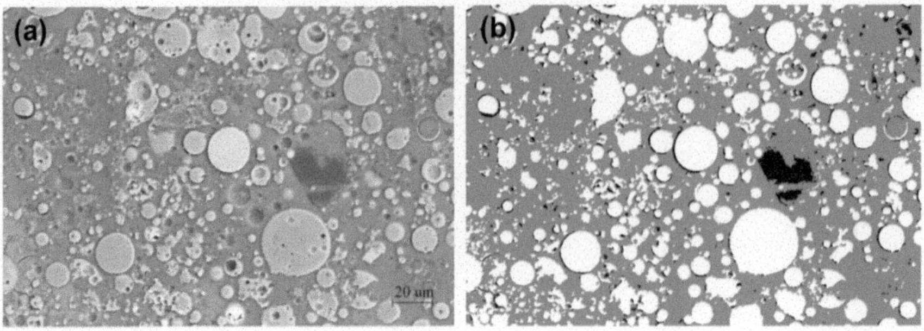

Fig. 5.19 Microstructures and phase segments of AAFA based on BSE images: **a** BSE image (500×); **b** binary image

The presence of crystals can then help to explain why the N–A–S–H gel detected in this study is less than results from other techniques. Typically, when segmenting of phases based on grey value [15, 21, 27] as shown in Fig. 5.19, the statistical amount of reacted products in AAFA-M1-H-8% is 50.38%. In fact, when referring to Table 5.7, for the upper-level mixed phases starting from 2 components identified by statistical nanoindentation, they all have relatively large proportions. When considering factors such as different resolutions between techniques and easy to encounter other phases (particles and crystals) in interaction zone even if the surface is identified as N–A–S–H by BSE, the final small proportion of pure N–A–S–H obtained by statistical nanoindentation is reasonable.

For micromechanical properties results in this study, when the alkali concertation increased from 6 to 10%, the results show that there is a trend of a slight increase of modulus and hardness. The micromechanical properties probability distribution of N–A–S–H was plotted in Fig. 5.17 for more intuitive observation. It shows that although there is an increasing trend it is in a small range and distributions are similar for AAFA with different concentrations. For the study of Portland cement paste with different water to cement ratios, regarding the modulus and hardness as intrinsic material properties is also not based on fully consistent results, but on a small range of variation, such as smaller mean properties difference than the standard deviation, and so on [42]. This is a reasonable consideration as errors are unavoidable in any real experiments, but it also cannot eliminate the possibility that the micromechanical properties themselves just change in a very small range and the increase trend also seems plausible. Therefore, further studies are needed for understanding whether the modulus and hardness are intrinsic material properties of N–A–S–H.

5.7 Correlation and Difference of Different Statistical Techniques Based on Alkali-Activated Fly Ash Paste

For the LSE method that generally used to fit the frequency density histogram for deconvolution analysis, it is known to depend on the bin size. The correlation and difference of the LSE and MLE methods are discussed by the hardness histogram of AAFA-M1-S. The possible N–A–S–H gel phase of AAFA-M1-S starts to appear in the model with 5 components and decompose into two phases in the model with 10 components. The possible gel would exist in models with 5–9 components. However, as displayed in Fig. 5.20, any model could match the histogram if suitable bin size is set for the histogram. If the bin size is more than 0.28, it is not possible to obtain the hardness of N–A–S–H gel, denoted by the green peak in Fig. 5.20b, c, as there is no corresponding peak in the histogram. The large green peak in Fig. 5.20a would be misidentified as the N–A–S–H. Generally, the bin size set in exiting studies is more than 0.28 and the number of components adopted is 4. In that case, the final results should be from a mixed phase instead of N–A–S–H gel. When the bin size is small to around 0.12, as can be clearly observed from Fig. 5.20c, d, the model with 10 components can even better match the histogram than the model with 9 components. Consequently, the properties of N–A–S–H would be erroneously derived from the 10 components model.

For the GMM estimated by the MLE method, reasonably increasing the number of components enables the model to reflect more details of the distribution of the collected data. Bin size plays a similar role in the histogram. Smaller bin size contributes to revealing more information on the collected data. Therefore, for the models with increased phases, there are histograms with decreased bin sizes that can match them. However, for both of the number of components and bin size, the inappropriate parameters would lead to either mixed phase ($k = 4$) or sub-phases ($k = 10$). For the MLE method adopted, the gel phase could be determined by empirical knowledge of multiple parameters within a few models with a different number of components. For LSE method, the bin size is hard to set reasonably, since it is a continuous number with an infinite range of values. Besides, the suitable bin size would be affected by lots of factors, such as features of the sample and the number of test data. For the same bin size, a small number of test data would be no more able to form a neat and identifiable histogram for fitting.

5.8 Mechanism of Characteristics of Nano/Micromechanical Properties Based on Alkali-Activated Fly Ash Paste

The nano/micromechanical properties of N–A–S–H gels obtained in this study are similar but slightly lower than that of LD C–S–H gel [30]. Besides, the proportion of it is also lower than reported for both N–A–S–H [19, 20] and LD C–S–H [54].

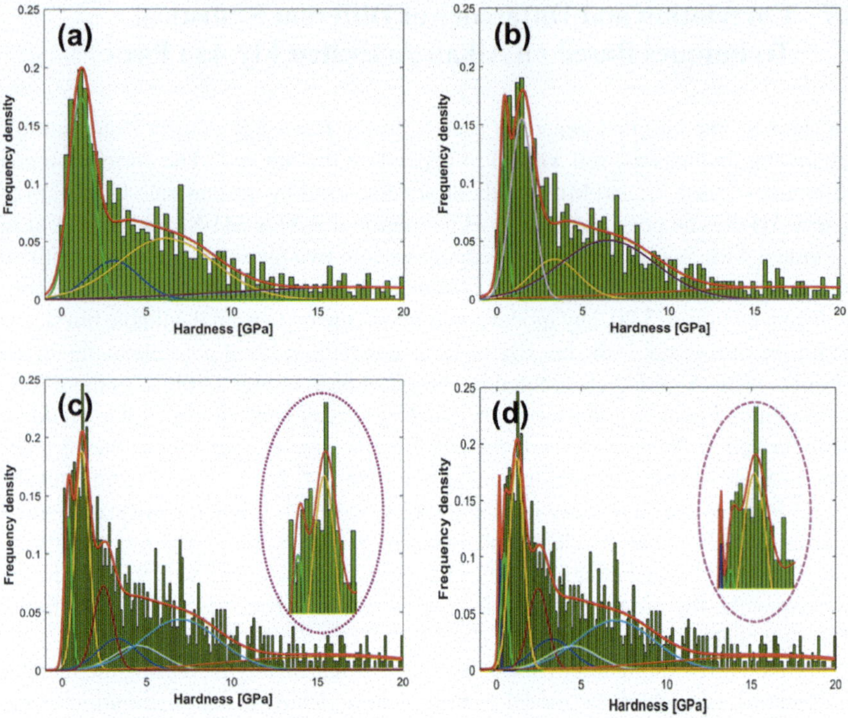

Fig. 5.20 Compared of PDF with hardness frequency histogram for AAFA-M1-S: **a** k = 4 (Bin = 0.28 GPa); **b** k = 5 (Bin = 0.24 GPa); **c** k = 9 (Bin = 0.12 GPa); **d** k = 10 (Bin = 0.12 GPa)

In fact, there is an important factor that would influence the test results but has not been considered in existing nanomechanical test research of N–A–S–H gel, namely, the presence of crystals. Although the study by the XRD shown in Fig. 5.21 does not detect any new crystals when compared with raw material fly ash, it can't eliminate the possibility that the crystals significantly affect the results. In the activation process, crystals are hard to react which would then be surrounded by the generated gel. Evidence can be found by the micro to nanoscale observation of N–A–S–H gel displayed in Fig. 5.22. Slender crystals can be observed clearly in Fig. 5.22a. Additionally, for the gel near the clustered crystals, the magnified photo in Fig. 5.22b indicates that there are also lots of crystals in the gel. Higher magnification observations were conducted at lots of other locations of the sample. Although there is gel without recognizable crystal as shown in Fig. 5.22c, nano to sub-micron crystals can be found to embed in gel in some other locations. The presence of crystals in the involved range of nanoindentation would significantly enhance the mechanical properties and then identified as the mixed phase in deconvolution analysis. It should be the main reason responsible for the low proportion of the gel obtained. For almost all the samples, there is a phase with elastic modulus slightly more than 20 GPa and hardness slightly more than 1 GPa. It can be typically found in Fig. 5.2f for AAFA-M1-S

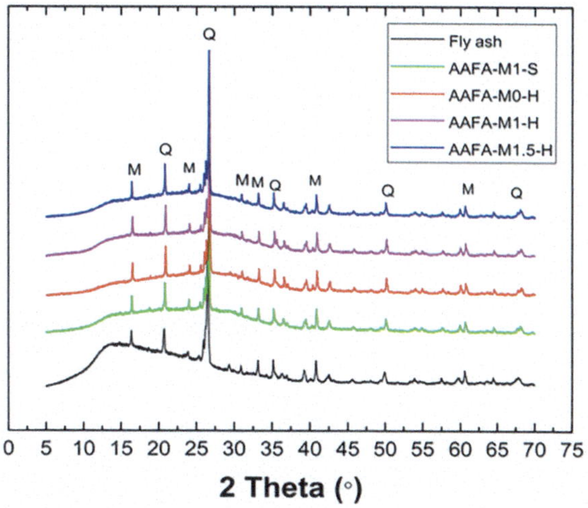

Fig. 5.21 X-ray diffractogram of geopolymer pastes

(yellow phase), Fig. 5.3b for AAFA-M0-H (yellow phase), Fig. 5.6d for AAFA-M1-H (blue phase), Fig. 5.7b for AAFA-M1.5-H (blue phase, 21.43 GPa, 1.52 GPa). For AAFA-M0-H based on 18 grids, the majority of the new test points from the additional nanoindentation grids are intentionally not put on unreacted fly ash. Hence, the test data would concentrate on the region with smaller nano/micromechanical properties. The part of this mixed phase with large mechanical properties would be lost in the deconvolution due to the small proportion, resulting in the slightly smaller of the mechanical properties of it. The same consideration could possibly also be applied for AAFA-M1.5-H. The maximum proportion of it may be ascribed to the grids. Some of them happen to be in gel-rich areas. This mixed phase also shows small properties when k is large.

The final N–A–S–H gel phase is separated from the part of this kind of mixed phase where it has the lowest mechanical properties. This phase may be mainly a mixture of crystals and gels, and have a small fraction of other inclusions. Firstly, the "big phase" can be found in all samples with similar properties which mean it may be a stable phase that exists in AAFA rather than a mixture of gel phase with random inclusions. Besides, the nano/micromechanical properties of this kind of mixed phase are close and a bit lower than the agglomeration of C–S–H and CH crystal identified in Portland cement paste [23], which is plausible to be the mixture of N–A–S–H and crystals as the nano/micromechanical properties of N–A–S–H is also a bit lower than the LD C–S–H. This kind of mixed phase has lower mechanical properties than other phases. This may be caused by the small scale of the crystals and different mechanical properties of it compared with other unreacted particles that come from fly ash.

As shown in Fig. 5.22, the microstructure of N–A–S–H is found to be looser than that normally known for C–S–H gel, which may be the reason for lower nano/micromechanical properties of N–A–S–H gel, and then the generally lower macro performance of fly ash-based geopolymer. In terms of the nano/micromechanical

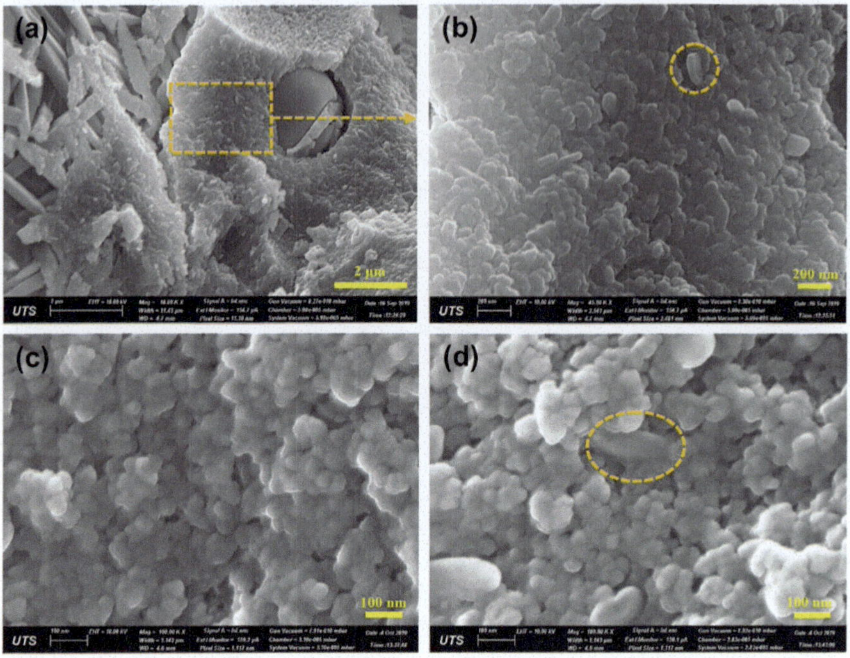

Fig. 5.22 Microstructure of N–A–S–H gel in AAFA: **a** 10,000×; **b** 45,000×; **c** 100,000×; **d** 100,000×

properties in different geopolymers, the variation may be affected by both the nanomechanical properties of the gel particles and the nanoscale pores. For the study of C–S–H in Portland cement paste, Jennings [50] proposed a simplified model to represent the microstructure of C–S–H. Basic spherical blocks with size about 2 nm cluster together to form globules and then the globules pack together to form LD and HD C–S–H structures. This model is further modified using packing structure with around 5 nm globules as the basic unit [29] to help the understanding of the microstructural changes associated with drying and heat curing. The globules are considered to have intrinsic parking density. The geopolymer gel is also formed by the packing of primary globular polymeric entities of several nanometers [49], which can't be observed in Fig. 5.22 due to the limited resolution of SEM. The globular polymeric entities form gel particles with the sizes of dozens of nanometers, while the gel particles are not always tightly integrated as shown in Fig. 5.22c, d. The defects and pores between gel particles would be a factor affecting the nano/micromechanical properties detected. The gel particles are governed by the globular polymeric entities and their parking density, while the stiffness of polymeric entities may be affected at least by its maturity and chemical structure.

The different maturity, and the different composition and chemical structure of the N–A–S–H gel are considered by Ma et al. [27] as the main reasons that account for the constant elastic's modulus in the study by Nemecek et al. [30] but varied elastic modulus in the study by Ma et al. [27]. The polymerization process of AAFA is very slow and heat curing is usually used to accelerate it. For ambient cured AAFA and heat cured AAFA with different curing temperature, the difference between their 28-day compressive strength and longer time compressive strength is quite different [54–56] and larger than the difference in the Portland cement paste, which means that at 28 days, the samples have significantly different reaction degree under different curing conditions. The different maturity would induce different stiffness of polymeric entities, the gel particles and then the nano/micromechanical properties of the N–A–S–H gel. This is considered as the dominant reason for the significantly smaller properties of AAFA-M1-S than AAFA-M1-H-8%. As for the influence of the composition and chemical structure of the gel phase, Georgios and Ulm [6] found that the C–S–H decalcification would cause remarkable degradation of elastic modulus of C–S–H gel. Additionally, the indentation modulus and hardness of the synthetic C–S–H would change with the Ca/Si molar ratio [58]. For geopolymer, within a suitable range, the increase of silica content may result in the increase of the fully condensed tetrahedral aluminosilicate network structures and then the strength of the geopolymer due to the higher strength of Si–O–Si bonds than the Si–O–Al bonds [51]. These results or theoretical basis indicate that the globular polymeric entities themselves may have different stiffness when chemical structure changed significantly. NaOH activated fly ash normally has lower macro strength than the alkali-silicate activated fly ash with similar mixture and condition. Provis et al. [34] proposed that larger crystals in NaOH activated fly ash are hard to pack densely within the binder phase, which could be one of the reasons for its low macro strength. The lowest nanoindentation results for AAFA-M0-H can provide another reason that the N–A–S–H gel in NaOH activated fly ash would have lower micro-strength than others. These two factors work together to make the lower macro strength of NaOH activated fly ash.

5.9 Conclusion

In this chapter, the nano/micromechanical properties of sodium aluminosilicate hydrate (N–A–S–H) gel in alkali-activated fly ash (AAFA) geopolymer under different silica modulus and curing conditions are investigated by deconvolution of grid nanoindentation data with maximum likelihood estimation method. The following conclusions can be drawn up:

(1) When using the MLE method, the process of separating stable N–A–S–H gel from mixed phases can be intuitively observed. Moreover, nanoindentation data can be clustered to corresponding components with a good match, which allows for further analysis. MLE also shows the advantages in independent on distribution histogram, more reasonable determination of component numbers, and less sensitive to the number of test points than LSE.

(2) Mixed phases and sub-phases are typical spurious phases in geopolymer. It is virtually impossible to reconcile these two kinds of spurious phases to obtain the accurate nano/micromechanical properties of each real component by statistical nanoindentation. A compromise approach was proposed, which ensures the accuracy of the gel phase by intentionally introducing spurious phases for other components.

(3) The errors generated from the analytical technique itself were estimated from the two aspects of the average method adopted, and the number of the experimental data. The average deviation of elastic modulus and hardness introduced by the average method are 0.44 GPa, 0.04 GPa, respectively. Correspondingly, the number of test data would bring the errors of

(4) Bayesian Information Criterion is not necessary to achieve the optimum value when just focusing on the N–A–S–H phase in the highly heterogeneous AAFA reaction system. When the number of components is large, sparse data in some locations leads to misidentification of new phases, which may also affect reaching the optimum value of BIC and increase the complexity of the model.

(5) The number of components needed to obtain the pure N–A–S–H phase is usually more than the typically used value of 4, and even the second deconvolution is sometimes needed. Although this is partially caused by different deconvolution method used, it is still necessary that the number of components should be enough to avoid mixed phases.

(6) GMM model obtained by MLE can match histogram well when bin size is reasonably small. A sufficiently small bin size is needed to present the actual and detailed micromechanical properties distribution of AAFA which calls for enough testing points to construct a regular and ordered histogram when using the LSE method. The N–A–S–H phase does not always show as a distinct peak in the histogram and is easy to be mixed into identifiable large peaks, which means that even if the bin size used is small enough it is still hard to obtain the properties of N–A–S–H phase by the LSE method.

(7) The nano/micromechanical properties obtained for geopolymers vary in a minor range of 10.50–14.30 GPa for elastic modulus and 0.40–0.57 GPa for hardness. The highest nano/micromechanical properties were achieved by AAFA-M1-H. Both AAFA-M0-H and AAFA-M1-S show significantly lower properties. The variation may be affected by both the nanomechanical properties of the gel particles and the nanoscale pores between them. The formal one would be determined at least by its maturity and chemical structure.

(8) The proportion, average modulus and average hardness of pure N–A–S–H phase in AAFA with different alkali concentrations vary in small ranges of 6.28% to 10.67%, 12.70 to 15.46 GPa and 0.59 to 0.73 GPa, respectively. The main reason for the small proportion is supposed to be the presence of mixed phases, such as mixing of crystals, partly-activated and un-activated small particles with the gel phase in the highly heterogeneous AAFA system, which largely decreases the volume of pure N–A–S–H gel detected by SNT.

(9) The intermixing of crystal gels was observed under high resolution SEM. The interaction of them in the involved range results in a mixed phase in deconvolution, which largely decreased the proportion of N–A–S–H gel obtained. Compared to C–S–H, the looser structure of N–A–S–H should be one of the reasons for its inferior nano/micromechanical properties.

References

1. P.G. Allison, C.A. Weiss, R.D. Moser, A.J. Diaz, O.G. Rivera, S.S. Holton, Nanoindentation and SEM/EDX characterization of the geopolymer-to-steel interfacial transition zone for a reactive porcelain enamel coating. Compos. B Eng. **78**, 131–137 (2015)
2. C.M. Bishop, Pattern recognition and machine learning. Springer **2**, 1122–1128 (2006)
3. A.R. Brough, A. Atkinson, Automated identification of the aggregate-paste interfacial transition zone in mortars of silica sand with Portland or alkali-activated slag cement paste. Cem. Concr. Res. **30**(6), 849–854 (2000)
4. D.K. Chanda, S.R. Chowdhury, M. Bhattacharya, A.K. Mandal, N. Dey, A.K. Mukhopadhyay, Intelligently designed fly-ash based hybrid composites with very high hardness and Young's modulus. Constr. Build. Mater. **158**, 516–534 (2018)
5. J.J. Chen, L. Sorelli, M. Vandamme, F.J. Ulm, G. Chanvillard, A coupled nanoindentation/SEM-EDS study on low water/cement ratio Portland cement paste: evidence for C–S–H/$Ca(OH)_2$ nanocomposites. J. Am. Ceram. Soc. **93**(5), 1484–1493 (2010)
6. G. Constantinides, F.-J. Ulm, The effect of two types of C–S–H on the elasticity of cement-based materials: results from nanoindentation and micromechanical modeling. Cem. Concr. Res. **34**(1), 67–80 (2004)
7. G. Constantinides, F.J. Ulm, The nanogranular nature of C–S–H. J. Mech. Phys. Solids **55**(1), 64–90 (2007)
8. S. Das, P. Yang, S.S. Singh, J.C.E. Mertens, X. Xiao, N. Chawla, N. Neithalath, Effective properties of a fly ash geopolymer: synergistic application of X-ray synchrotron tomography, nanoindentation, and homogenization models. Cem. Concr. Res. **78**, 252–262 (2015)
9. D. Davydov, M. Jirásek, L. Kopecký, Critical aspects of nano-indentation technique in application to hardened cement paste. Cem. Concr. Res. **41**(1), 20–29 (2011)
10. G. Fang, M. Zhang, Multiscale micromechanical analysis of alkali-activated fly ash-slag paste. Cem. Concr. Res. **135**, 106141 (2020)
11. C. Hu, Z. Li, Property investigation of individual phases in cementitious composites containing silica fume and fly ash. Cement Concr. Compos. **57**, 17–26 (2015)
12. C. Hu, Z. Li, A review on the mechanical properties of cement-based materials measured by nanoindentation. Constr. Build. Mater. **90**, 80–90 (2015)
13. D. Khale, R. Chaudhary, Mechanism of geopolymerization and factors influencing its development: a review. J. Mater. Sci. **42**(3), 729–746 (2007)
14. F.-J. Ulm, M. Vandamme, H.M. Jennings, J. Vanzo, M. Bentivegna, K.J. Krakowiak, G. Constantinides, C.P. Bobko, K.J. Van Vliet, Does microstructure matter for statistical nanoindentation techniques? Cement Concr. Compos. **32**(1), 92–99 (2010)
15. M. Khedmati, Y.R. Kim, J.A. Turner, H. Alanazi, C. Nguyen, An integrated microstructural-nanomechanical-chemical approach to examine material-specific characteristics of cementitious interphase regions. Mater Charact **138**, 154–164 (2018)
16. H. Lee, V. Vimonsatit, P. Chindaprasirt, Mechanical and micromechanical properties of alkali activated fly-ash cement based on nano-indentation. Constr. Build. Mater. **107**, 95–102 (2016)
17. H. Lee, V. Vimonsatit, P. Chindaprasirt, T. Ngo, P. Mendis, Creep properties of cement and alkali activated fly ash materials using nanoindentation technique. Constr. Build. Mater. **168**, 547–555 (2018)

18. N. Li, C. Shi, Z. Zhang, H. Wang, Y. Liu, A review on mixture design methods for geopolymer concrete. Compos. B Eng. **178**, 107490 (2019)
19. W. Li, S. Kawashima, J. Xiao, D.J. Corr, C. Shi, S.P. Shah, Comparative investigation on nanomechanical properties of hardened cement paste. Mater. Struct./Materiaux et Constructions **49**(5), 1591–1604 (2016)
20. W. Li, J. Xiao, S. Kawashima, G.S. Shekhawat, S.P. Shah, Experimental investigation on quantitative nanomechanical properties of cement paste. ACI Mater. J. **112**(2) (2015)
21. W. Li, J. Xiao, Z. Sun, S. Kawashima, S.P. Shah, Interfacial transition zones in recycled aggregate concrete with different mixing approaches. Constr. Build. Mater. **35**, 1045–1055 (2012)
22. Y. Ling, K. Wang, W. Li, G. Shi, P. Lu, Effect of slag on the mechanical properties and bond strength of fly ash-based engineered geopolymer composites. Compos. B Eng. **164**, 747–757 (2019)
23. R.R. Lloyd, J.L. Provis, K.J. Smeaton, J.S.J. van Deventer, Spatial distribution of pores in fly ash-based inorganic polymer gels visualised by Wood's metal intrusion. Microporous Mesoporous Mater. **126**(1–2), 32–39 (2009)
24. Z. Luo, W. Li, Y. Gan, K. Mendu, S.P. Shah, Applying grid nanoindentation and maximum likelihood estimation for N–A–S–H gel in geopolymer paste: Investigation and discussion. Cem. Concr. Res. **135**, 106112 (2020)
25. Z. Luo, W. Li, K. Wang, S.P. Shah, Research progress in advanced nanomechanical characterization of cement-based materials. Cement Concr. Compos. **94**, 277–295 (2018)
26. P. Lura, P. Trtik, B. Münch, Validity of recent approaches for statistical nanoindentation of cement pastes. Cement Concr. Compos. **33**(4), 457–465 (2011)
27. Y. Ma, G. Ye, J. Hu, Micro-mechanical properties of alkali-activated fly ash evaluated by nanoindentation. Constr. Build. Mater. **147**, 407–416 (2017)
28. M. Miller, C. Bobko, M. Vandamme, F.J. Ulm, Surface roughness criteria for cement paste nanoindentation. Cem. Concr. Res. **38**(4), 467–476 (2008)
29. M. Nedeljković, B. Šavija, Y. Zuo, M. Luković, G. Ye, Effect of natural carbonation on the pore structure and elastic modulus of the alkali-activated fly ash and slag pastes. Constr. Build. Mater. **161**, 687–704 (2018)
30. J. Němeček, V. Šmilauer, L. Kopecký, Nanoindentation characteristics of alkali-activated aluminosilicate materials. Cement Concr. Compos. **33**(2), 163–170 (2011)
31. V. Nikolić, M. Komljenović, N. Džunuzović, T. Ivanović, Z. Miladinović, Immobilization of hexavalent chromium by fly ash-based geopolymers. Compos. B Eng. **112**, 213–223 (2017)
32. W.C. Oliver, G.M. Pharr, An improved technique for determining hardness and elastic modulus using load and displacement sensing indentation experiments. J. Mater. Res. **7**(6), 1564–1583 (1992)
33. J.L. Provis, G.C. Lukey, J.S.J. Van Deventer, Do geopolymers actually contain nanocrystalline zeolites? a reexamination of existing results. Chem. Mater. **17**(12), 3075–3085 (2005)
34. J.L. Provis, R.J. Myers, C.E. White, V. Rose, J.S.J. Van Deventer, X-ray microtomography shows pore structure and tortuosity in alkali-activated binders. Cem. Concr. Res. **42**(6), 855–864 (2012)
35. J.L. Provis, A. Palomo, C. Shi, Advances in understanding alkali-activated materials. Cem. Concr. Res. **78**, 110–125 (2015)
36. J.L. Provis, J.S. Van Deventer, *Alkali Activated Materials: State-of-the-Art Report, RILEM TC 224-AAM* (Springer Science & Business Media, 2013)
37. A.M. Rashad, A comprehensive overview about the influence of different additives on the properties of alkali-activated slag—a guide for Civil Engineer. Constr. Build. Mater. **47**, 29–55 (2013)
38. A.M. Rashad, A comprehensive overview about the influence of different admixtures and additives on the properties of alkali-activated fly ash. Mater. Des. **53**, 1005–1025 (2014)
39. M.S. Reddy, P. Dinakar, B.H. Rao, A review of the influence of source material's oxide composition on the compressive strength of geopolymer concrete. Microporous Mesoporous Mater. **234**, 12–23 (2016)

40. G. Schwarz, Estimating the dimension of a model. Ann. Stat., 461–464 (1978)
41. K.L. Scrivener, Backscattered electron imaging of cementitious microstructures: Understanding and quantification. Cement Concr. Compos. **26**(8), 935–945 (2004)
42. K.L. Scrivener, R.J. Kirkpatrick, Innovation in use and research on cementitious material. Cem. Concr. Res. **38**(2), 128–136 (2008)
43. C. Shi, D. Roy, P. Krivenko, in *Alkali-Activated Cements and Concretes* (CRC Press, 2003)
44. F. Škvára, L. Kopecký, V. Šmilauer, Z. Bittnar, Material and structural characterization of alkali activated low-calcium brown coal fly ash. J. Hazard. Mater. **168**(2–3), 711–720 (2009)
45. L. Sorelli, G. Constantinides, F.J. Ulm, F. Toutlemonde, The nano-mechanical signature of ultra high performance concrete by statistical nanoindentation techniques. Cem. Concr. Res. **38**(12), 1447–1456 (2008)
46. Z. Tang, W. Li, Y. Hu, J.L. Zhou, V.W.Y. Tam, Review on designs and properties of multifunctional alkali-activated materials (AAMs). Constr. Build. Mater. **200**, 474–489 (2019)
47. M. Vandamme, F.J. Ulm, P. Fonollosa, Nanogranular packing of C–S–H at substochiometric conditions. Cem. Concr. Res. **40**(1), 14–26 (2010)
48. J. Xiao, W. Li, D.J. Corr, S.P. Shah, Effects of interfacial transition zones on the stress-strain behavior of modeled recycled aggregate concrete. Cem. Concr. Res. **52**, 82–99 (2013)
49. J. Xiao, W. Li, Z. Sun, D.A. Lange, S.P. Shah, Properties of interfacial transition zones in recycled aggregate concrete tested by nanoindentation. Cement Concr. Compos. **37**(1), 276–292 (2013)
50. L. Xiao, H. Huang, Equilibrium trip scheduling in bottleneck model with stochastic capacity, in *Proceedings of the 2012 5th International Joint Conference on Computational Sciences and Optimization, CSO 2012* (2012), pp. 504–508
51. K. Yan, Y. Guo, Z. Ma, Z. Zhao, F. Cheng, Quantitative analysis of crystalline and amorphous phases in pulverized coal fly ash based on the Rietveld method. J. Non-Cryst. Solids **483**, 37–42 (2018)
52. L. Yan, B. Kasal, L. Huang, A review of recent research on the use of cellulosic fibres, their fibre fabric reinforced cementitious, geo-polymer and polymer composites in civil engineering. Compos. B Eng. **92**, 94–132 (2016)
53. C. Zhai, Y. Gan, D. Hanaor, G. Proust, D. Retraint, The Role of Surface Structure in Normal Contact Stiffness. Exp. Mech. **56**(3), 359–368 (2016)
54. Z. Zhang, J.L. Provis, A. Reid, H. Wang, Geopolymer foam concrete: An emerging material for sustainable construction. Constr. Build. Mater. **56**, 113–127 (2014)
55. Z. Zhang, H. Wang, Y. Zhu, A. Reid, J.L. Provis, F. Bullen, Using fly ash to partially substitute metakaolin in geopolymer synthesis. Appl. Clay Sci. **88–89**, 194–201 (2014)
56. P. Zhao, X. Liu, A. De La Torre, L. Lu, K. Sobolev, Assessment of the quantitative accuracy of Rietveld/XRD analysis of crystalline and amorphous phases in fly ash. Anal. Methods **9**(16), 2415–2424 (2017)
57. W. Zhu, T. Howind, S. Barbhuiya, R. Stark, B. Kümmerling, J.S. Dolado, Nanomechanical study of cement pastes by statistical nanoindentation and peakforce QNM, in *Proceedings of the 2nd International RILEM Conference on Microstructure-related Durability of Cementitious Composites* (2012), p. 6
58. W. Zhu, J.J. Hughes, N. Bicanic, C.J. Pearce, Nanoindentation mapping of mechanical properties of cement paste and natural rocks. Mater. Charact. **58**(11-12 Special Issue), 1189–1198 (2007)

Chapter 6
Nanocharacterization on Micromechanical Properties and Microstructure of Geopolymer with Nano Particles

6.1 Introduction

Presently, there is an increasing trend to modify and understand materials from the "bottom" scale of nano and even molecular and atomic scale [1–3]. One of the typical applications in construction materials is nanotechnology, which is defined from two aspects of nanoengineering and nanoscience by Sanchez [4], with the keywords of "manipulation", "develop" for the former one and "understanding" for the later one. Typical cases for them include modification with nanomaterials and investigation by nanomechanical techniques. Nanomaterials have been widely studied in Portland cement materials [5–7]. Positive effects including filler effect and nucleation effect of nanoparticles, crack arrest and interlocking effects of two-dimensional nanomaterials, and pozzolanic effect of pozzolanic active nanomaterials have been identified [8, 9], which lead to significant improvements in material properties. Moreover, various nanomechanical testing techniques such as nanoindentation, nanoscratch and modulus mapping have contributed a lot to the understanding of Portland cement materials. There is an increasing trend to combine nanoengineering and nanoscience approaches in research [10–14]. For the application of nanotechnology in geopolymer, relevant studies are insufficient. In addition, the nanomaterial modification and nanomechanical characterization are separate in existing studies. In these limited studies, the benefits of incorporating nanomaterials in geopolymer have also been recognized [15]. Typically, geopolymer strength improvement can be realized by the addition of nano-SiO_2 [16–18], nano-Al_2O_3 [19], nano-TiO_2 [20], nano-clay [21] and carbon nanotubes [22]. Higher durability such as enhanced sulfuric acid attack resistance of nano-SiO_2 modified geopolymer [15], and lower water absorption of nano-clay modified geopolymer [21] were found. In addition to excellent engineering properties, nanomaterials can help geopolymer gain special properties simultaneously, such as anti-bacterial properties [23] and self-sensing function [24].

W. Li et al., *Nanocharacterization of Cementitious Composites*,
https://doi.org/10.1007/978-981-96-2917-6_6

With the deepening of the research on the modification of geopolymer by nanoengineering, it is necessary to promote nanoscience study to help understand the relevant mechanism and lay a basis for further properties improvement.

The most commonly used nanomaterials in cementitious materials are nano-SiO_2 (NS) and nano-TiO_2 (NT) [25], which were adopted in this chapter to reinforce geopolymer. A high dosage makes it difficult to disperse the nanoparticles, thereby reducing the modification effect and even bringing adverse effects. The dispersion performance of nano particles was discussed in this study. The synthesized nano-geopolymer composites were investigated by statistical nanoindentation technique (SNT) to reveal the variation in micromechanical properties of the most crucial phase, sodium aluminosilicate hydrate (N–A–S–H) gel. Besides, XRD test, SEM observation, thermogravimetry analysis (TGA) and hydration heat test were combined conducted for properties investigation and mechanism analysis.

6.2 The Influence of Nanosilica Dispersion on the Properties and Microstructure of Fly-Ash Based Geopolymer Composite

6.2.1 *Raw Materials*

Alkali silicate solution was prepared by mixing of commercial sodium silicate and sodium hydroxide solution. The commercial sodium silicate consists of 14.7% of Na_2O, 29.4% of SiO_2 and 55.9% of H_2O. Sodium hydroxide solution was prepared by dissolving sodium hydroxide pellets into water and cooled down before incorporated into sodium silicate solution. The precursor material used was Class F low calcium fly ash with the chemical composition given in Table 6.1. The mix design of the samples is shown in Table 6.2, where the nanoparticles were incorporated by the way of replacing fly ash. Table 6.3 lists the properties of nanosilica particles. A comparison between nanosilica particles and fly ash particles is provided in Fig. 6.1. Based on the same weight, the nanosilica has far more particles than fly ash and occupies a larger volume. The micro and nano-scale morphology of the nanosilica particles were observed under SEM and provided in Fig. 6.2. It clearly reveals that the nanoparticles are easy to combine with each other to form aggregates from sub-micron to larger sizes.

Table 6.1 Chemical composition of fly ash

Oxide	Al_2O_3	CaO	SiO_2	Fe_2O_3	K_2O	MgO	Na_2O	MnO	P_2O_5	TiO_2	LOI
Weight (%)	25.21	1.73	64.55	2.85	1.47	0.41	0.48	0.07	0.19	0.91	1.54

Table 6.2 Mix proportions and curing, dispersion conditions of geopolymer paste and mortar

Samples	Ms	Na_2O/ash	w/s	Nanoparticles	Sonication duration (min)	Curing condition
Reference	1	10%	0.32	–	120	65 °C for 24 h
Geo-NS	1	10%	0.32	2% Nano-SiO_2	0; 20; 120; 180	65 °C for 24 h

Table 6.3 Properties of nanoparticles

Material	Appearance	Average particle size (nm)	Purity (%)	Type
Nano-SiO_2	White powder	20	99.9%	Amorphous

Fig. 6.1 Appearance of nanosilica particles: **a** comparison of particles; **b** nanosilica particles

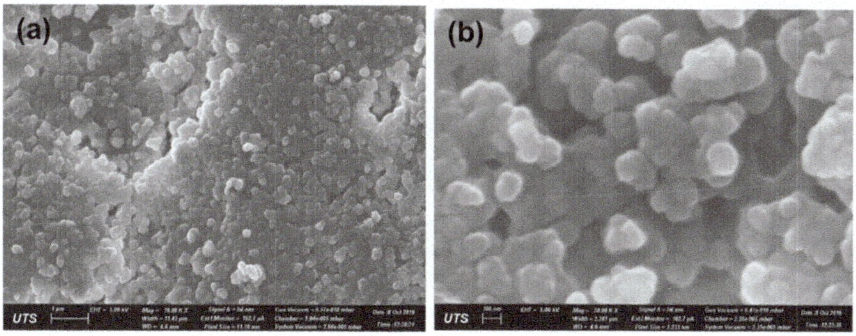

Fig. 6.2 Morphology of nanosilica particles: **a** 10,000×; **b** 50,000×

Previous studies [26, 27] indicate that the compressive strength of geopolymer would increase up to 2% of nanosilica is incorporated and a further increase in content would result in decreased strength. Besides, nanosilica is reported to be able to cover cement particles even if the percentage of it is lower than 2% [28]. Fly ash particles have a similar scale as the cement grains. Therefore, 2% of nanosilica by weight of fly ash is selected for the dispersion study. As suggested in the previous study [29], Polycarboxylate superplasticizers (SP) is superior to any other surfactants in dispersing and stabilizing the dispersed suspensions, and is beneficial to reduce the viscosity and air bubbles of fresh mixture. It was therefore adopted in the high viscosity alkali silicate system. After incorporating and mixing 0.5% SP into the prepared alkaline solution evenly, nanoparticles were added and mixed manually for 3 min. Then, the suspensions were subjected to different dispersion procedures.

6.2.2 Nanoparticle Dispersion and Sample Preparation

Nanoparticles are easy to aggregate into large particles due to high Van der Waals forces on their surfaces [11]. As a result, the advantages such as filling sub-micron to nanoscale pores and high activity would be diminished. It would be followed by properties change of modified material from micro to macroscale. A related comparison is silica fume and nanosilica, research found that nanosilica with smaller size is able to generate compacter gel structure and poorer crystallinity [30, 31]. Existing studies also revealed that the agglomeration of nanoparticles would account for the decreased modification effect and even negative effect when a high dosage of nanoparticles was incorporated [32–34]. Different from the aqueous solution in Portland cement paste, alkali silicate solution for the synthesis of geopolymer is characterized by high viscosity, especially when high silica modulus is applied. It would increase the difficulty to disperse, although may be also accompanied by the benefit of harder to re-agglomeration of nanoparticles. This feature once again highlights the issue of dispersion.

Mechanical stirring and sonication are the most used two methods for the dispersion of nanoparticles. In this study, one prepared suspension was manually stirred for 20 min, while another suspension was subjected to sonication (120 W, 40 kHz). Twenty minutes is roughly the comfortable limit of manual stirring but longer times are easily reached by sonication. Therefore, the sonication period was extended to 180 min for further dispersion. For the sonicated suspensions, due to the nature of sodium silicate solution and the presents of another solute of sodium hydroxide, it is not possible to evaporate water and then observe the dispersion by approaches such as SEM. The appearance change of suspensions was used as an indicator for the dispersion effect, which is considered as a convenient and also feasible method that suits for engineering applications. The sonicated suspensions were placed in a transparent cup with constant content. A small piece of white paper was pasted on the table under the cup. Then, the appearance of 20, 40, 60, 90, 120 and 180 min sonicated suspensions were recorded. In order to prevent heat generated during sonication, the water in the sonication bath was replaced every 3 min and always kept the same content.

After judging the dispersion by appearance observation, alkali suspensions were re-prepared and dispersed to different degrees for the heat of reaction testing and preparing of paste and mortar samples. The time interval for replacing water was reduced to 2 min and 1 min (final 10 min) in the last 30 min. Besides, alkaline

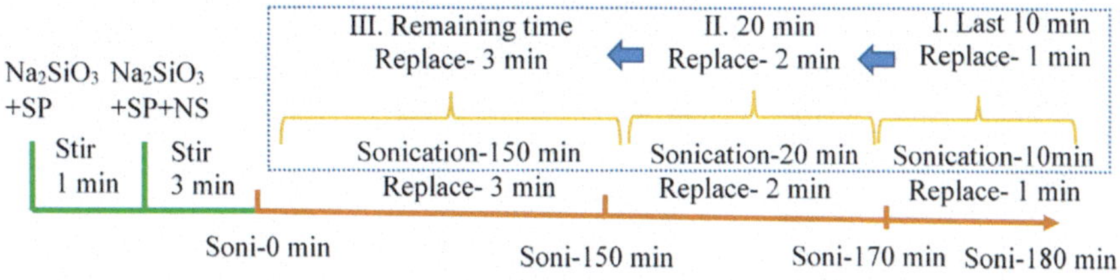

Fig. 6.3 Sonication procedure

solution used for preparing the reference sample was also sonicated in the bath to eliminate temperature differences. The detailed sonication process is summarized in the dotted box of Fig. 6.3 and illustrated by the case of 180 min of sonication. The dispersed suspensions were mixed with fly ash in Hobart mixer for 5 min and then cast in 50 mm × 50 mm × 50 mm cubic plastic mould to make paste sample. Mortar samples were prepared correspondingly with the sand to fly ash weight ratio of 2.0. The prepared samples were sealed with plastic film and cured in an oven at 65 °C for 24 h, followed by standard curing until 28 days. Afterward, 10 mm × 10 mm × 5 mm small pieces were taken from the centre part of the cubic paste samples and cast in epoxy resin. The surface of the samples was well prepared for the following nanoindentation and BSE observation.

The preparation procedure consists of grinding on 320, 600 and 1200 grits of abrasive paper (10 min each) and polishing with alumina of 0.3 μm for 40 min and cerium oxide slurry of 0.05 μm for 40 min. For the heat of reaction test, 30 g fly ash was mixed with alkali-nanosilica dispersions at the same liquid to solid ratio as paste and mortar sample.

The alkali suspensions after different dispersion processes are shown in Fig. 6.4. After 20 min of manual stirring, the appearance of the suspension is almost the same as that before dispersion. For the 20 min sonication counterpart, the suspension becomes clearer, and the small white paper is faintly visible. It means that although the relatively long period of 20 min was used for manual mixing, it does not contribute too much for the dispersion of nanosilica particles in alkali suspensions. Sonication is a more efficient dispersion method. The increase of sonication time is accompanied by cleaner alkali suspensions until it reaches 120 min. A longer time of sonication would no longer bring perceptible appearance change. By the method of appearance discrimination, the time needed in this study for the suspensions to realize good dispersion is more than 90 min. However, what unknown is whether the suspensions are already dispersed well by 120 min of sonication. It is possible that there is invisible dispersion when further sonicating the suspensions that still contributes to the performance improvement of geopolymer. Based on the results of appearance observation, the suspensions without dispersion, and the suspensions subjected to 20, 120 and 180 min of sonication were used for making samples, which represent the situations of without dispersion, poor dispersion, good dispersion and potential better dispersion for the following compressive strength, flowability and heat of reaction test.

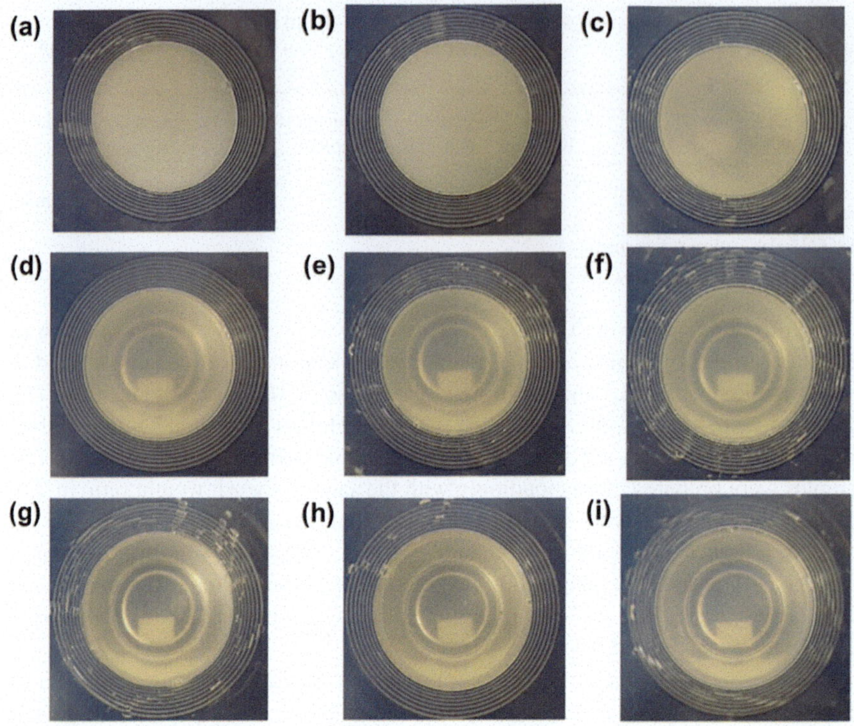

Fig. 6.4 Appearance changes of alkali-nanosilica suspensions: **a** before; **b** manual stirring for 20 min; **c** sonication for 20 min; **d** sonication for 40 min; **e** sonication for 60 min; **f** sonication for 90 min; **g** sonication for 120 min; **h** sonication for 180 min; **i** alkaline solution without NS

6.2.3 Effects of Nanosilica Dispersion on Heat of Reaction

TAM Air Isothermal Calorimeter was adopted for the test of heat of reaction during the heat curing of geopolymers. The ambient temperature in the machine was set as 65 °C, the same temperature as heat curing. The heat of reaction of samples is presented in Fig. 6.5. For Portland cement paste, the process of the hydration is commonly subdivided into four periods including the initial period, the induction period, the acceleration period and the retardation period [32]. However, the heat flow curves of heat cured geopolymers did not present similar reaction features. Only a single peak that presents in the early stage of reaction can be found. As can be seen in the detailed early stage heat flow curves in Fig. 6.5a, after reaching the peak, the heat flow curves gradually drop to almost coincide with each other. For easy comparison of differences between curves, the heat flow curves can be understood from the two stages of the peak period and the almost coincidence period. In the peak period, all of the nano-geopolymer composites reveal higher heat flow peak than the reference sample, and there is an obvious trend of increasing the heat flow peak with the sonication time. However, when the curves enter the almost coincidence period, as shown in the dotted box, it is the reference sample that has a higher heat flow than others. The above phenomenon indicates that similar to that in Portland cement paste [33], nanosilica would also accelerate the early reaction in geopolymer. The well dispersed nanosilica has a more evident effect. The accelerating effect of nanosilica in Portland cement paste is understood to benefit from Pozzolanic effect [34], ion adsorption [35], and nucleation effect (e.g. providing nucleation sites by itself [36, 37] or generating C–S–H seed by reaction [32]). Besides, for the investigation of the filler effect of mineral additions, the interparticle distance is identified as the dominant factor [38]. In this study for geopolymer, the solution used is sodium silicate and a small part of sodium hydroxide, which is not as powerful as some high concentration of strong base for the dissolving of nanosilica. At least for the Geo-0 min sample, the dissolution of nanosilica particles should be very low, as there is a very short time between adding nanosilica in solution and doing heat of reaction test. However, it still significantly enhances the peak. Additionally, if there is a significant solubility of the nanosilica particles during sonication, there should be evidently different heat flow curves. The different dispersion would lead to different degrees of dissolution and different amount of residual nanoparticles. Heat curing may increase the release of soluble silica with time, leading to different heat flow curves of geopolymers in the almost coincidence period. Therefore, for geopolymers in this study, it is more plausible that the physical presence of nanosilica is the main reason that accelerates the reaction, instead of its reacted production. It also can be supported by previous research that the inner nano-Al_2O_3 can eliminate the induction period in hydroxide activation geopolymer [39]. Therefore, the nanomodified geopolymer should be in the form of nanosilica-geopolymer composites. For the cumulative heat in Fig. 6.5b, the geopolymers with nanosilica all show higher cumulative heat than the reference sample, especially for samples with properly dispersed nanoparticles. The maximum and minimum cumulative heat after 24 h of curing are 181.16 J/g for Geo-120 min, and 172.29 J/g for the reference sample, respectively. The more obvious difference is present at the early stage.

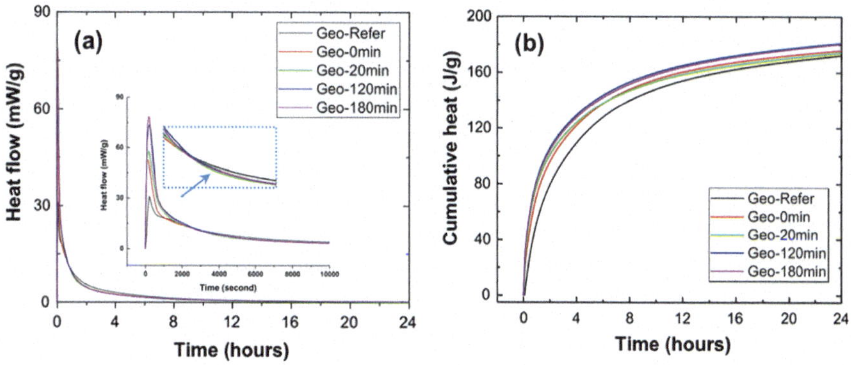

Fig. 6.5 Effect of dispersion on the heat of reaction during 24 h of heat curing: **a** heat flow; **b** cumulative heat

Combining with the results of compressive strength and flowability, it is clear that there is no significant difference in properties between Geo-180 min and Geo-120 min sample. It indicates that by the sonication method here, extending the sonication time would not bring further improvement to the engineering properties of nano-geopolymer, but increasing the waiting time for construction. Therefore, the Geo-120 min sample can represent the well dispersed sample, although it doesn't mean that the nanosilica particles are all sufficiently dispersed to the form of single particles. In fact, knowledge of the differences in appearance characteristics and related particle size ranges of suspension, colloid and solution has suggested that most particles in transparent colloids after 120 min of sonication should be nanoscale.

6.2.4 Effects of Nanosilica Dispersion on Morphology of and Properties of N–A–S–H Gel Phase

Microstructure observation was conducted on a high-resolution SEM, Zeiss Supra 55VP, under the accelerating voltages of 5 or 10 kV. For the matrix generated, as shown in Fig. 6.6a, the gel phase surrounds the partially reacted or unreacted fly ash particles. Crystals can be found on the surface of the remnant fly ash cavity. When observing the gel phase at high magnification, it shows that the N–A–S–H gel is present in the form of agglomerates of small gel particles with a size of dozens of nanometers, while other phases such as the Mullite crystals are embedded in the gel. The appearance and scale of the small gel particles are similar to the nanosilica particles as shown in Fig. 6.2. In addition to the needle-like Mullite, there are also more rounded Mullite clusters [40] as shown in Fig. 6.7. The more rounded Mullite has an average size larger than the small gel particles and nanoparticles but also has a similar appearance. Therefore, it is difficult to determine the dispersion of nanosilica in the gel and whether there are a small number of inclusions in the gel through microstructure observation. However, these particles can be detected by the nanoindentation test due to their effects on the final micromechanical properties of the gel phase.

The modulus and hardness of N–A–S–H gel were tested by nanoindentation. In the test procedure, a 20 nm Berkovich indenter was loaded to a peak force of 1.5 mN in 7.5 s. The force was kept for 5 s and then unloaded to 10% of the peak force in 7.5 s. For the highly heterogeneous geopolymer, even if a large number of 900 test points were adopted, the test data are still scattered and not able to provide reliable estimation for phases with high mechanical properties such as fly ash and partially reacted fly ash. Therefore, this research aims to the gel phase only. Grid nanoindentations with 10 μm space were applied in the regions rich in gel to provide more abundant data for the estimation of gel properties. The modulus and hardness of the indented phase are calculated by Eqs. 6.1 to 6.3. Constant Poisson's ratio of 0.2 was set for testing.

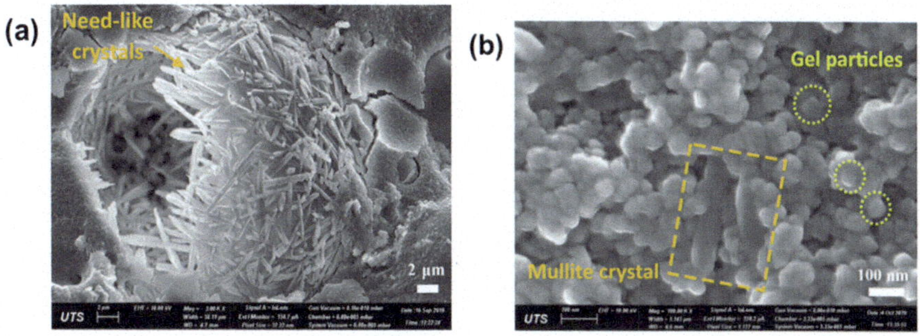

Fig. 6.6 Morphology of matrix in geopolymer: **a** 3000×; **b** 100,000×

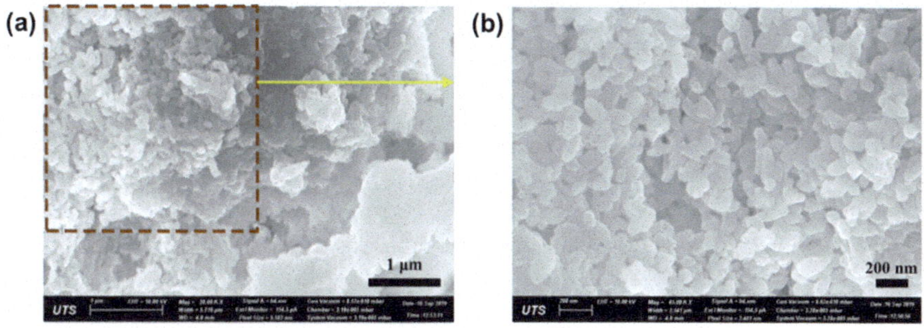

Fig. 6.7 Mullite in geopolymer matrix: **a** 20,000×; **b** 4500×

$$H = \frac{P_{\max}}{A} \tag{6.1}$$

$$S = \frac{\mathrm{d}P}{\mathrm{d}h}|_{h=h\max} = \frac{2}{\sqrt{\pi}} E_r \sqrt{A} \tag{6.2}$$

$$\frac{1}{E_r} = \frac{1 - v^2}{E} + \frac{1 - v_i^2}{E_i} \tag{6.3}$$

where P and h are the indentation load and indentation depth, respectively; S is the initial unloading stiffness; E_r is the reduced elastic modulus; E_i and v_i are the elastic modulus and Poisson's ratio of the indenter, respectively; E and v are the corresponding parameters for sample; A is the projected contact area.

Two-dimensional Gaussian mixture model (GMM) given in Eqs. 6.4 to 6.8 was used to describe the micromechanical properties distribution of geopolymers. The parameters in the model were estimated by maximum likelihood estimation (MLE) shown in Eq. 6.7. Expectation Maximization (EM) algorithm [31] was used for solving the log-likelihood function. In order to prevent the overfitting errors, Bayesian Information Criterion (BIC) was adopted as given in Eq. 6.8. After completing the parameter estimation for the model, each raw nanoindentation test point was assigned to a phase in the model, based on the principle of achieving the maximum posterior probability.

$$p(x) = \sum_{k=1}^{K} \pi_k \mathcal{N}\left(x|\mu_k, \sum_k\right) \quad (6.4)$$

$$\sum_{k=1}^{K} \pi_k = 1 \quad (6.5)$$

$$\mathcal{N}\left(x|\mu, \sum\right) = \frac{1}{\sqrt{\det(2\pi \sum)}} \exp\left(-\frac{1}{2}(x-\mu)^T \sum^{-1}(x-\mu)\right) \quad (6.6)$$

$$\ln L = \sum_{n=1}^{N} \ln\left(\sum_{k=1}^{K} \pi_k \mathcal{N}\left(x_n|\mu_k, \sum_k\right)\right) \quad (6.7)$$

$$BIC = k \ln n = 2 \ln L \quad (6.8)$$

where $x = (M, H)^T$, π_k, μ_k and Σ_k are the corresponding weighting coefficient, mean value and covariance of the kth component, respectively. N is the number of nanoindentation data.

The properties variation of the gel phase is generally in a small range. In order to increase the accuracy of the result, representative samples including the reference sample, Geo-20 min (poor dispersion) and Geo-120 min (good dispersion) were chosen. Then, sufficient tests were conducted on them to investigate the effect of nanosilica particles and their dispersion on the gel properties. When the MLE deconvolution method is used, the number of components needed to generate the gel phase is not a constant value [41]. The gel phase can be determined by comprehensively considering the micromechanical properties and variance [42].

For the Geo-120 min sample, the phase that conforms to the characteristics of the gel phase appears when the number of components k is three. This phase can be observed in Table 6.4 and Fig. 6.8a, b. The properties of the gel phase are almost

Table 6.4 Deconvolution results (clustered sand brown points)

k	Geo-120 min			Geo-20 min			Geo-Refer		
	$k_1 = 3$	$k_2 = 4$	$k_3 = 5$	$k_1 = 2$	$k_2 = 3$	$k_3 = 4$	$k_1 = 3$	$k_2 = 4$	$k_3 = 5$
M [GPa]	12.34	12.18	12.20	11.76	10.85	9.94	11.03	8.63	8.61
H [GPa]	0.51	0.51	0.50	0.59	0.53	0.47	0.58	0.47	0.47
f	54.49%	53.07%	53.09%	49.46%	41.90%	30.93%	29.50%	18.77%	18.65%

Note k is the number of components in a model. The data in the column of k_1 is the results of the first model where the gel phase starts to appear, while the data in the column of k_2 and k_3 are the results of the next two models with an increased number of components. M, H and f are the elastic modulus, hardness and fraction of the possible gel phase in the model

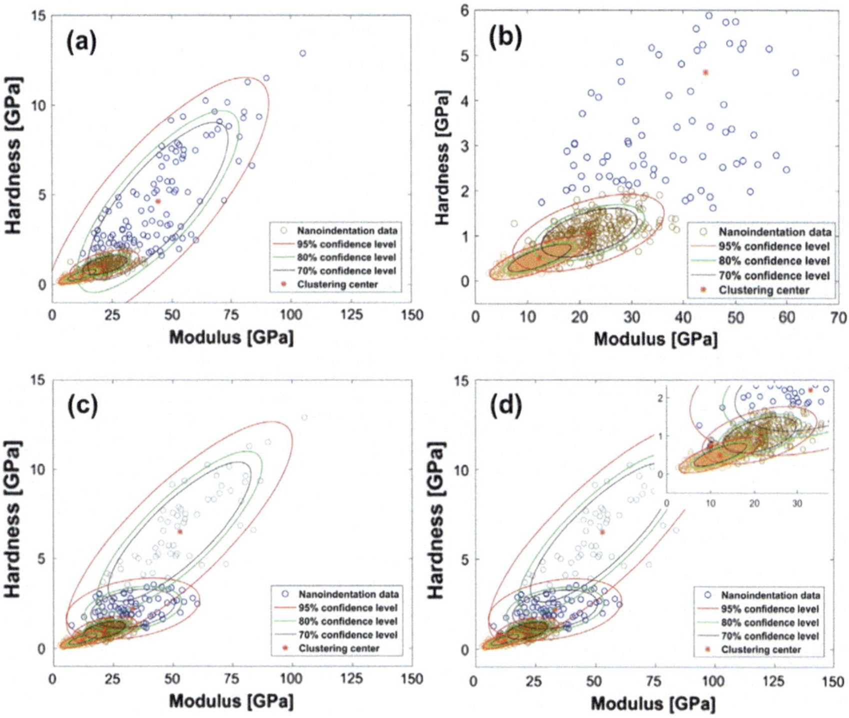

Fig. 6.8 Deconvolution results for Geo-120 min: **a** k = 3; **b** k = 4_local detail; **c** k = 4; **d** k = 5 (new phase: $M = 10.01, H = 0.75$)

invariant in models with increased components. Compared with the four-component model, the new phase in the five-component model is a spurious phase with a negligible proportion. As displayed in Table 6.4 and Figs. 6.9 and 6.10, the first model to generate the possible gel phase has a number of components of two and three for the Geo-20 min sample and the reference sample, respectively. The properties of the gel phase experience change with the increase of components. In fact, the properties of the gel phase in both samples become almost stable when the number of components reached to 4. Further increase in components doesn't lead to a significant variation of the results. The previous study indicated that there are mixed phases and sub-phases during the deconvolution analysis [43]. It is hard to distinguish whether the first gel results or the stable gel results are the real properties of the gel phase. However, previous studies indicate that the test and deconvolution itself just cause very small errors for elastic modulus [44, 45]. Even if based on the first gel results, after taking into account the errors, it can still know that the dispersion contributes to the enhancement of elastic modulus. This trend becomes more significant when the comparison is based on the results from the final stable gel phase. To decrease the errors caused by uncertainty, the final micromechanical properties of the gel phase are determined by the average value from the first three models where the phase that conforms to

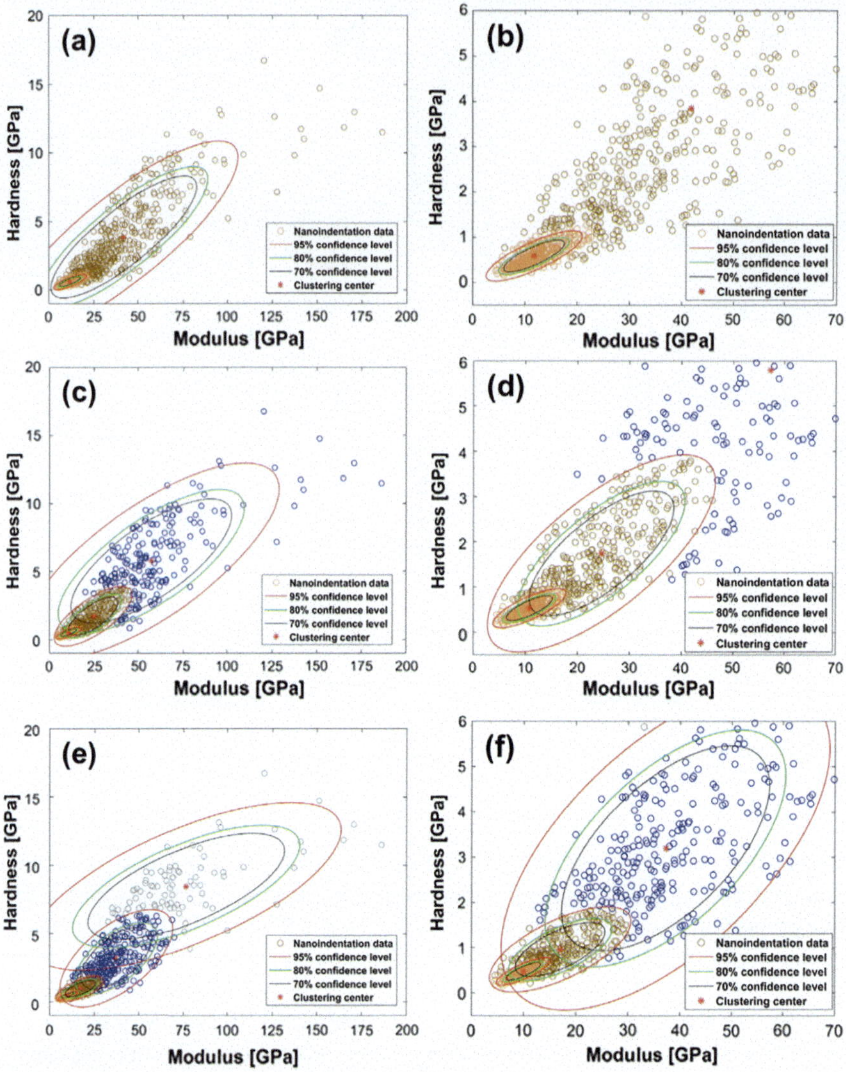

Fig. 6.9 Deconvolution results for Geo-20 min: **a** k = 2; **b** k = 2_local detail; **c** k = 3; **d** k = 3_local detail; **e** k = 4; **f** k = 4_local detail

the characteristics of the gel phase has appeared. In this study, the nanoindentation tests are on the gel rich regions. Therefore, the small phase in the models containing only two components was also examined whether it is the gel phase.

The results for all the three kinds of samples are summarized in Table 6.5. It can conclude that the geopolymer would achieve higher micro modulus after incorporating of nanosilica particles. For the part of hardness, the maximum hardness difference obtained between the three kinds of samples is 0.03 GPa. It means that

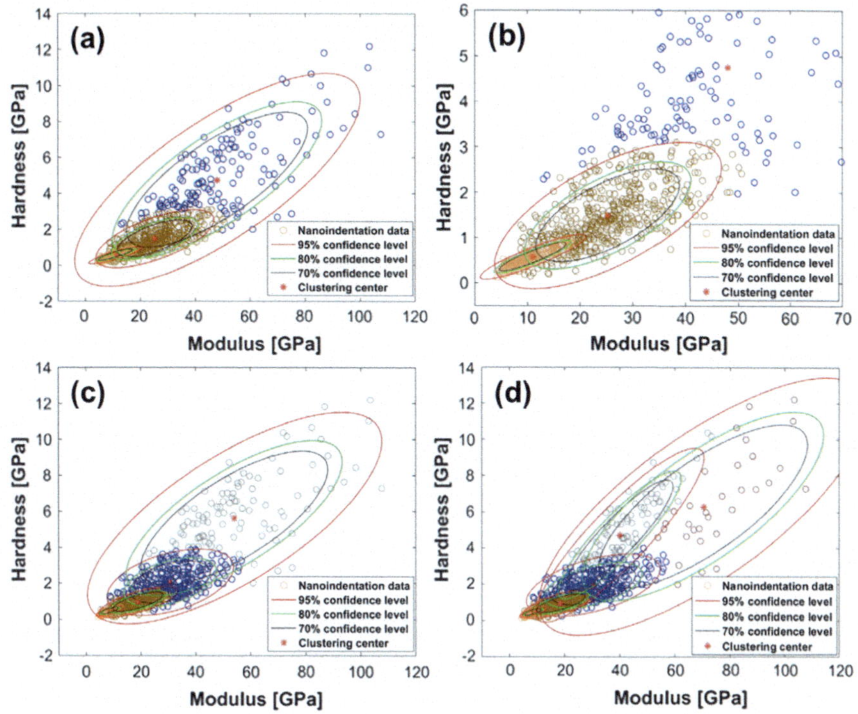

Fig. 6.10 Deconvolution results for Geo-Refer: **a** k = 3; **b** k = 3_local detail; **c** k = 4; **d** k = 5

Table 6.5 Micromechanical properties of gel in nano-geopolymers with different dispersion

Sample	Modulus (GPa)	Hardness (GPa)	Proportion from SNT (%)	Proportion from BSE (%)
Geo-120 min	12.24	0.51	53.55	54.18
Geo-20 min	10.85	0.53	40.76	52.38
Geo-Refer	9.42	0.50	22.31	49.03

the hardness of them are close. The errors generated from the testing, deconvolution and average process cover up the real difference. In addition to the modulus, a significantly lower proportion of gel phase is obtained for the reference sample by SNT analysis. The limited test points of the SNT technique on the heterogeneous geopolymer is not considered to be able to reflect the content of gel accurately. Moreover, the set of the grid on the gel rich areas would also influence the results. The proportion of gel was then evaluated by the BSE analysis method as illustrated in Fig. 6.11. Twenty 500× backscattered electron (BSE) images were taken at random locations of samples to statistic the proportion of gel phase, which were operated on Zeiss EVO LS15 with accelerating voltage of 15 kV. As shown in Table 6.5, the

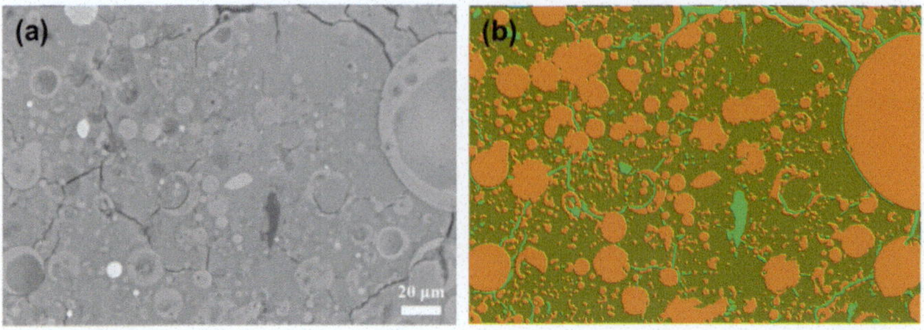

Fig. 6.11 Determination of the proportion of N–A–S–H gel based on BSE images: **a** BSE image; **b** segmented image

proportion of gel increases with the incorporation of nanoparticles but which is not as significant as the increase in the micromechanical properties.

In terms of the effect of dispersion, the benefits of the dispersion on the micro properties and content of gel can be identified from Table 6.5. Based on the modulus result of 9.42 GPa for the reference sample, the micro modulus increases by 15.18% for the Geo-20 min sample. The increased proportion reaches a high value of 29.93% for the Geo-120 min with just longer time of dispersion. The enhancement of micromechanical properties by dispersion is more pronounced than that of macro compressive strength compared with the reference sample.

For the nanosilica particles, the agglomerations with very large size are known to have dramatically decreased mechanical properties [46]. Hence, the bulk agglomerations with a loose appearance in Fig. 6.2 would be easily broken down by sonication. The translucent appearance of the Geo-20 min in Fig. 6.4 shows the process of transforming the suspension to colloid. It would have a considerable number of nanoscale particles, but also a large number of larger particles. For the colloidal dispersion of Geo-120 min, most of the particles should be at the nanoscale. When the dispersions are further mixed with fly ash immediately, the state of particles would not change significantly. These nanoparticles are far smaller than the fly ash particles, which can play the micro-filling effect to help the geopolymer achieve denser microstructure. Therefore, Geo-20 min and especially Geo-120 min have improved mechanical properties at both macro and micro scale. In fact, the macro strength is also related to the reaction degree and bonding, whereas the micromechanical properties of gel are affected by the micromechanical properties of nanosilica.

For the Geo-20 min and Geo-120 min sample, the final gel generated would be the composites of gel particles and nanoparticles. The elastic modulus of the single nanosilica particle obtained by nanoindentation test is reported to be 68.9 ± 9.6 GPa [42]. For the gel particles, no research has reported the mechanical properties of it. The LD C–S–H gel could be treated as an analogue of the NA–S–H gel, as they both consist of small globules of around 5 nm [47, 48]. Besides, the micromechanical properties of N–A–S–H are also known to be similar to the LD C–S–H gel. The modulus of the C–S–H globules is normally found to be in the range of around

53.55–63.5 GPa [27]. For the N–A–S–H gel particles packed by globules and with pore inside it, its modulus would be lower than the nanosilica. In the nanoindentation test, the microvolume probed would reach to about 3–5 times of the penetration depth, namely, slightly larger than 1 mm. For the well dispersed nanosilica particles, they have a similar size as the gel particles but with much less proportion. These features make the composite with gel particles and well dispersed nanosilica particles be detected by nanoindentation as the final gel phase. For the poorly dispersed nanosilica particles, the uneven distribution or present of agglomerates that cause significantly high micromechanical properties on test points would not affect the final gel properties significantly, because these points would not be clustered to the gel phase in the deconvolution process. The same situation is also for inclusions such as Mullite, which has high mechanical properties (modulus of 230 GPa [49]) and large size.

Among the three kinds of samples, the Geo-120 min would have the most uniform and dense structure of the nanoparticles-gel composites. The dense structure and the present of high modulus nanoparticles in gel making it achieve the highest micro modulus value. However, for the Geo-20 min, the nanoparticles-gel composites would be less uniform and have some nanosilica agglomerates inside it. The test points that have an abnormally high amount of nanoparticle in the microvolume probed would be excluded during deconvolution. For the remaining part of the test points on gel where the nanoparticles have relatively better dispersion and are embedded more evenly, because of the decreased content of the high modulus nanoparticles and the less effective filling effect, the micromechanical properties obtained would be lower than the Geo-120 min. Thus, the presence of dense gel structure by the filling effect of small particles and the presence of high modulus nanosilica itself are considered as the two important factors for the improved micro modulus. However, for the macro level performance, in addition to the dense gel structure and slightly improved proportion of gel, the bonding between gel and nanoparticles would also be an important aspect. The large size agglomerates would have low tensile strength and would be hard to pack densely within the gel phase [39]. The better bonding between gel and small size nanoparticles resulted from dispersion could also be a reason for the high compressive strength of Geo-120 min.

6.3 Micromechanical Properties and Microstructure of Geopolymer with nano-SiO_2 and nano-TiO_2

6.3.1 Material and Sample

Class F low calcium Fly ash with the chemical composition listed in Table 6.1 was adopted as the precursor material. The alkali solution used was a mixture of NaOH solution and Na_2SiO_3 solution. The NaOH solution was prepared by dissolving NaOH pellets in water. The commercial Na_2SiO_3 adopted has a silica modulus of

Fig. 6.12 Morphology of nanoparticles: **a** NT particles (10,000×); **b** NT particles (20,000×); **c** NS particles (10,000×); **d** NS particles (20,000×)

Table 6.6 Basic properties of nanoparticles

Material	Appearance	Average particle size (nm)	Purity (%)	Type
Nano-TiO_2	White powder	20	99.9	Anatase
Nano-SiO_2	White powder	20	99.9	Amorphous

2.07 (14.7% of Na_2O and 29.4% of SiO_2). Alkali solutions and fly ash were mixed based on a silica modulus (SiO_2/Na_2O) of 1, Na_2O to ash ratio of 10%, and water to solid ratio of 0.32. The morphology and properties of NS and anatase NT are shown in Fig. 6.12 and Table 6.6. They have the same average particle size of 20 nm. Nanoparticles were incorporated by the way of replacing 2% of fly ash while other factors were kept the same as the reference sample. Nanoparticles are easy to combine with each other owing to the high van der Waals force [50]. In this research, 0.5% of Polycarboxylate superplasticizer was incorporated into the alkali solution (also for the reference sample) to help disperse nanoparticles. After adding nanoparticles to the alkali solution, the solution was manually stirred for 3 min and then dispersed in an ultrasonic bath (120 W, 40 kHz) for 2 h. In order to avoid the heat generated in the dispersion process, the water in the bath was replaced every 3 min in the first

1.5 h, every 2 min in the following 20 min and every 1 min in the last 10 min. Alkali solution for the reference sample was also sonicated in the bath.

Geopolymer paste and mortar were mixed by Hobart mixer and cast in 50 mm × 50 mm × 50 mm cubic plastic moulds. In addition to 2 times of sand by weight of fly ash, the mortar samples had the same mix proportions as paste samples. After vibrated for 3 min, samples were sealed by plastic film and cured in a 65 °C oven for 48 h, followed by standard curing until 28 days. Then, small samples with a size of about 10 mm × 10 mm × 5 mm were cut from the core part of geopolymer pastes and embedded in epoxy resin. They were ground by 320, 600 and 1200 grits abrasive papers with each grade lasted for 10 min and polished with a small force by 0.3 μm alumina (for 20 min) and 0.05 μm cerium oxide slurry (for 20 min) to achieve a satisfactory surface. The polished samples were used for the nanoindentation test and BSE image analysis. Besides, paste samples were ground into powders for XRD and TGA analysis. Thereafter, samples were soaked in isopropanol for 3 days and then put in 50 °C vacuum oven drying for 3 days. Prepared samples were stored in a vacuum desiccator.

6.3.2 Heat of Reaction of Geopolymer Paste with Nanoparticles

The heat of reaction was measured by TAM Air Isothermal Calorimeter. In order to simulate the heat curing of geopolymer, the temperature condition in the isothermal calorimeter was pre-set and kept at 65 °C for testing. The heat of reaction of samples during the first 48 h is given in Fig. 6.13. As presented in Fig. 6.13a, the heat flow curves are similar for these three samples. When compared with Portland cement paste [51], much higher heat flow is found for the heat cured geopolymers in the initial stage, and no peak was obtained in the later stage. The detailed figure in Fig. 6.13a indicates that the highest heat release rate occurs at a similar time for samples. The nanomodified samples show significantly higher heat flow than the reference sample in the initial 2000s, with the highest value achieved by Geo-NS. It indicates that NT and especially NS have accelerated the reaction rate, quite similar to the phenomenon observed in Portland cement paste [25, 48, 52]. It would be attributed to increased interparticle space [39], as well as the possible nucleation effect from the ultra-small size of particles [46, 47, 53, 54]. Increased thermal conductivity could also contribute to the acceleration of reaction. However, considering the fact that NT with high thermal conductivity than NS brings a less significant increase in the heat flow of geopolymer, this factor should be less dominant than others in the initial stage. In terms of the cumulative heat generated, similar values were achieved for Geo-NS (193.56 J) and Geo-NT (193.95 J) samples, all larger than the reference sample (186.64 J). The higher cumulative heat in the nanomodified samples implies the higher reaction degree promoted by both nanoparticles during the first 48 h. A previous study for nano-Al_2O_3 indicated that it can avoid the induction period in

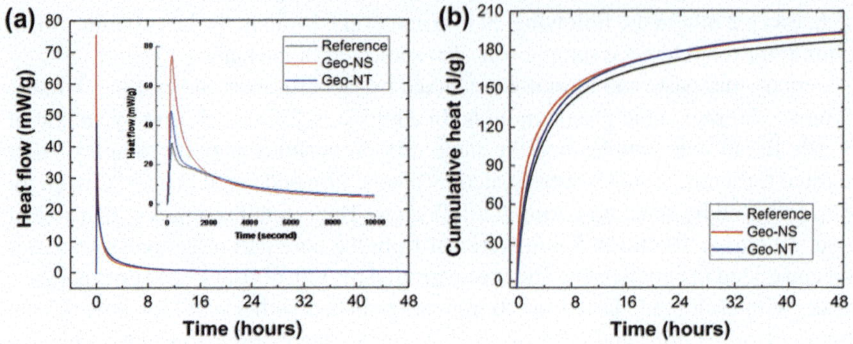

Fig. 6.13 Effect of nanoparticle type on the heat of reaction during the first 48 h: **a** heat flow over time; **b** cumulative heat over time

NaOH activated fly ash, generating gel much earlier [48]. Thus, the results for NS and NT again reveal the potential of nanoparticles for improving the early performance of geopolymer, which is in agreement with macro-mechanical results [19, 55].

6.3.3 Microstructure and Mineralogical Phases of Geopolymer Paste

Zeiss Supra 55VP was adopted for scanning electron (SE) image observation of geopolymer paste samples to the nanoscale. The voltage for SE observation was 5 or 10 kV. The micro and nanoscale structure of geopolymers is shown in Fig. 6.14. At the magnification of 5000×, Fig. 6.14a shows that geopolymer is mainly composed of residual fly ash and matrix. Defects such as pores and cracks are found to present in both fly ash and matrix. At the higher magnification of 10,000×, the local microstructure observation of matrix in Fig. 6.14b indicates that in addition to the main reacted product of N–A–S–H gel, sometimes there are also microcrystals. At this magnification, the image starts to show the feature that the matrix generated has a looser microstructure than the raw materials fly ash. The nanoscale's images can more clearly reveal the difference. In Fig. 6.14c, the N–A–S–H gel in matrix is found to be a loose structure packed by different particles, while the appearance of the fly ash (upper right corner) is still smooth and dense at this magnification. Nanoscale's details of the matrix are displayed in Fig. 6.14d. The particles pile up and entangle with each other. Most of them display in the granular form. Some typical particles displayed are found to have a size of slightly more than 30 nm. It should be noted that these particles actually consist of smaller globules as the basic unit. Due to the resolution limitation of SEM techniques, the smaller globules can be just seen faintly in the enlarged picture of 300,000×, which are reported to have a size of around 5 nm [40].

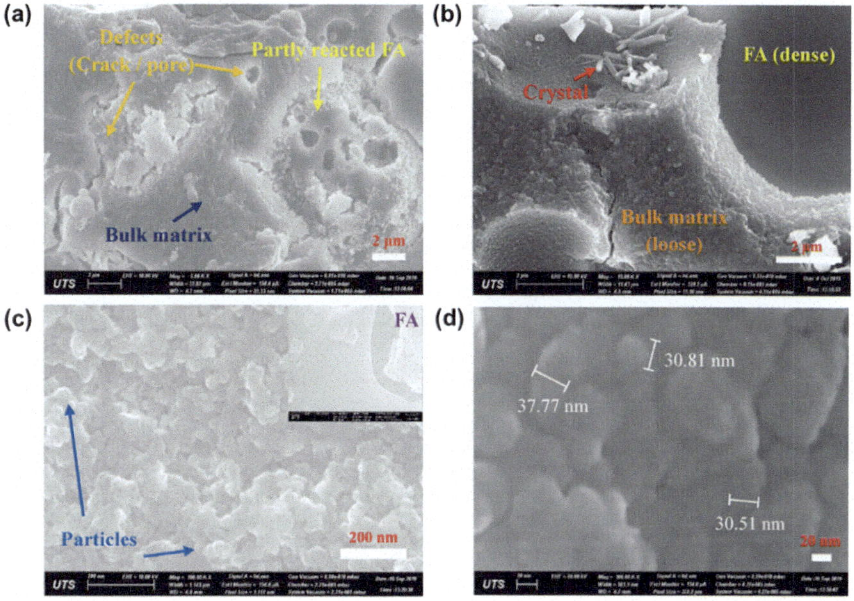

Fig. 6.14 Microstructure observation of geopolymer paste: **a** microscale of geopolymer (5000×); **b** microscale of geopolymer (10,000×); **c** nanoscale of N–A–S–H gel (100,000×); **d** nanoscale of N–A–S–H gel (300,000×)

The crystals in raw material fly ash and reacted geopolymer were detected by Bruker D8 Discover diffractometer. The step size and 2θ range were 0.02° and 5–70°, respectively. The crystals detected by XRD are presented in Fig. 6.15. Quartz and Mullite are the main crystals in geopolymer. Compared with the raw material fly ash, it shows that there is no new kind of crystal generated in the activated system, which is in line with some previous studies [56]. The geopolymer aluminosilicate gel has been proposed to be related to the precursor gel for the formation of zeolite [56]. The new zeolite phase Na-F was formed in the nano-Al_2O_3 seeded geopolymer [39]. However, a previous study for geopolymer incorporated with Zirconia showed that it is not able to act as a nucleation germ for zeolite formation [55]. The XRD analysis in this research reveals that the NS and NT also fail to promote the evolution of the zeolite phase.

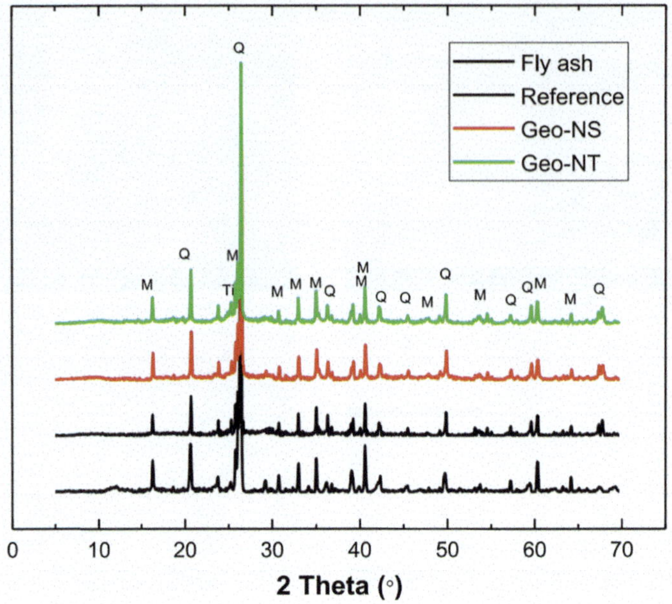

Fig. 6.15 XRD results of nanoparticles reinforced geopolymer paste

6.3.4 *Micromechanical Properties and Proportion of Nano-Reinforced N–A–S–H Gel*

6.3.4.1 Spurious Phases and a Compromise Strategy for Deconvolution

Nine of 10 × 10 grids with a grid spacing of 15 μm were performed on samples. In an individual nanoindentation point test, the maximum force was set as 1.5 mN. The loading and unloading time (unloading to 10% of peak force) was 7.5 s. When reaching the maximum force, it was maintained for 5 s to remove creep. Constant Poisson's ratio of 0.2 was set for testing. The modulus and hardness of each point could then be obtained [57]. After deleting abnormal test curves as shown in Fig. 6.16, the nanoindentation data were subjected to deconvolution analysis.

The distribution of micromechanical properties of geopolymer was described by a two-dimensional Gaussian mixture model (GMM) in Eqs. (6.4–6.8), Maximum likelihood estimation (MLE) [23, 53, 58] is adopted as a deconvolution technique in this study. The parameters of the model including the weighting coefficient, mean value of modulus and hardness and corresponding covariance of phases that can make the log-likelihood function ln L given in Eq. (6.7) achieve the maximum value are the final estimated results. Expectation Maximization (EM) algorithm was used to find the maximum likelihood solution. 1000 times of repeated calculation with random initial input values were made to find the optimum value. After obtaining the estimated model, the testing data was clustered to the corresponding component

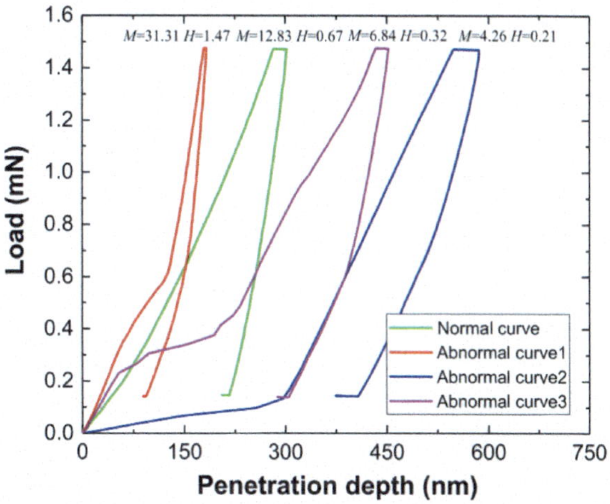

Fig. 6.16 Typical abnormal nanoindentation test points on sample

(phases) based on the maximum posterior probability. In addition, Bayesian Information Criterion (BIC) as shown in Eq. (6.8) was used to penalize the overfitting errors and find a suitable number of components (phases) to describe the model.

Geopolymer is mainly composed of N–A–S–H gel, unreacted fly ash, pores/cracks and crystals. However, the deconvolution of the Gaussian mixture model would not be able to accurately obtain the distribution of micromechanical properties of all phases [59]. Spurious phases such as mixed phase and sub-phases would be generated owing to the combined effects of phase characteristics, test factors, deconvolution parameters, etc. In the nanoindentation test, the involved range of gel is 3–4 times of the penetration depth [60]. Thus, the test results of nanoindentation sometimes come from multiphase interactions, such as interactions between gel and microcrystalline, gel and fly ash. These test points are one of the sources of the spurious phases. Typically, because of the small size of the crystal phase, most of the test results for it actually come from the composite response of the crystal and gel.

For the nanoindentation data collected, the deconvolution analysis would play a very important role in presenting micromechanical information of phases. When the number of phases specified is small, the deconvolution result of phases would be a mixture of multiple phases (mixed phases caused by deconvolution). However, increasing the number of phases in the model would sometimes lead to sub-phases. For instance, Phase like fly ash has a wide range of micromechanical distribution especially when pores are presented. Its micromechanical distribution is sparse at some locations and would sometimes overlap with other phases (crystals and mixed phases from multiphase interactions). As a consequence, some fly ash test data would be misidentified as individual phases (sub-phases) or incorporated into other phases in deconvolution. The mixed phases and sub-phases would exist in the deconvolution model simultaneously. In lots of the case of deconvolution, the sub-phases of fly ash occur before obtaining the gel phase from the mixture of multiple phases. Therefore, the micromechanical properties of the gel phase and fly ash cannot be accurately obtained at the same time. Additionally, the small amount of the real test points on the crystal phase makes it very easy to be combined with others with similar micromechanical properties in deconvolution instead of presence as an individual peak. The real micromechanical properties of the crystal phase are almost impossible to be obtained from the deconvolution analysis. Considering the above issues, a compromise strategy [56, 61] is adopted. The micromechanical investigation in this study just focuses on the most crucial phase, N–A–S–H gel.

6.3.4.2 Micromechanical Properties of Nano-Reinforced N–A–S–H Gel

The deconvolution process is briefly illustrated based on the case of Geo-NT. As shown in Table 6.7 and Fig. 6.17, for the model of Geo-NT with 3 phases, it is the sand brown phase that has the minimum micromechanical properties, with the modulus of 19.25 GPa, hardness of 1.15 GPa and covariance matrix of (C11 = 31.31, C12 = C21 = 1.53, C22 = 0.14), respectively. When increasing the number of phases to 4 and 5, there is no significant change for the sand brown phase. For this phase, it has high micromechanical properties and especially large variance, which does not conform to the characteristics of the gel phase [34, 39, 48, 52]. The N–A–S–H gel phase with small micromechanical properties ($M = 15.44$ GPa, $H = 0.90$ GPa) and small covariance matrix (C11 = 11.43, C12 = C21 = 0.49, C22 = 0.06) appears when the number of phases was set as 6, which changes just slightly with the increase in the number of phases. The decomposition of the above sand brown phase to the new smaller sand brown phase (N–A–S–H) can be observed clearly in Fig. 6.17b, c and e, f. However, the BIC does not reach the optimal value (minimum value) even if 7 components are assigned to the model as shown in Table 6.7. As mentioned above, there are two kinds of spurious phases, mixed phase and sub-phase. In the deconvolution analysis, with the increase in the number of phases, the mixed phases would decrease while the sub-phases would increase. It is obvious that sub-phases (e.g. phase with M of 112.79 GPa and H of 10.39 GPa) already present in the model with 5 components while the sand brown phase is still the mixture of gel and other phases. The scattered test points of fly ash make the spurious sub-phases evaluated as reasonable by the mathematical method of BIC but does not conform to the real distribution. Therefore, it is not feasible to obtain the optimum model by BIC to analyze the properties of N–A–S–H gel because the excessive separation of phases is wrongly evaluated as reasonable. Besides, the normally used 4 components in deconvolution would make the result actually the mixture of gel and other phases. The compromise strategy of studying the N–A–S–H gel merely instead of all of the phases can avoid these problems. The micromechanical properties of N–A–S–H are determined by the average value of the N–A–S–H phase in the first three models as shown in Table 6.7. The elastic modulus, hardness and proportion of N–A–S–H gel in Geo-NT are 15.17 GPa, 0.89 GPa and 22.74%, respectively.

For reference and Geo-NS sample, the N–A–S–H gel appears in model with 4 components, and behaviours as a stable phase with the increase in the number of components. The model and SNT results for Geo-NS and reference with 4 and 6

Table 6.7 Micromechanical properties of Geo-NT

k	M [GPa]	H [GPa]	*F* (%)	BIC	C		
					C11	C12 = C21	C22
3	19.25	1.15	47.42	9151.100	31.31	1.53	0.14
4	19.30	1.15	47.76	9094.102	31.77	1.54	0.14
5	18.73	1.13	44.78	9055.36	27.58	1.37	0.13
6	15.44	0.90	24.67	9029.64	11.43	0.49	0.06
7	14.68	0.86	18.93	9026.98	9.33	0.41	0.05
8	15.39	0.90	24.63	9019.36	11.20	0.48	0.06
Average 6–8	15.17	0.89	22.74	–	10.65	0.46	0.06

Note Geo-NT is 2% NT modified geopolymer with 65 °C heat curing of 48 h. k is the number of phases. "*M*" and "*H*" refer to mean value of elastic modulus and hardness, respectively. *f* is the proportion of the phase. C11 is the variance of elastic modulus, C22 is the variance of hardness and C12 is the covariance of modulus and hardness

components are shown in Fig. 6.18 as well as Tables 6.8 and 6.9, respectively. The average elastic modulus, hardness and proportion of N–A–S–H gel in Geo-NS are 13.30 GPa, 0.75 GPa and 23.78%, respectively. The corresponding values in the reference sample are 11.03 GPa, 0.64 GPa and 12.16%, respectively.

6.3.4.3 Repeatability and Validity of Gel Proportion and Micromechanical Properties

According to the above results, it is clear that the nanoparticles would increase the micromechanical properties of the gel, and the Geo-NT achieves the highest value. Besides, the proportion of gel for the reference sample is just about 50% of the Geo-NT and Geo-NS samples. Before using the above results to analyze the micro mechanism of the macro properties, verification was conducted to understand the repeatability and validity of the results.

Since the reference sample has the lowest gel proportion, SNT test was conducted on the sample again to investigate if the results were affected by limited test points and also the repeatability of SNT results. For the above results of reference, Geo-NT and Geo-NS, they were based on the random principle of the SNT, namely, each grid was selected randomly. Since the research focusing on the gel phase only, grids nanoindentation were performed on gel rich areas to provide richer test data and more accurate test results for gel. In the repeated test procedure, six locations were selected randomly under a microscope and then grids were intentionally set on the gel rich area within the field of vision. The test result is provided in Fig. 6.19 and Table 6.10. Regarding the verification of proportion, BSE images were taken. The BSE images were taken at the low magnification of 500× to balance the requirement between details and overall information of samples [62, 63]. A total of twenty figures were

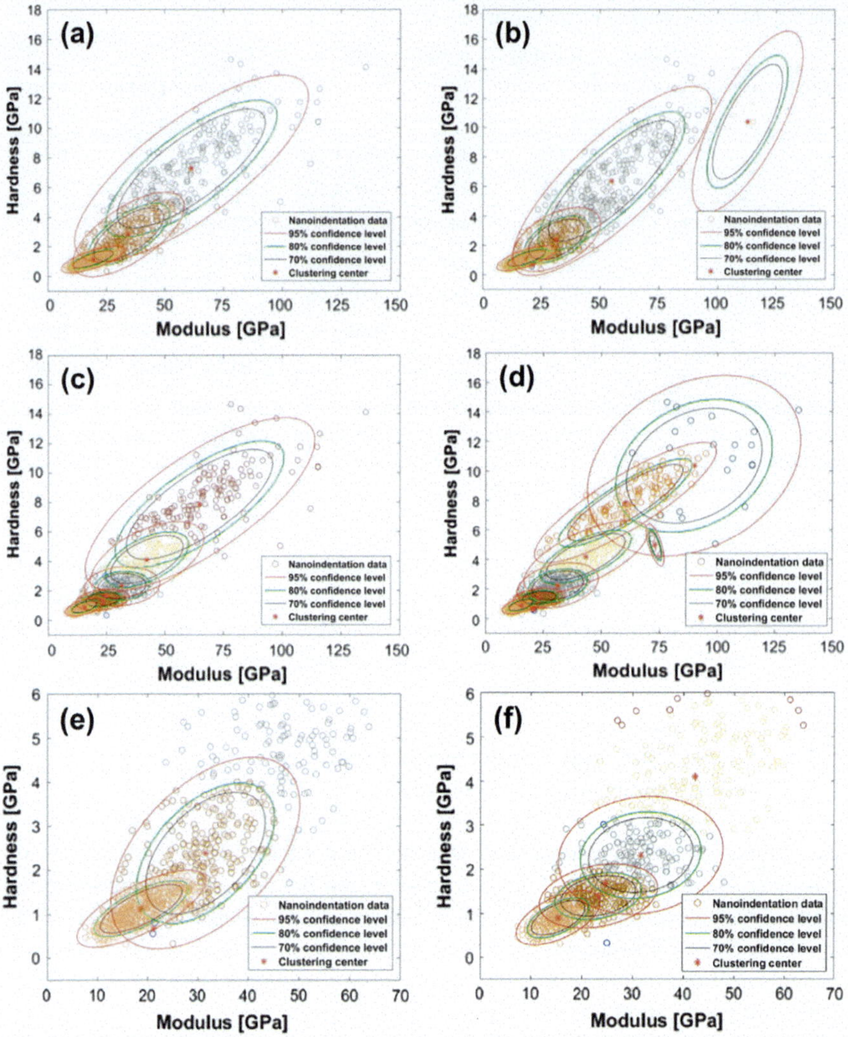

Fig. 6.17 Statistical nanoindentation results for Geo-NT: **a** k = 3; **b** k = 5 (blue dots also represent a phase); **c** k = 6 (blue dots also represent a phase); **d** k = 8 (blue dots also represent a phase); **e** k = 5_local detail; **f** k = 6_local detail

taken for statistical analysis of each sample. BSE was operated under the accelerating voltage of 15 kV on Zeiss EVO LS15. These images were segmented into three phases of pores/cracks, gel and unreacted fly ash as shown in Fig. 6.20. The proportions of the gel phase determined by BSE and by nanoindentation are summarized in Table 6.11.

As shown in Table 6.11, the SNT results based on 12.16% gel and 31.01% gel data set in different tests are quite similar to each other. It means that the accuracy of the test method would not be a significant obstacle for the comparison of micromechanical

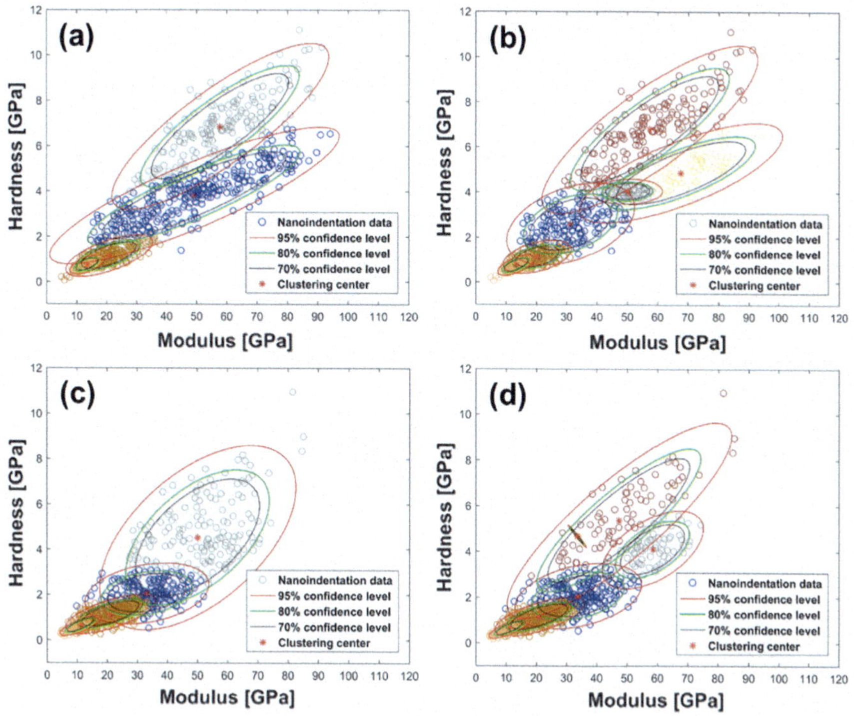

Fig. 6.18 Statistical nanoindentation results for Geo-NS and reference sample: **a** Geo-NS, k = 4; **b** Geo-NS, k = 6; **c** reference, k = 4; **d** reference, k = 6

Table 6.8 Micromechanical properties of Geo-NS

k	M [GPa]	H [GPa]	*F* (%)	BIC	C		
					C11	C12 = C21	C22
4	13.38	0.76	22.80	9051.71	5.47	0.25	0.03
5	13.27	0.75	24.36	9020.92	6.27	0.31	0.04
6	13.26	0.75	24.19	9001.09	6.23	0.3	0.04
Average	13.30	0.75	23.78	–	5.99	0.29	0.04

Note: Geo-NS is 2% NS modified geopolymer with 65 °C heat curing of 48 h. k is the number of phases. "*M*" and "*H*" refer to mean value of elastic modulus and hardness, respectively. *f* is the proportion of the phase. C11 is the variance of elastic modulus, C22 is the variance of hardness and C12 is the covariance of modulus and hardness

properties. However, the gel proportion obtained by statistical nanoindentation is obviously different from that obtained by the generally used BSE technique. In this study, epoxy resin was not used to impregnate the surface to avoid its effect on micromechanical testing. Thus, crystals are basically dropped, making the gel

Table 6.9 Micromechanical properties of the reference sample

k	M [GPa]	H [GPa]	*F* (%)	BIC	C		
					C11	C12 = C21	C22
4	11.04	0.64	12.20	8341.59	7.78	0.40	0.03
5	11.02	0.64	12.15	8297.51	7.74	0.40	0.03
6	11.02	0.64	12.13	8278.39	7.74	0.40	0.03
Average	11.03	0.64	12.16	–	7.75	0.40	0.03

Note k is the number of phases. "*M*" and "*H*" refer to mean value of elastic modulus and hardness, respectively. *f* is the proportion of the phase. C11 is the variance of elastic modulus, C22 is the variance of hardness and C12 is the covariance of modulus and hardness

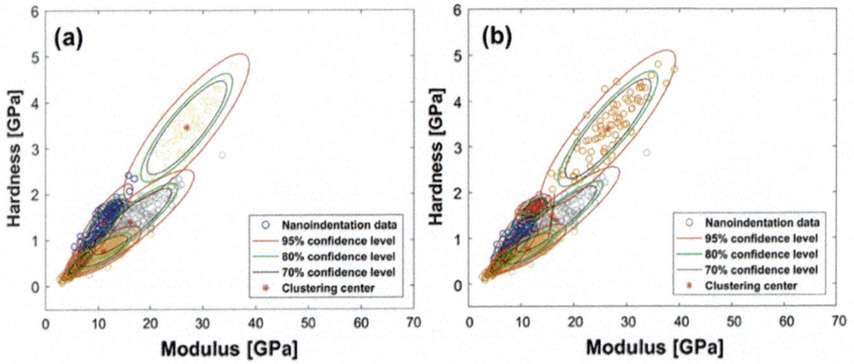

Fig. 6.19 Statistical nanoindentation results for Reference-repeated; **a** k = 5; **b** k = 7

Table 6.10 Micromechanical properties of reference sample-repeated

k	M [GPa]	H [GPa]	*F* (%)	BIC	C		
					C11	C12 = C21	C22
4	11.07	0.78	33.42	4991.91	8.88	0.52	0.05
5	11.30	0.77	29.89	4972.91	9.03	0.54	0.05
6	11.32	0.77	29.71	4960.76	9.10	0.54	0.05
Average	11.23	0.77	31.01	–	9.00	0.53	0.05

Note k is the number of phases. "*M*" and "*H*" refer to mean value of elastic modulus and hardness, respectively. *f* is the proportion of the phase. C11 is the variance of elastic modulus, C22 is the variance of hardness and C12 is the covariance of modulus and hardness

proportion obtained by BSE greater than the true value. However, since crystals in matrix may be mainly from raw material fly ash, the interference would not impact the relative trend of gel proportion significantly. BSE would still be a more accurate method than SNT on phase proportion evaluation as analyzed in the next section.

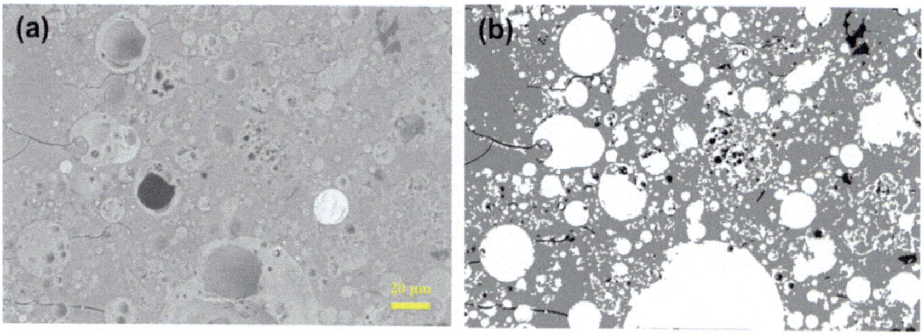

Fig. 6.20 Segment of phases based on the grey value of BSE image: **a** BSE image (500×); **b** segmented image

Table 6.11 Comparison of nano-mechanical properties and test results from SNT and BSE of geopolymer containing different types of nanoparticles

Samples	Modulus (GPa)	Hardness (GPa)	Proportion from SNT (%)	Proportion from BSE (%)
Geo-NT	15.17	0.89	22.74	54.02
Geo-NS	13.30	0.75	23.78	55.69
Reference	11.03 (11.23)	0.64 (0.77)	12.16 (31.01)	49.16

Note Samples were heat cured at 65 °C for 48 h. The results in brackets are from the repeated test on the reference sample where some grids are intentionally set on gel rich area

Based on the result of BSE, and considering that nanoparticles are added in a way of replacing FA, the difference of gel content in samples should be small.

6.3.5 Microscale Mechanism of Macro Performance

As opposed to the phenomenon at the macro scale where Geo-NS displays better performance than others, it is the gel in Geo-NT that has the highest modulus and hardness. Besides, for random nanoindentation results, the reference sample presents only 12.16% of gel while Geo-NT and Geo-NS have 22.74% and 23.78% gel, respectively. If based on the SNT results solely, it is easy to consider that the nanoparticles have increased the content of the gel/reaction degree and then better macro performance. In fact, the number of nanoindentation test points is very few when compared with pixels in BSE images, which is hard to accurately reflect the overall information of the highly heterogeneous geopolymer. Besides, the interaction of gel with other phases in the involved volume would also decrease the proportion of gel detected by deconvolution analysis. Therefore, the proportion obtained from SNT should be less reliable than that from BSE. More evidence is the TGA result as shown in Fig. 6.21. TGA test was conducted by STA449 F5 JUPITER. Vacuum dried powder was heated

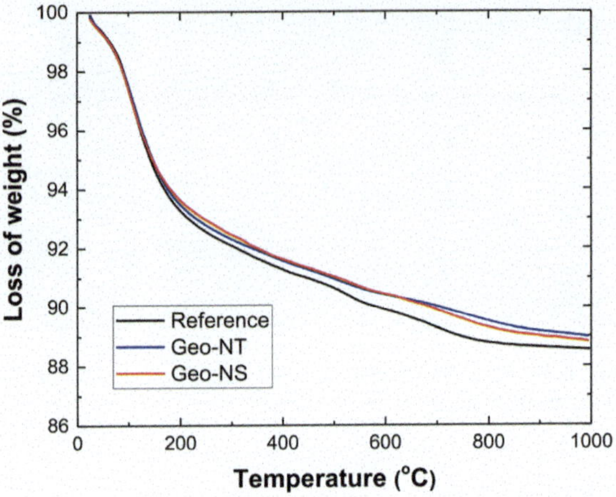

Fig. 6.21 Thermogravimetric analysis of geopolymer paste

from 25 °C to 1000 °C with a heating rate of 10 °C/min. The remaining weights for reference, Geo-NS and Geo-NT are 88.55%, 88.98% and 88.82%, respectively. The weight loss between 25 and 300 °C is the free water and loosely bound water, while the weight loss in the later period is caused by the loss of structure water and carbonaceous substances [46]. When the influence of free water is eliminated, the TGA results are almost the same for all the specimens, indicating that there would be no significant difference between the content of the gel phase.

As analyzed through the heat of reaction, the nanoparticles have an accelerating effect in the early stage. It would lead to a higher amount of gel in nano-geopolymers, which plays a significant role in the early performance of geopolymer. In fact, both NT and NS particles in this study could be treated as unreactive particles due to the limited solubility of NS in the sodium silicate solution. These particles promote the growth of gel in a more physical way, just as the interparticle distance is demonstrated as the most important parameter for the mineral additions to accelerate the clinker hydration [40]. Therefore, the relative difference of gel content (not the absolute difference) between geopolymer and nanomodified geopolymer would narrow over time. For the later stage, instead of the gel proportion, the differences in macro-mechanical properties between samples would be more related to the strength of the gel, as well as the coordination (e.g. bonding) between gel and other phases.

For nano-geopolymer, the nanoparticles could act as filler which contributes to the formation of a denser gel structure and then improved macro behavior. However, the NS would be more compatible with both sodium silicate solution and N–A–S–H gel particles due to the similar chemical composition. The NS may achieve better dispersion and also integrate with the N–A–S–H gel particles better, resulting in more prominent macro mechanical properties. For micromechanical properties, the denser gel structure would also result in higher modulus and hardness. However, there is another factor that would influence the micromechanical results significantly.

In the nanoindentation test, the involved depth is usually considered as 3–4 times of the indentation depth [64], reaching around 1 μm on N–A–S–H gel. When the nanoparticles are well dispersed (single particle, very small aggregations), they have a similar size to the gel particles shown in Fig. 6.14d, and works together with the gel particles as a composite. Owing to the large involved range, the well-dispersed nanoparticles are able to be detected by the nanoindentation test. The micromechanical properties of N–A–S–H in nano-geopolymers come from the interaction of gel particles and nanoparticles. Single nanosilica is reported to have elastic modulus and hardness of 68.9 ± 9.6 GPa and 2.8 ± 0.4 GPa, respectively in nanoindentation test [65]. For nano-titania (anatase), the modulus and hardness are as high as around 170 GPa and 8 GPa, respectively [4, 24]. N–A–S–H gel is actually similar to the counterpart LD C–S–H gel. Both are reported to consist of globules of about 5 nm [64, 66]. The nanoindentation results for them are also quite close. The globules of C–S–H are reported to have an elastic modulus of 59.7 ± 1.9 GPa [22]. Therefore, the N–A–S–H gel particles with a size larger than 30 nm would have a smaller elastic modulus due to the existence of nanopores between the packed globules. The higher mechanical properties of NS and especially NT particles than gel particles contribute to the high micromechanical properties of nano-reinforced N–A–S–H gel detected. Meanwhile, because of the small size and small proportion of the nanoparticles in each test point, the SNT results of the nano-reinforced gel would be improved but within a small range. For some of the nanoparticles that happen to present densely in the nanoindentation test points (poor dispersion), these test points with significantly high mechanical properties would not be identified and clustered to the gel phase in the deconvolution process. The above reasons lead to the results in Table 6.11 for the gel particles-nanoparticles composites where Geo-NT and reference sample achieve the highest and the lowest value, respectively. Thus, the properties of the particles themselves cause the different reinforcement effect of NS and NT on macro compressive strength and micro modulus and hardness.

6.4 Conclusions

In this chapter, the effects of dispersion of nanosilica particles using sonication on the compressive strength, flowability and microstructure properties of geopolymer were reviewed. The effects of different types of nanoparticles (NS and NT) on the microscale properties of geopolymer, especially the most important component N–A–S–H gel, were discussed. The related conclusions can be drawn as follows:

(1) Proper dispersion of nanoparticles is important to optimize microstructure and micromechanical properties of nanoparticle modified geopolymers. Sufficient dispersion of nanoparticles would make the sample a more uniform nanoparticles-gel composite. Smaller particles after dispersion play the filling effect better to achieve denser microstructure and would have a better bonding with the gel.

(2) The presence of both NS and NT particles increase the early reaction rate in geopolymer, especially for the samples with better dispersion, while the reaction degree of different samples in the later age is not vastly different. The contents of gel in 28 days geopolymer and nano-geopolymers vary in a small range of 49.16–54.02%.
(3) SNT results due to the limited test points on the highly heterogeneous geopolymer are considered less reliable than the results from BSE. The multiple phase interaction would be responsible for the significantly lower amount of gel detected by SNT.
(4) Gel particles packed by around 5 nm globules are typically observed to have a size of more than 30 nm, similar to the size of well-dispersed nanoparticles. Nanoparticles integrate with gel particles to form a composite with higher mechanical properties. The elastic modulus of N–A–S–H gel is 11.03 GPa, 13.30 GPa and 15.17 GPa, respectively, for reference, Geo-NS and Geo-NT sample.
(5) Nanoparticles have higher mechanical properties (highest for NT) but a much lower proportion than the gel particles, leading to slightly higher micromechanical properties of gel in nanogeopolymers obtained by SNT. Nano-SiO_2 is more compatible with sodium silicate solution and gel particles, resulting in better dispersion and bonding, then a higher macro strength of Geo-NS than Geo-NT.

References

1. Z. Tang, Y. Hu, V.W.Y. Tam, W. Li, Uniaxial compressive behaviors of fly ash/slag-based geopolymeric concrete with recycled aggregates. Cement Concr. Compos. **104** (2019)
2. E. Berodier, K. Scrivener, Understanding the filler effect on the nucleation and growth of C–S–H. J. Am. Ceram. Soc. **97**(12), 3764–3773 (2014)
3. F. Matalkah, P. Soroushian, Synthesis and characterization of alkali aluminosilicate hydraulic cement that meets standard requirements for general use. Constr. Build. Mater. **158**, 42–49 (2018)
4. F. Sanchez, K. Sobolev, Nanotechnology in concrete—a review. Constr. Build. Mater. **24**(11), 2060–2071 (2010)
5. B. Lothenbach, K. Scrivener, R.D. Hooton, Supplementary cementitious materials. Cem. Concr. Res. **41**(12), 1244–1256 (2011)
6. O. Zywitzki, T. Modes, H. Sahm, P. Frach, K. Goedicke, D. Glöß, Structure and properties of crystalline titanium oxide layers deposited by reactive pulse magnetron sputtering. Surf. Coat. Technol. **180–181**, 538–543 (2004)
7. D. Adak, M. Sarkar, S. Mandal, Structural performance of nano-silica modified fly-ash based geopolymer concrete. Constr. Build. Mater. **135**, 430–439 (2017)
8. P.S. Deb, P.K. Sarker, S. Barbhuiya, Sorptivity and acid resistance of ambient-cured geopolymer mortars containing nano-silica. Cement Concr. Compos. **72**, 235–245 (2016)
9. Y. Wang, X.-H. Xu, J. Han, Y.-S. Yan, Separation/enrichment of trace tetracycline antibiotics in water by [Bmim]BF_4–$(NH_4)_2SO_4$ aqueous two-phase solvent sublation. Desalination **266**(1), 114–118 (2011)
10. C. Hu, Z. Li, A review on the mechanical properties of cement-based materials measured by nanoindentation. Constr. Build. Mater. **90**, 80–90 (2015)

11. A.H. Korayem, N. Tourani, M. Zakertabrizi, A.M. Sabziparvar, W.H. Duan, A review of dispersion of nanoparticles in cementitious matrices: Nanoparticle geometry perspective. Constr. Build. Mater. **153**, 346–357 (2017)
12. S. Riahi, A. Nazari, Retracted: The Effects of Nanoparticles on Early Age Compressive Strength of Ash-Based Geopolymers (Elsevier, 2012)
13. R. Madandoust, E. Mohseni, S.Y. Mousavi, M. Namnevis, Retracted: An Experimental Investigation on the Durability of Self-Compacting Mortar Containing Nano-SiO_2, nano-Fe_2O_3 and nano-CuO (Elsevier, 2015)
14. Z. Luo, W. Li, K. Wang, S.P. Shah, Research progress in advanced nanomechanical characterization of cement-based materials. Cement Concr. Compos. **94**, 277–295 (2018)
15. C. Hu, Z. Li, Property investigation of individual phases in cementitious composites containing silica fume and fly ash. Cement Concr. Compos. **57**, 17–26 (2015)
16. J. Xiao, W. Li, Z. Sun, D.A. Lange, S.P. Shah, Properties of interfacial transition zones in recycled aggregate concrete tested by nanoindentation. Cement Concr. Compos. **37**(1), 276–292 (2013)
17. E. John, T. Matschei, D. Stephan, Nucleation seeding with calcium silicate hydrate—a review. Cem. Concr. Res. **113**, 74–85 (2018)
18. F. Pacheco-Torgal, S. Jalali, Nanotechnology: advantages and drawbacks in the field of construction and building materials. Constr. Build. Mater. **25**(2), 582–590 (2011)
19. T. Phoo-ngernkham, P. Chindaprasirt, V. Sata, S. Hanjitsuwan, S. Hatanaka, The effect of adding nano-SiO_2 and nano-Al_2O_3 on properties of high calcium fly ash geopolymer cured at ambient temperature. Mater. Des. **55**, 58–65 (2014)
20. P. Duan, C. Yan, W. Luo, W. Zhou, Effects of adding nano-TiO_2 on compressive strength, drying shrinkage, carbonation and microstructure of fluidized bed fly ash based geopolymer paste. Constr. Build. Mater. **106**, 115–125 (2016)
21. H. Assaedi, F.U.A. Shaikh, I.M. Low, Characterizations of flax fabric reinforced nanoclay-geopolymer composites. Compos. B Eng. **95**, 412–422 (2016)
22. S.M. Abbasi, H. Ahmadi, G. Khalaj, B. Ghasemi, Microstructure and mechanical properties of a metakaolinite-based geopolymer nanocomposite reinforced with carbon nanotubes. Ceram. Int. **42**(14), 15171–15176 (2016)
23. M. Zou, D. Yang, Nanoindentation of silica nanoparticles attached to a silicon substrate. Tribol. Lett. **22**(2), 189–196 (2006)
24. M. Saafi, K. Andrew, P.L. Tang, D. McGhon, S. Taylor, M. Rahman, S. Yang, X. Zhou, Multifunctional properties of carbon nanotube/fly ash geopolymeric nanocomposites. Constr. Build. Mater. **49**, 46–55 (2013)
25. L. Senff, D. Hotza, S. Lucas, V.M. Ferreira, J.A. Labrincha, Effect of nano-SiO_2 and nano-TiO_2 addition on the rheological behavior and the hardened properties of cement mortars. Mater. Sci. Eng. A **532**, 354–361 (2012)
26. J.J. Thomas, H.M. Jennings, J.J. Chen, Influence of nucleation seeding on the hydration mechanisms of tricalcium silicate and cement. J. Phys. Chem. C **113**(11), 4327–4334 (2009)
27. A. Naji Givi, S. Abdul Rashid, F.N.A. Aziz, M.A.M. Salleh, Experimental investigation of the size effects of SiO_2 nano-particles on the mechanical properties of binary blended concrete. Composites Part B: Eng. **41**(8), 673–677 (2010)
28. J. Xu, B. Wang, J. Zuo, Modification effects of nanosilica on the interfacial transition zone in concrete: a multiscale approach. Cement Concr. Compos. **81**, 1–10 (2017)
29. C.A. Rees, J.L. Provis, G.C. Lukey, J.S.J. van Deventer, The mechanism of geopolymer gel formation investigated through seeded nucleation. Colloids Surf. A **318**(1–3), 97–105 (2008)
30. C.J.W.C. Astm, PA: American Society for Testing, M. International, 1437, Standard test method for flow of hydraulic cement mortar (2001)
31. D. Khale, R. Chaudhary, Mechanism of geopolymerization and factors influencing its development: a review. J. Mater. Sci. **42**(3), 729–746 (2007)
32. M. Sumesh, U.J. Alengaram, M.Z. Jumaat, K.H. Mo, M.F. Alnahhal, Incorporation of nanomaterials in cement composite and geopolymer based paste and mortar—a review. Constr. Build. Mater. **148**, 62–84 (2017)

33. W. Li, Z. Huang, T. Zu, C. Shi, W.H. Duan, S.P. Shah, Influence of nanolimestone on the hydration, mechanical strength, and autogenous shrinkage of ultrahigh-performance concrete. J. Mater. Civil Eng. **28**(1) (2016)
34. E. García-Taengua, M. Sonebi, K.M.A. Hossain, M. Lachemi, J. Khatib, Effects of the addition of nanosilica on the rheology, hydration and development of the compressive strength of cement mortars. Compos. B Eng. **81**, 120–129 (2015)
35. X. Zhu, Y. Gao, Z. Dai, D.J. Corr, S.P. Shah, Effect of interfacial transition zone on the Young's modulus of carbon nanofiber reinforced cement concrete. Cem. Concr. Res. **107**, 49–63 (2018)
36. P. Hou, J. Qian, X. Cheng, S.P. Shah, Effects of the pozzolanic reactivity of nanoSiO_2 on cement-based materials. Cement Concr. Compos. **55**, 250–258 (2015)
37. P.S. Deb, P.K. Sarker, S. Barbhuiya, Effects of nano-silica on the strength development of geopolymer cured at room temperature. Constr. Build. Mater. **101**, 675–683 (2015)
38. Z. Zhang, J.L. Provis, X. Ma, A. Reid, H. Wang, Efflorescence and subflorescence induced microstructural and mechanical evolution in fly ash-based geopolymers. Cement Concr. Compos. **92**, 165–177 (2018)
39. T. Alomayri, Experimental study of the microstructural and mechanical properties of geopolymer paste with nano material (Al_2O_3). J. Build. Eng. **25** (2019)
40. S. Diamond, Considerations in image analysis as applied to investigations of the ITZ in concrete. Cement Concr. Compos. **23**(2–3), 171–178 (2001)
41. S.A. Bernal, J.L. Provis, Durability of alkali-activated materials: Progress and perspectives. J. Am. Ceram. Soc. **97**(4), 997–1008 (2014)
42. S. Kawashima, J.W.T. Seo, D. Corr, M.C. Hersam, S.P. Shah, Dispersion of $CaCO_3$ nanoparticles by sonication and surfactant treatment for application in fly ash-cement systems. Mater. Struct./Materiaux et Constr. **47**(6), 1011–1023 (2014)
43. W. Li, Z. Luo, C. Long, C. Wu, W.H. Duan, S.P. Shah, Effects of nanoparticle on the dynamic behaviors of recycled aggregate concrete under impact loading. Mater. Des. **112**, 58–66 (2016)
44. J.L. Provis, G.C. Lukey, J.S.J. Van Deventer, Do geopolymers actually contain nanocrystalline zeolites? a reexamination of existing results. Chem. Mater. **17**(12), 3075–3085 (2005)
45. J.J. Chen, L. Sorelli, M. Vandamme, F.J. Ulm, G. Chanvillard, A coupled nanoindentation/SEM-EDS study on low water/cement ratio portland cement paste: evidence for C–S–H/$Ca(OH)_2$ nanocomposites. J. Am. Ceram. Soc. **93**(5), 1484–1493 (2010)
46. J. Xu, F. Kong, S. Song, Q. Cao, T. Huang, Y. Cui, Effect of Fenton pre-oxidation on mobilization of nutrients and efficient subsequent bioremediation of crude oil-contaminated soil. Chemosphere **180**, 1–10 (2017)
47. W. Li, X. Li, S.J. Chen, G. Long, Y.M. Liu, W.H. Duan, Effects of nanoalumina and graphene oxide on early-age hydration and mechanical properties of cement paste. J. Mater. Civil Eng. **29**(9) (2017)
48. D. Davydov, M. Jirásek, L. Kopecký, Critical aspects of nano-indentation technique in application to hardened cement paste. Cem. Concr. Res. **41**(1), 20–29 (2011)
49. Z. Zhang, J.L. Provis, A. Reid, H. Wang, Fly ash-based geopolymers: the relationship between composition, pore structure and efflorescence. Cem. Concr. Res. **64**, 30–41 (2014)
50. M. Stefanidou, I. Papayianni, Influence of nano-SiO_2 on the Portland cement pastes. Compos. Part B: Eng. 2706–2710 (2012)
51. D. Jansen, F. Goetz-Neunhoeffer, B. Lothenbach, J. Neubauer, The early hydration of ordinary Portland cement (OPC): an approach comparing measured heat flow with calculated heat flow from QXRD. Cem. Concr. Res. **42**(1), 134–138 (2012)
52. K. Sobolev, I. Flores, L. Torres-Martinez, P. Valdez, E. Zarazua, E. Cuellar, Engineering of SiO_2 nanoparticles for optimal performance in nano cement-based materials. In *Nanotechnology in Construction 3: Proceedings of the NICOM3* (Springer, 2009), pp. 139–148
53. J.L. Provis, A. Palomo, C. Shi, Advances in understanding alkali-activated materials. Cem. Concr. Res. **78**, 110–125 (2015)
54. W. Li, S. Kawashima, J. Xiao, D.J. Corr, C. Shi, S.P. Shah, Comparative investigation on nanomechanical properties of hardened cement paste. Mater. Struct./Materiaux et Constr. **49**(5), 1591–1604 (2016)

55. R. Madandoust, E. Mohseni, S.Y. Mousavi, M. Namnevis, An experimental investigation on the durability of self-compacting mortar containing nano-SiO_2, nano-Fe_2O_3 and nano-CuO. Constr. Build. Mater. **86**, 44–50 (2015)
56. S. Riahi, A. Nazari, The effects of nanoparticles on early age compressive strength of ash-based geopolymers. Ceram. Int. **38**(6), 4467–4476 (2012)
57. Z. Luo, W. Li, Y. Gan, K. Mendu, S.P. Shah, Applying grid nanoindentation and maximum likelihood estimation for N–A–S–H gel in geopolymer paste: investigation and discussion. Cement Concr. Res. **135** (2020)
58. J. Björnström, A. Martinelli, A. Matic, L. Börjesson, I. Panas, Accelerating effects of colloidal nano-silica for beneficial calcium-silicate-hydrate formation in cement. Chem. Phys. Lett. **392**(1–3), 242–248 (2004)
59. J. Chen, S.C. Kou, C.S. Poon, Hydration and properties of nano-TiO_2 blended cement composites. Cement Concr. Compos. **34**(5), 642–649 (2012)
60. Z. Luo, W. Li, Y. Gan, K. Mendu, S.P. Shah, Maximum likelihood estimation for nanoindentation on sodium aluminosilicate hydrate gel of geopolymer under different silica modulus and curing conditions. Compos. Part B: Eng. **198** (2020)
61. R.R. Lloyd, J.L. Provis, J.S.J. Van Deventer, Microscopy and microanalysis of inorganic polymer cements. 1: Remnant fly ash particles. J. Mater. Sci. **44**(2), 608–619 (2009)
62. K.L. Scrivener, Backscattered electron imaging of cementitious microstructures: understanding and quantification. Cement Concr. Compos. **26**(8), 935–945 (2004)
63. P.J.M. Monteiro, G. Geng, D. Marchon, J. Li, P. Alapati, K.E. Kurtis, M.J.A. Qomi, Advances in characterizing and understanding the microstructure of cementitious materials. Cement Concr. Res. **124** (2019)
64. I. Gheewala, R. Smith, S.D. Kenny, Nanoindentation and nanoscratching of rutile and anatase TiO_2 studied using molecular dynamics simulations. J. Phys. Condensed Matter **20**(35) (2008)
65. S. Nath, A. Dey, A.K. Mukhopadhyay, B. Basu, Nanoindentation response of novel hydroxyapatite-mullite composites. Mater. Sci. Eng.: A 513–514(C), 197–201 (2009)
66. G. Constantinides, F.J. Ulm, The nanogranular nature of C–S–H. J. Mech. Phys. Solids **55**(1), 64–90 (2007)

Chapter 7
Nano/Micromechanical Properties of Interfacial Transition Zones in Recycled Aggregate Concrete Tested by Nanoindentation

7.1 Introduction

In views of environmental protection, conservation of natural resources and shortage of lands for waste disposal, it is necessary to promote recycled aggregates for effective utilization of waste concrete. Researchers have tried to relate the quality of Recycled Aggregate Concrete (RAC) to the properties of the original waste concrete, the new mix composition, the mixing approach and the deterioration condition of the Recycled Coarse Aggregate (RCA) [1–5]. It is generally accepted that the Interfacial Transition Zones (ITZs), the quality of the old mortar, and the old mortar content of the original concrete influence the properties of the RAC. In recent years, recycling of waste concrete to produce RAC has been proven to be commercially and technically sound for both non-structural and structural applications [6–9].

Although the recycling of demolished concrete can benefit the environment, there are concerns about the structural safety and durability of the RAC structure. To improve the quality of RAC, a new mixing method: Two-Stage Mixing Approach (TSMA) has been developed by Tam et al. [10, 11]. Using this method, the RCA was firstly mixed with cement paste, which was followed by adding natural sand to the mixture. The two-stage mixing can help to form a layer of cement slurry on the surface of RCA to fill up the initial damage such as micro-cracks and voids, leading to an improved interfacial zone at the pre-mix stage. Compared to Normal Mixing Approach (NMA), it has been demonstrated that TSMA leads to the improvement in strength and durability of RAC [12, 13].

Concrete is highly complex and non-homogeneous materials, and its properties depend on the properties of individual constitutes, i.e. coarse aggregate, paste matrix and ITZs [8, 9]. While researches on the strength and durability of RAC have been explored, it is still necessary to better understand the micro- and nanoscale structure and local mechanical properties of the ITZs in RAC. Compared to conventional concrete, RAC contains more types of ITZs [14, 15]. The old ITZ is between the original aggregate and the old paste matrix, while the new ITZ is between the old paste

W. Li et al., *Nanocharacterization of Cementitious Composites*,
https://doi.org/10.1007/978-981-96-2917-6_7

matrix and the new paste matrix. One of the major reasons for the lower strength exhibited by RAC is the weak ITZs. Improving the mixing approach of RAC to enhance its ITZ properties can help improve the mechanical properties of RAC. Although some investigators have found that TSMA can significantly improve the mechanical properties and durability of RAC, to the best knowledge of the authors, very few in-deep studies related to the improvement mechanism have been done. Therefore, there is a great need to investigate the influences of mixing approaches on microstructure and nanomechanical properties of ITZs in RAC.

With the emergence of nanoindentation technology, it is now possible to directly measure the mechanical properties of the ITZs [16–19], and SEM is typically applied to investigate the morphological characteristic of the ITZs [20, 21]. This chapter will focus on evaluating the microstructure and nanomechanical properties of ITZs in RAC prepared with different mixing approaches using nanoindention, Atomic Force Microscopy (AFM) and SEM techniques. Investigating the mechanical properties of ITZs in RAC at multiple scales will provide a means for correlating micro- and nanoscale properties to macroscale application. The overall objective of this study is to improve the mechanical behavior of RAC in order to encourage and broaden the application of RAC [22, 23].

7.2 Recycled Aggregate Concrete Sample Preparation for Nano/Micromechanical Property Tests

The RCA in this study was from Chicago O'Hare International Airport provided by Rossi Contractors Inc. (Fig. 7.1). The density and the absorption capacity were 2410 kg/m^3 and 5.73%, respectively. Most of natural aggregates in the recycled coarse aggregate were limestone, except a few of them were gravel. Due to the high water absorption, the RCAs used for casting RAC specimens were presoaked before mixing. The water amount used to presoak the RCA was calculated according to the water absorption capacity of RCA. Both RCA and natural coarse aggregate (NCA) were sieved to have the same size graduation. The maximum size of coarse aggregates was 20 mm. During casting RAC with 100% replacement of RCA, two different mixing approaches (TSMA and NMA) were used. With the TSMA approach, the RCA and cement were firstly mixed for one minute, and subsequently mixed for another one minute with half of the mixing water. Secondly, sand and the remaining water were added and mixed for two minutes. The TSMA was compared with the NMA in which the RAC, sand, cement and water were added altogether and mixed for two minutes. A mixture of normal concrete with limestone aggregate (here is called as NAC) with the same mix proportions was also designed. The mixture proportions of RAC and NAC are listed in Table 7.1. Three different RACs were mixed and investigated in this study. One was prepared with a water-to-binder ratio (w/b) of 0.42. The Type I Portland cement content was 246 kg/m^3 and the Class C fly ash content was 61 kg/m^3. The other RACs used in this study had a water-to-cement ratio (w/c) of 0.45, where the Type I Portland cement, water, and river sand were applied. Compared to the first RAC, no fly ash was used in the latter two RAC mixtures. The difference is that one RAC mixture was prepared through TSMA and the other was prepared through NMA. Throughout the whole process of this study, all the specimens were cured under 25 °C and 55% humidity. The compressive strength was obtained by testing with 100 mm by 200 mm cylindrical specimens at 7, 28 and 90 days. The reported compressive strength value is the average of three measurements. The compressive strength of NAC at 7, 28 and 90 days is 38 MPa, 46 MPa and 50.5 MPa, respectively; for RAC with NMA, the value is 36.4 MPa, 42 MPa and 45.7 MPa, respectively; for RAC with TSMA, the value is 39.6 MPa, 45 MPa and 50.2 MPa, respectively. Obviously, the compressive strength of RAC with NMA was lower than that of NAC, especially at the later ages. However, the compressive strength of RAC with TSMA was comparable to that of NMA. These results again indicate that TSMA is an effective method for enhancing the mechanical properties of RAC. The strength discrepancy between the two kinds of RACs respectively prepared with NMA and TSMA can be due to the influences of mixing approaches on microstructure and nanomechanical properties of ITZ, which will be investigated in the following details.

Fig. 7.1 Recycled coarse aggregate from Rossi Contractors Inc. in Chicago

Table 7.1 Mixture proportion of RAC and NAC

	w/c	kg/m^3						Mixing approach
		Water	Cement	Sand	Fly ash	Coarse aggregate		
						RCA	NCA	
RAC I	0.42	129	246	854	61	978	0	TSMA
RAC II	0.45	200	444	702	0	1054	0	TSMA
RAC	0.45	200	444	702	0	1054	0	NMA
NAC	0.45	200	444	702	0	0	1054	NMA

7.3 RAC Sample Preparation for Nanoindentation and AFM Tests

For nanoindentation test, 20 mm thick RAC slices were first cut from RAC prisms cured for 7, 28 and 90 days. The 20 mm thick slices were then cut into approximate 5 mm thick slices of 10 × 5 mm (cross section) with a diamond saw. The cut slices were immersed in ethanol for 24 h. The slices were then removed from the ethanol and were subjected to a heat treatment (drying) of 50 °C for 12 h [24]. For the impregnation, the concrete slice was first placed in a 38.1 mm diameter blue rubber mould. Epoxy called Struers provided by Cytec Industries Inc. was then used. The vacuum to the chamber was then released, allowing atmospheric pressure to facilitate impregnation. The mould was then removed from the vacuum chamber and cured for 12 h in an oven at 70 °C.

The biggest challenge and critical step in preparing RAC slices for nanoindentation is to polish the sample to provide a smooth and flat surface, while minimizing the sample disturbance [25–28]. This is mainly because RAC contains more multi-hardness phases (e.g., two types of ITZs). Hard phases resist polishing, while softer phases may wear down quickly, thus leading to uneven surfaces. In addition, the old mortar of RCA has high porosity, soft cement paste may smear across pores or cracks during polishing. Over-polishing can lead to the loss of important microstructural details, such as hydration products or aggregate and interface boundaries, making them less distinguishable in electronic images. Therefore, appropriate polishing speeds, pressures, and abrasives should be used to prevent pull-outs. Gentle polishing with more progressively finer abrasives is recommended. Fine abrasives like diamond paste (down to 1200 or 1500 grit) or colloidal silica can be employed to achieve the desired surface smoothness (Arithmetic Mean Roughness (R_a) < 100 nm). In this study, a sample was mounted on a metal holder using adhesive for grinding and polishing. And the sample holder was always placed within an annular metal collar. This configuration prevented any wobbling of the sample holder and helped to maintain parallel top and bottom surfaces while grinding and polishing. A sample after grinding and polishing is shown in Fig. 7.2.

Fig. 7.2 RAC sample after grinding and polishing

The degree to which the RAC can be polished for nanoindentation studies is associated with the microstructure heterogeneity. On the micro-scale, RAC is composed of several relatively hard phases (natural aggregate, quartz and unhydrated cement particles) surround by old paste matrix and new paste matrix. At higher magnification, many different phases of RAC can be observed by optical microscopy (Fig. 7.3).

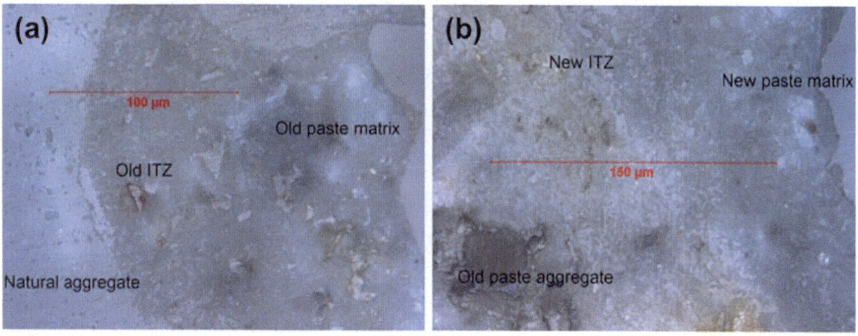

Fig. 7.3 Optical microscopy images of RAC ITZs after polishing: **a** old ITZ; **b** new ITZ

7.4 Nanoindentation and AFM Details

7.4.1 Atomic Force Microscopy Details

In order to guarantee the surface quality of polishing and grinding, the polished samples were further evaluated by AFM (Bruker AXS Dimension ICON) to obtain the surface topographic information. In AFM, a silicon tip was used with a cantilever arm. All the samples were scanned by using the tapping mode to obtain a topographic map around the interfaces in RAC, and the scanning was performed with an amplitude set point of 250 mV and a scan rate of 0.75 Hz. The scan was concentrated to observe a region of the paste matrix at the interfaces in RAC. The scanning sizes performed at both old and new ITZs were 50 × 50 μm. A total of 10 scans were conducted for each kind of ITZ. The results presented in this paper consisted of typical results obtained from these AFM scans, such as the surface topography and quantitative roughness. The Root Mean-Squared (RMS) average of the topography of the surface was chosen to measure the surface roughness.

Following the AFM imaging procedure, each image was digitally analyzed in order to extract a roughness value. Before calculating the roughness, a linear slop correction was performed to account for an alignment difference between the reference plan of AFM imaging and the overall slope of the sample surface. The preferred parameter for roughness was a RMS average of the topography of the surface (R_q). The average RMS values of 10 scans at each location (old ITZ and new ITZ) of the sample are given in Table 7.2.

The AFM results of roughness at different locations suggest roughness measurement exhibit variation. A possible explanation for this variation is that the polished surface is inherently heterogeneous. The old ITZs exhibited higher roughness than that of the new ITZs. A comparison of the 3D topography images obtained by the AFM at old and new ITZs is presented in Fig. 7.4. In the figures, the bright areas represent the natural aggregate or unhydrated cement particles, while the dark areas are considered to be the pores and cracks. Figure 7.5 shows the comparison of old ITZ and new ITZ cross-section height distribution (surface topography) along different sections. According to the three height distribution curves, it can be found that both the old ITZ and new ITZ regions are relatively flat in RAC.

Table 7.2 RMS roughness measurement for interfaces in RAC I

Scanning location	Scanning size (μm)	Average RMS roughness (nm)	RMS roughness deviation (nm)
Old ITZ	50 × 50	97.2	25.3
New ITZ	50 × 50	90.2	21.6

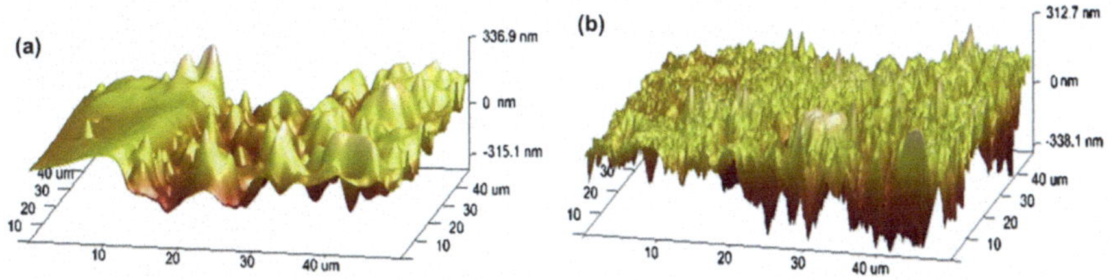

Fig. 7.4 Typical AFM 3D topography image of ITZs in RAC

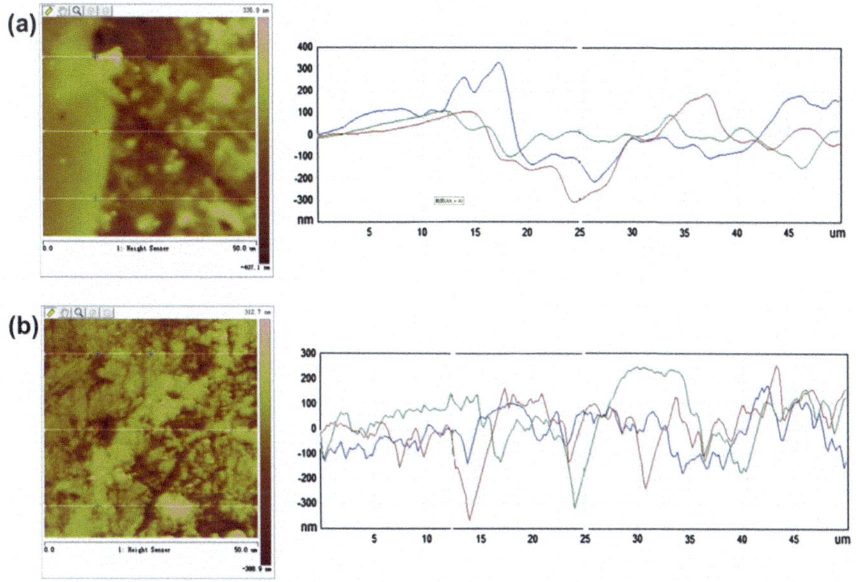

Fig. 7.5 Typical surface topography of RAC ITZs by AFM: **a** old ITZ region; **b** new ITZ region

7.4.2 Nanoindentation Details

Hysitron Triboindenter fitted with a Berkovitch tip (tip radius of 0.6 μm, angle of 142.3°) was used to determine the nanomechanical properties of the ITZs in RAC. A quartz standard was indented during each test to ensure accuracy. To account for possible surface roughness, head repositioning was carried out before each indent.

Researchers have shown that the indentation depth should be larger than the RMS of roughness, but how much larger remains an open argument for cement-based materials [10, 29]. In this study, the nanoindentation was performed to an average depth of 350 nm for the ITZ phase with a sufficient surface roughness lower than 100 nm. The average indentation depth, which is 3 times more than the surface roughness, should be sufficient to avoid effects of roughness. Each load/displacement diagram was plotted and inspected for signs of problems due to surface roughness,

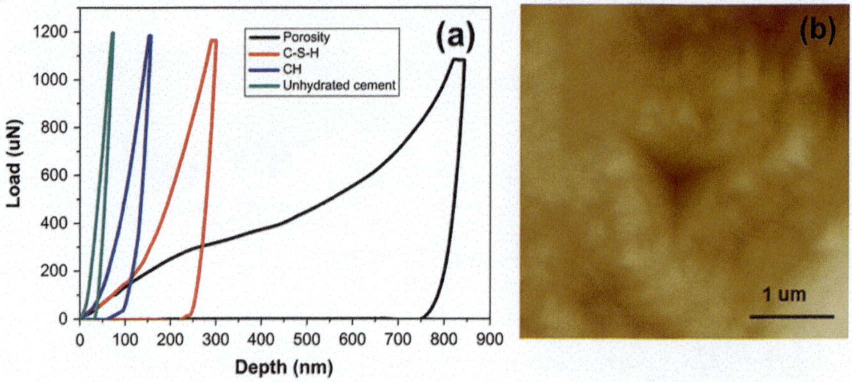

Fig. 7.6 Typical indentation in ITZs of RAC: **a** load-depth diagrams of different phases; **b** indent mark on C–S–H phase

such as abnormal or discontinuous shapes. Figure 7.6 presents load-depth curves of different phase in ITZ and the image of indent mark on the calcium silicate hydrate (C–S–H) gel after nanoindentation captured with indenter tip of the Triboindenter. The trapezoidal load functions which included single-cycle and multiple-cycle loads were used in this study. It is found that multiple cycles of partial loading and unloading lead to minimizing of the short-term creep and size effect [30].

All tests were programmed in a multi-cycle loading way that the indenter encountered the sample surface with the load increased at a constant rate of 120 μN and the maximum load was 1200 μN. Normally, grid indentation is a method to perform large number of indents on the sample. Statistical analysis of these data obtained from grid indentation provides information about the overall nanomechanical properties of the sample and volume fractions of different phases involved. In this study, indentation with in-situ imaging was used to determine the local nanomechanical properties at specific locations of the sample.

A typical indentation load and indentation displacement curve includes peak indentation load ($P_{\max}$), indentation depth at peak load ($h_{\max}$), and final depth of the contact impression after unloading (h_f). As to the penetration of elastic half space by an axisymmetric indenter of arbitrary smooth profile, by applying a continuum scale model, the initial unloading stiffness S = dp/dh, is given as follows [31, 32]:

$$S = \frac{\mathrm{d}p}{\mathrm{d}h} = \frac{2}{\sqrt{\pi}} E_r \sqrt{A} \tag{7.1}$$

where A is the projected contact area at the peak load; E_r is the reduced elastic modulus, which can be determined by nanoindentation. The relationship between E_r and E_i can be given by:

$$\frac{1}{E_r} = \frac{\left(1 - v^2\right)}{E} + \frac{\left(1 - v_i^2\right)}{E_i} \tag{7.2}$$

where E and v are Young's modulus and Poisson's ratio of the sample respectively, and E_i and v_i are the corresponding parameters of the indenter. For the indenter used in the present experiment, the elastic modulus $E_i = 1140$ GPa and the Poisson's ratio $v_i = 0.07$. Then, the elastic modulus E of the sample can be calculated by:

$$E = \left(1 - v^2\right) \times \left[\frac{1}{E_r} - \frac{\left(1 - v_i^2\right)}{E_i}\right]^{-1} \quad (7.3)$$

The suggested Poisson's ratio for RAC is $v = 0.2$, which was used in this study to calculate E in Eq. (7.3) [33, 34].

The hardness H has the definition as follows:

$$H = \frac{P_{\max}}{A} \quad (7.4)$$

The studied areas were taken randomly within old ITZs and new ITZs in RAC that were around aggregates with similar sizes. The spacing between the indents was 5 μm and 10 μm in the lateral and vertical directions, respectively. The indent areas for old and new ITZs had the dimensions of 100 × 100 μm and 150 × 100 μm, respectively, as shown in Fig. 7.7. As for the old ITZs, 231 indents were made in an array at each studied area. For new ITZs, 341 indents were made in an array. For each old and new ITZ, four studied areas across interfaces were randomly chosen for nanoindentation. These four indented areas obtained for ITZs were based on the statistical consideration, which was used to calculate the mean and standard deviation values for the nanomechanical properties of ITZs. As the spacing between indents was 5 μm in lateral direction, the strips of 5 μm wide were successively taken from the indented areas in RAC. For each trip, 11 indentation values with 10 μm spacing in one trip were then averaged. The distributions of nanomechanical properties were obtained with the distance starting at natural aggregate surrounded by old ITZ or old paste aggregate surrounded by new ITZ. In this study, the distributions of indentation modulus were used as a basis for the characterizations of old ITZ and new ITZ respectively, such as ITZ thickness.

7.5 Grid Nanoindentation Test Results of RAC ITZ with and Without Fly Ash

For the RAC samples with fly ash at 90 days, areas on the samples that contained old and new ITZs were chosen for indentation (Fig. 7.8a, b). The ITZs around coarse limestone aggregates with around 20 mm size were chosen for indentation. Figure 7.8c, d shows the contour maps of indentation modulus of old and new ITZs, respectively. The deep blue and black areas have the lowest moduli, which indicate relatively weaker zones in the RAC. The results also show that these blue and black areas are

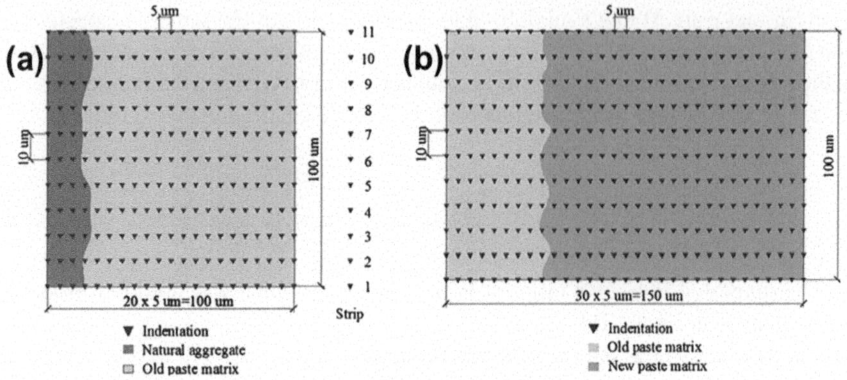

Fig. 7.7 Schematic showing indented area of old ITZ and new ITZ in RAC: **a** across old ITZ; **b** across new ITZ

distributed in the regions between natural aggregates and old paste matrix, as well as between old paste matrix and new paste matrix. This is an important indication of the existence of old ITZ and new ITZ in RAC.

Figure 7.8e, f shows the indentation modulus distributions with the distances across old ITZ and new ITZ, respectively. According to other researchers [35, 36], the indentation modulus of calcium hydroxide (CH) is higher than that of calcium silicate hydrate (C–S–H). The porosity and CH can be concentrated when approaching the aggregate surface and old paste matrix in both old and new ITZs due to the wall effect.

In this study, the ITZ thickness was estimated by locating the place where there is little variation in indentation modulus with the distance from the natural aggregate (for old ITZ) or old paste aggregate surface (for new ITZ), and the indentation modulus distribution of ITZ seems to become close to those of the paste matrix. For the old ITZ, the thickness of the interfacial zone ranges between 40 and 50 μm. This supports the ITZ thickness of conventional concrete previously found by other researchers to be from 20 to 50 μm by analyzing the lateral variation in properties around aggregate [37–41]. The average modulus of the old ITZ was found to be 70–80% of that of the old paste matrix. However, the indentation modulus values obtained around the aggregate does not increase consistently when moving away from the aggregate surface and the ITZ modulus is not necessary always lower than that of the paste matrix. This phenomenon may be due to the concentration of porosity and CH crystals in the interfacial zone.

For the new ITZ, the thickness was estimated to be in the range of 55–65 μm, which was greater than that of old ITZ. Moreover, the average indentation modulus of new ITZs was found to be approximately 80–90% of that of the new paste matrix. Similar to the old ITZ, there is no obvious trend for the modulus of new ITZ to increase with an increasing distance from the old paste aggregate surface. However, the indentation modulus of new ITZ appears to be higher than that of old ITZ in RAC. This may be because the improved mixing method TSMA coats the recycled

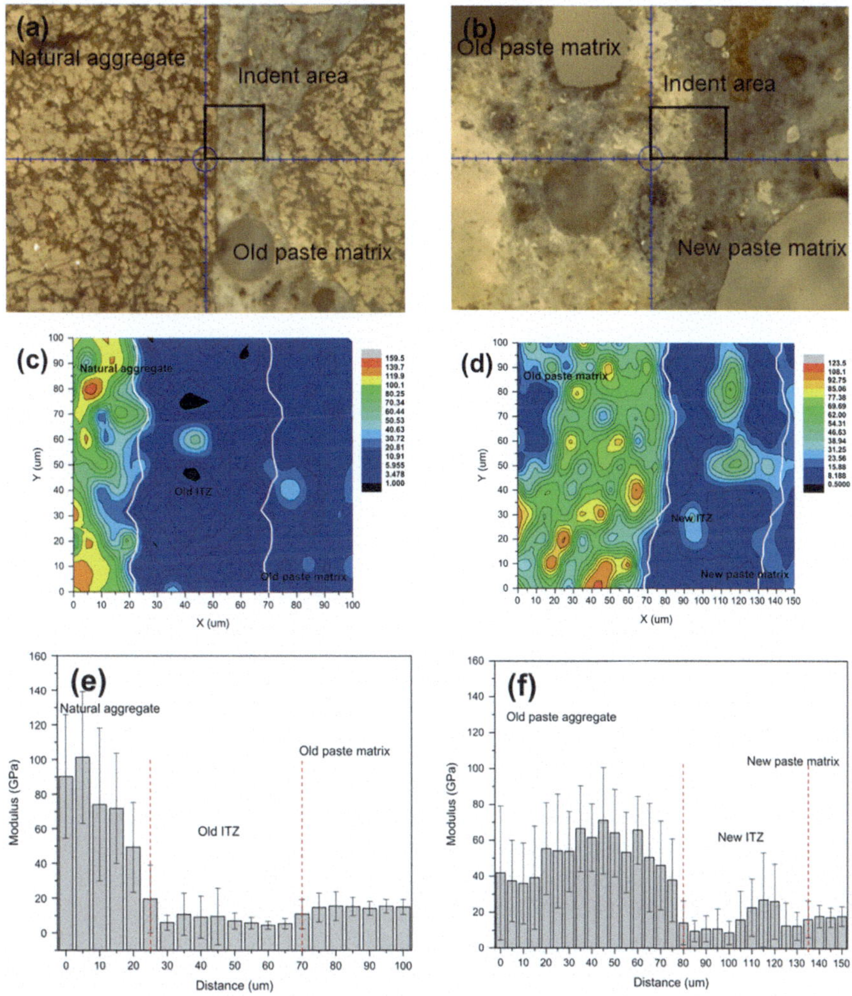

Fig. 7.8 Grid indentation modulus on ITZs of RAC with fly ash at 90 days: **a** old ITZ, indent area:100 × 100 μm; **b** new ITZ, indent area:150 × 100 μm; **c** contour map of modulus in GPa (old ITZ); **d** contour map of modulus in GPa (new ITZ); **e** modulus distribution across old ITZ; **f** modulus distribution across new ITZ

aggregate with cement, and further hydration takes place between old paste matrix and new paste matrix.

Figures 7.9a, b show the contour map of indentation hardness of the old and new ITZs in RAC with fly ash. It is found that the characterization of indentation hardness distribution was close to that of indentation modulus. The thicknesses of old ITZ and new ITZ obtained from the hardness distribution (Fig. 7.9c, d) were also in a good agreement with that from the modulus distribution (Fig. 7.8e, f). On the other hand, the indentation hardness of old ITZ was found to be roughly 85% of that of old

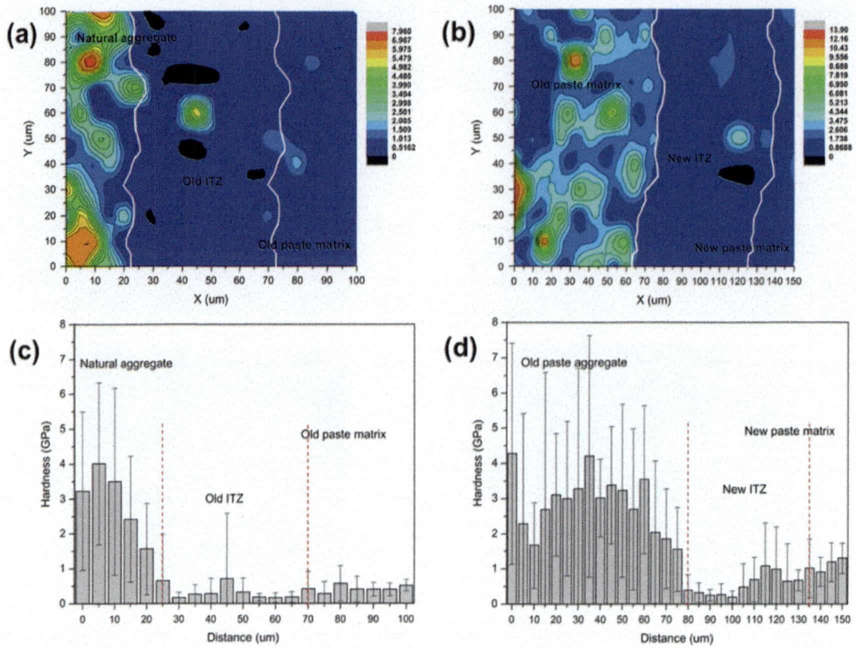

Fig. 7.9 Grid indentation hardness on ITZs of RAC with fly ash at 90 days: **a** contour map of hardness in GPa (old ITZ); **b** contour map of hardness in GPa (new ITZ); **c** hardness distribution across old ITZ; **d** hardness distribution across new ITZ

paste matrix, while the indentation hardness of new ITZ was found to be 90% that of new paste matrix. As the indentation hardness characteristics at old and new ITZs appeared to generally be similar to the indentation modulus characteristics, this study will mainly discuss the indentation modulus in the following sections.

This SEM images provides evidence that the ITZs in RAC with fly ash at 90 days have different microstructure characteristics than the paste matrix. It can be observed that there are voids and calcium hydroxide (CH) crystals in the ITZs, especially perpendicular to old ITZ on the old paste matrix side. In comparison with old ITZ, there are more C–S–H gels existing at new ITZ (Fig. 7.10).

The SEM micrographs of polished sample with respect to microstructure in RAC with fly ash at 90 days are presented in Fig. 7.11. It shows that heterogeneities of the microstructure exist in both old and new ITZs. It can be seen clearly that higher porosity appears in old and new ITZ compared to the corresponding paste matrix. A comparatively denser zone close to new ITZ was found at the old paste aggregate side. It can be seen that interactions at new ITZ is considered to contribute to reduce the high porosity and improve the properties of the old paste aggregate perpendicular to new ITZ [42]. The denser zone around the paste aggregate surface is probably due to the further hydration process of the old paste matrix in vicinity of the new ITZ.

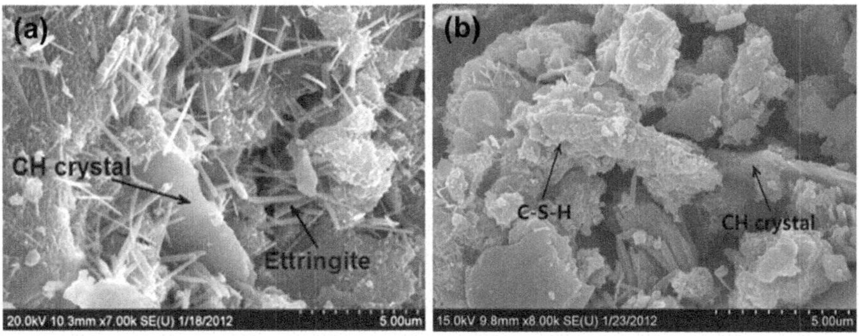

Fig. 7.10 SEM images of old ITZ and new ITZ in RAC at 90 days: **a** perpendicular to old ITZ, at old paste; **b** perpendicular to new ITZ, at new paste matrix side

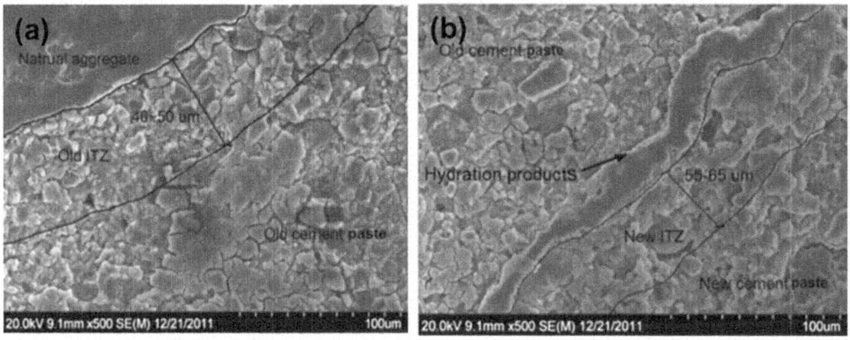

Fig. 7.11 SEM images of polished new ITZ and old ITZ of RAC at 90 days: **a** old ITZ; **b** new ITZ

7.6 Grid Nanoindentation Test Results of Paste Matrix in RAC with and Without Fly Ash

With the RAC samples with fly ash, nanoindentation was also performed at old paste matrix and new paste matrix (Fig. 7.12a, b). The chosen areas for indentation for both old and new paste matrix were 100 × 100 μm. And the spacing between indents is 10 μm in both lateral and vertical directions. Figure 7.12c, d show the contour map of indentation modulus at old paste matrix and new paste matrix, respectively. Some deep blue and black areas represented the relatively weak parts, such as voids and cracks. The red areas with maximum modulus typically represent the unhydrated cement particles [43]. Figure 7.12e, f show no obvious weak zones have been found in both old and new paste matrices. In comparison, the average indentation modulus of new paste matrix is slightly higher than that of the old paste matrix. This result is probably due to the different mix proportions that used when casting original concrete and the recycled concrete. But for the original concrete, the mix proportion record is usually difficult to obtain.

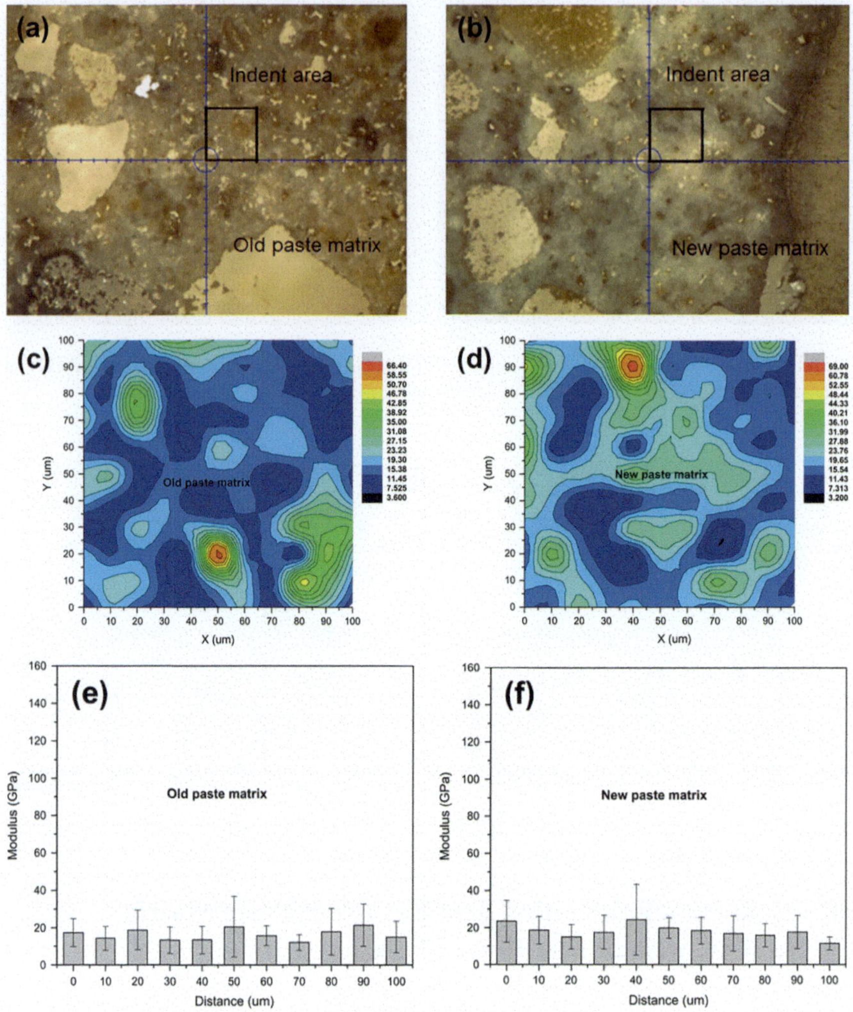

Fig. 7.12 Indentation modulus in old and new paste matrices in RAC with fly ash at 90 days: **a** old paste matrix: 100 × 100 μm; **b** new paste matrix: 100 × 100 μm; **c** contour map of indentation modulus in GPa; **d** contour map of indentation modulus in GPa; **e** modulus in old paste matrix; **f** modulus in new paste matrix

7.7 Imaging Nanoindentation Test Results of RAC with and Without Fly Ash

Hundreds of indentations can be performed on a grid to generate enough data without any bias for a statistical analysis. However, grid indentation occasionally misses some positions of specific interest for the statistical analysis. On the other hand, imaging indentation becomes preferred when the nanomechanical properties of a local area on

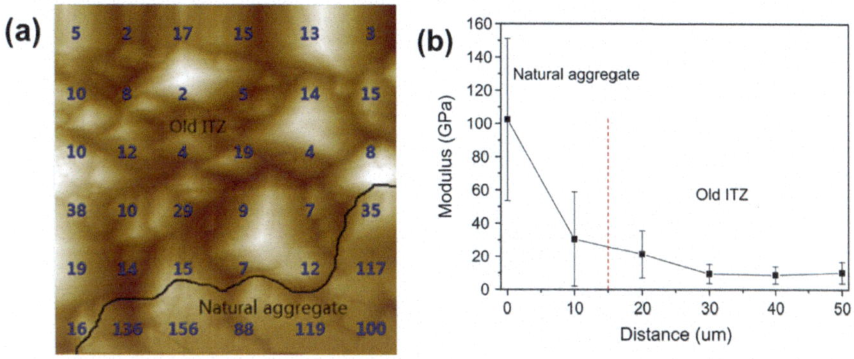

Fig. 7.13 Imaging indentation on old ITZ of RAC with fly ash at 90 days: **a** modulus on each location (60 × 60 μm); **b** modulus distribution across old ITZ

the sample are focused on. The image capability of the Triboindenter can also provide an effective method to examine the different phases in RAC. Figure 7.13a is a 60 × 60 μm Scanning Probe Microscopy (SPM) image of the old ITZ in RAC containing fly ash with the indentation modulus written on each point. The bright area near the right bottom of the image represents the old natural aggregate. There are areas very close to the natural aggregate that seems softener and more porous. Thirty-six indents were performed on selected points. The spacing between these indented points was 10 μm. It can be figured out that at the extreme vicinity of the natural aggregate, the old ITZ generally has a lower modulus (ranged from 2 to 9 GPa). It should be noted that the connectivity of the weaker areas such as large voids and cracks along the interface influences the mechanical properties of RAC. Figure 7.13b shows that the relatively lower modulus in the vicinity of natural aggregate (around 20–30 μm from the aggregate surface) has been measured within the old ITZ. It shows that extreme heterogeneity and poor nanomechanical properties exist in old ITZ.

Higher resolution SPM image with the indentation modulus value obtained from each location shows some pores existing in the new ITZ (Fig. 7.14a). Indentation was performed on 36 selected points. It is observed that the new ITZ has higher porosity in the vicinity of old paste aggregate (dark areas in the figure). The modulus of the areas with large voids ranges from 3 to 9 GPa. It is also seen that the relatively lower modulus of new ITZ exists at the vicinity of old paste aggregate (around 10–20 μm from the old paste aggregate surface), but no trend of increasing modulus with distance moving away from the old paste matrix surface was found (Fig. 7.14b). This can tell that new ITZ itself is not a uniform region and the vicinity of new interface at the old paste aggregate side is a weak part within the new ITZ region.

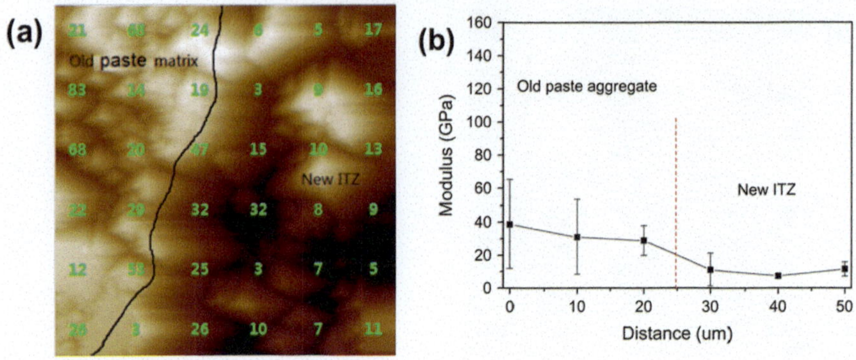

Fig. 7.14 Imaging indentation on new ITZ of RAC with fly ash at 90 days: **a** modulus on each location (60 × 60 μm); **b** modulus distribution across new ITZ

7.8 Nanoindentation Test Results of RAC Under Different Conditions

7.8.1 Old ITZ

Considering the old ITZ being between old mortar and natural aggregate, it is assumed that the old ITZ would not be influenced by the mixing approach. The following investigated old ITZ was from the TSMA. In order to obtain representative distributions of the mechanical properties in old ITZ, 4 grid areas of 231 indentations were performed, and then statistical analysis was carried out, correspondingly. Mechanical response distribution represented as frequency plots are employed to analyze indentation data on old ITZ. All the four grid areas in old ITZ give similar nanoindentation results. Figure 7.15 shows the nanoindentation results for one of the four grid areas in old ITZ shown as Fig. 7.15a. The indentation modulus contour and histogram seem to exhibit variations which are related to the modulus values of different phases.

Figure 7.15b represents the in situ 2D mapping of mechanical properties in old ITZ using nanoindentation for examining the properties of different phases independently of each other. In order to identify the indentation modulus distribution, statistical average and deviation value were applied to histograms of modulus across the old ITZ (Fig. 7.15c). To identify different phases, the bin-size for deconvolution of the indentation modulus is 5 GPa, and the frequency histogram was plotted to show the volume fraction distribution in the old ITZ (Fig. 7.15d). It can be found that the dominant phase in the old ITZ is OHP, especially the C–S–H gel.

The overall results from 4 grid areas are averaged and plotted in Fig. 7.16a. Results revealed that the indentation modulus increases with increasing distance from the original aggregate surface. The indentation modulus range of the different phases in the ITZ can be assumed to be as following: porosity ($\leq$ 7 GPa), other hydration

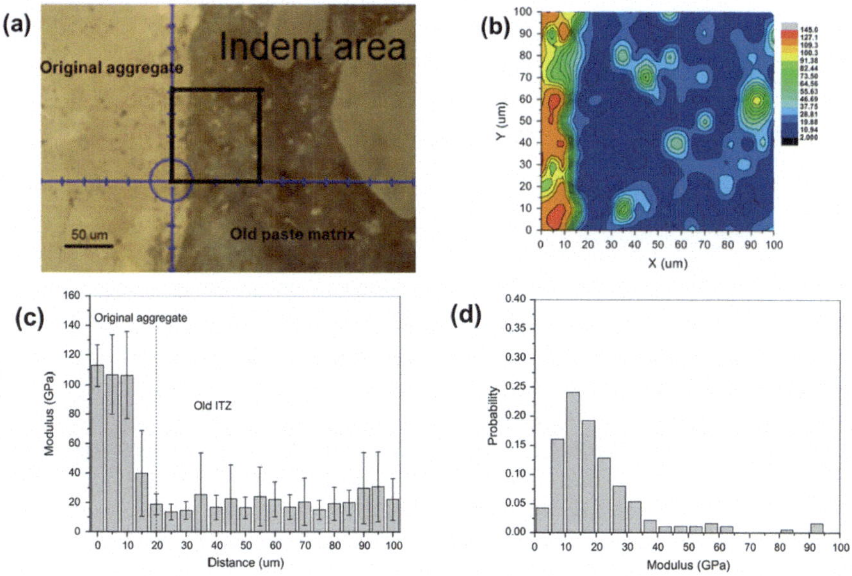

Fig. 7.15 Nanoindentation in old ITZ (one of four indent areas): **a** indent area for old ITZ in RAC; **b** contour map of indentation modulus [GPa]; **c** modulus distribution across old ITZ; **d** statistical analysis of modulus [bin-size = 5]

product (7–34 GPa), CH (34–50 GPa) and unhydrated cement (≥ 50 GPa) [44–47]. The results supplied by the normalized probability represent the mechanical properties of each phase as the surface volume fractions. Figure 7.16b shows the mean probability density of each phase in the old ITZ starting from the original aggregate surface with 5 μm spacing. It is apparent that the OHP (C–S–H gel) is the main compound in the old ITZ, exhibiting a volume fraction of about 69%.

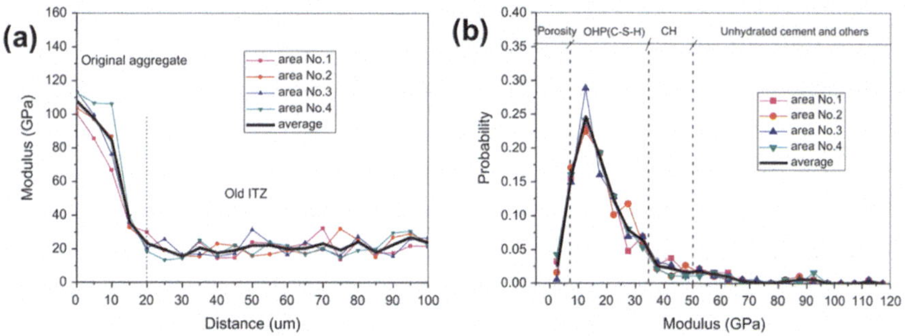

Fig. 7.16 Indentation modulus distribution in old ITZ: **a** general modulus distribution; **b** general modulus probability

7.8.2 New ITZ with TMSA

Nanoindentation was performed over 4 grid areas (341 indents each) on the new ITZ with TSMA (Fig. 7.17a). Figure 7.17b shows the in situ 2D mapping of one indentation grid across the new ITZ. It shows that the indentation modulus of the old paste matrix region adjacent to the new ITZ is obviously higher compared to the other regions. The distribution of statistic indentation modulus across the new ITZ is shown in Fig. 7.17c. The modulus gradient of the new ITZ is variable, which may be attributed to the greater porosity and heterogeneity of the new ITZ compared to the bulk cement paste. From the statistical indentation modulus of new ITZ with TSMA, it is apparent that the main compound in the new ITZ was also OHP (primarily C–S–H gel), and the content is close to that in old ITZ (Fig. 7.17d).

The results from all the four grid areas across the new ITZ with TSMA are plotted in Fig. 7.18a. The four indent areas yielded similar indentation modulus distributions with no obvious trend indicating that the indentation modulus increase or decrease with increasing distance from the old paste matrix surface. The statistic probability density of each phase in the new ITZ was measured starting from the old paste matrix surface with 5 μm spacing (Fig. 7.18b). The C–S–H gel is one of complicated materials in cement-based materials, which can be divided into low density C–S–H and high density C–S–H [48, 49]. The low density C–S–H has a dominant volume

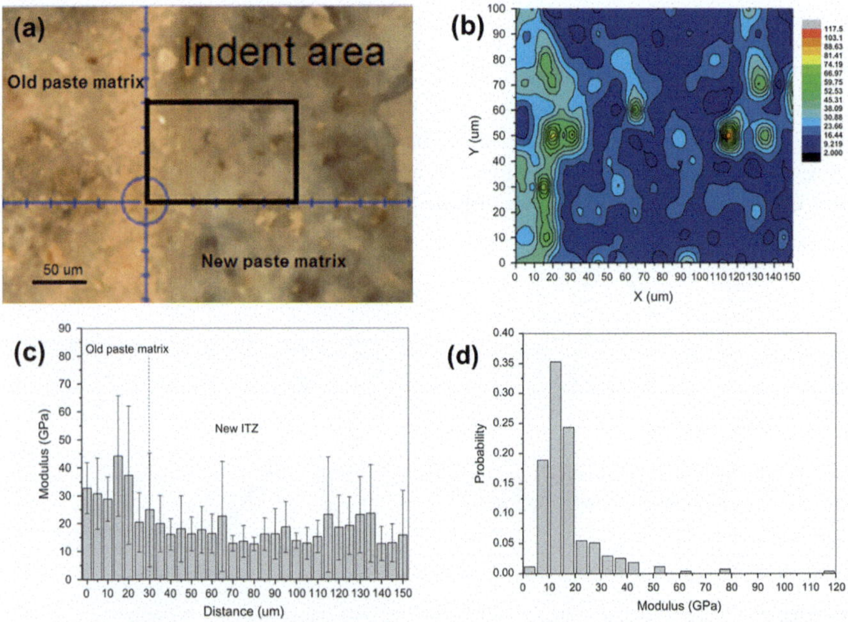

Fig. 7.17 Nanoindentation in new ITZ with TSMA (one of four indent areas): **a** indent area for new ITZ in RAC; **b** contour map of indentation modulus [GPa]; **c** modulus distribution across new ITZ; **d** statistical analysis of modulus [bin size = 5]

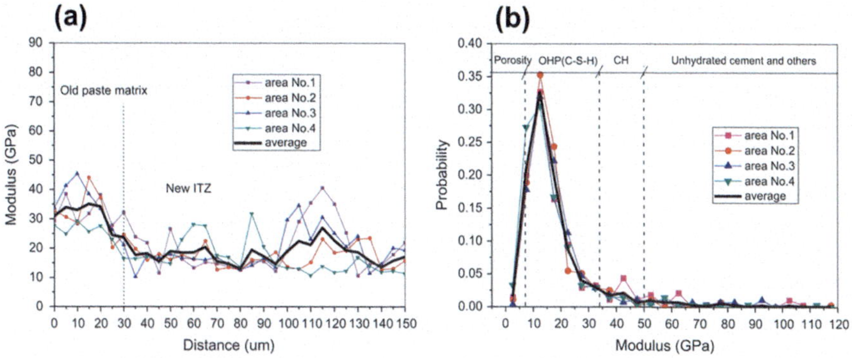

Fig. 7.18 Nanomechanical properties characteristics of new ITZ with TSMA: **a** general modulus distribution; **b** general modulus probability

fraction of the other hydration product, which suggests that the low density C–S–H volume content in the new ITZ with TSMA is higher compared to the old ITZ. In addition, the OHP (C–S–H) exhibits a volume fraction of about 68% in new ITZ with TSMA.

7.8.3 New ITZ with NMA

Nanoindentation was performed over 4 grid areas (341 indents each) in the new ITZ with NMA as shown in Fig. 7.19a. From Fig. 7.19b, it appears that the old paste matrix region adjacent to the new ITZ has a higher indentation modulus. The indentation modulus distribution of the new ITZ with NMA is plotted versus the distance from the old paste matrix to new paste matrix (Fig. 7.19c). Compared to the old ITZ and new ITZ with TSMA, the variation in modulus along the new ITZ with NMA is significantly greater. Similar to the old ITZ and new ITZ with TSMA, OHP phase is also the main component in new ITZ with NMA, but it appears to contain relatively greater amount of porosity, CH and unhydrated cement grain as displayed in Fig. 7.19d.

The average distribution of the indentation modulus in the new ITZ with NMA is shown in Fig. 7.20a. Results reveal that the modulus decreases with the increasing distance from the old paste matrix surface. This may be due to the presence of large deposits of CH in the vicinity of the old paste matrix surface. The discussion here is qualitative in nature and will be followed with quantitative microstructural analysis. Figure 7.20b shows the statistic probability density of each phase in the new ITZ with NMA. The high volume fraction of porosity, CH and unhydrated cement can help to explain the higher variation in the modulus distribution in the new ITZ with NMA. In addition, the OHP (C–S–H) phase exhibits a volume fraction of about 55%

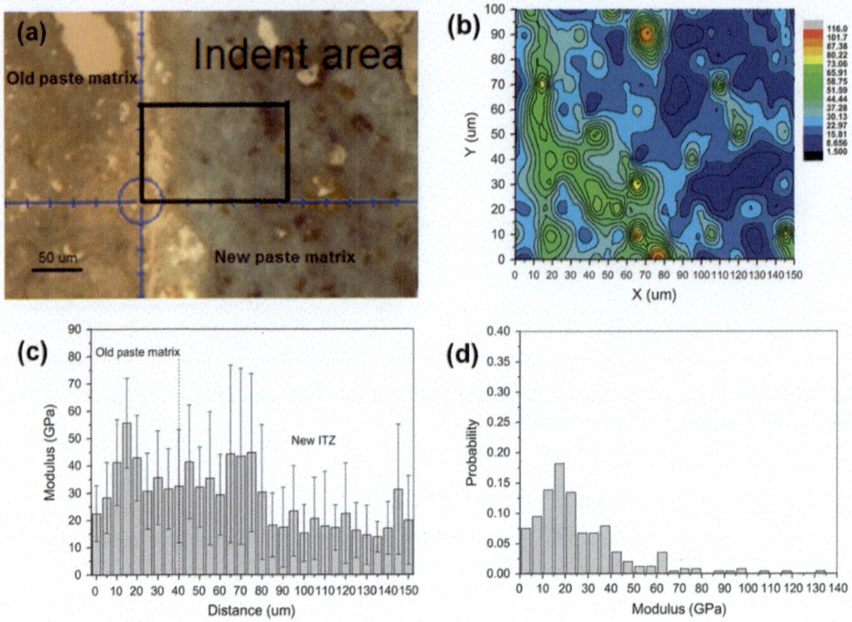

Fig. 7.19 Nanoindentation in new ITZ with NMA (one of four indent areas): **a** indent area for new ITZ in RAC; **b** contour map of indentation modulus [GPa]; **c** modulus distribution across new ITZ; **d** statistical analysis of modulus [bin-size = 5]

in the new ITZ with NMA, which is much lower than those in the old ITZ and new ITZ with TSMA.

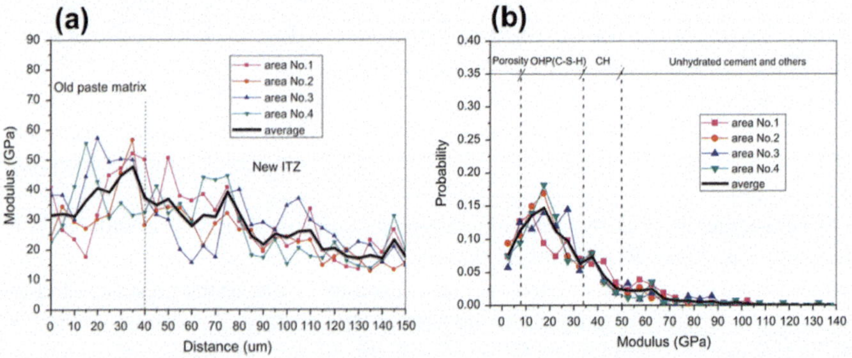

Fig. 7.20 Nanomechanical properties characteristics of new ITZ with NMA: **a** general modulus distribution of old ITZ; **b** general modulus probability in old ITZ

7.8.4 Nanoindentation on Paste Matrices

The old paste matrix and new paste matrix were also investigated by nanoindentation as well. The main components of the paste matrix in RAC can also be assumed to contain porosity, CH, OHP and unhydrated cement. Nanoindentation was performed in serials of grids of 121 indents covering a representative area of 100 × 100 μm on the surface of the paste matrix with the indent spacing of 10 μm. The indent areas of the paste matrices were examined using an optical microscopy.

Figure 7.21a, b show an optical image and contour map of the indent area in the old paste matrix, respectively. All the different grid areas on the old paste matrix showed similar results. They indicate that there is a less concentration of unhydrated cement and CH deposition than those seen in the ITZs. Results of mean values and standard deviation for the indentation modulus in the old paste matrix are shown in Fig. 7.21c. Compared to ITZs, no significant variations in the modulus distribution were found. This phenomenon is most likely due to a decrease in the amount of porosity, CH and unhydrated cement. Statistical probability of individual phases in the old paste matrix is shown in Fig. 7.21d, in which the C–S–H phase is the dominant, and the presence of unhydrated cement is limited. It is also apparent that the variation in modulus was smaller in the old paste matrix than in the ITZs.

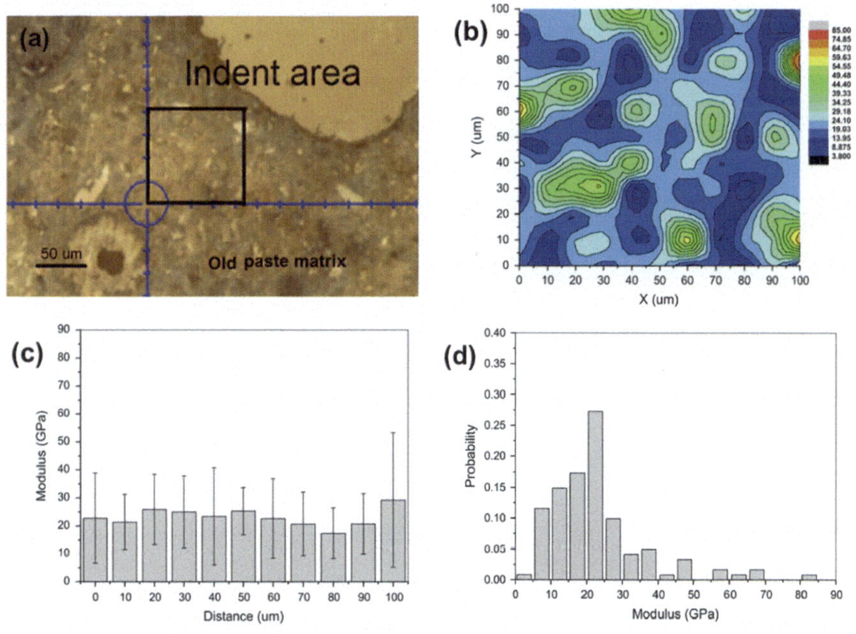

Fig. 7.21 Nanoindentation in old paste matrix: **a** indent area for old paste matrix; **b** contour map of indentation modulus [GPa]; **c** modulus distribution in old paste matrix; **d** statistical analysis of modulus [bin-size = 5]

For the new paste matrix, the optical image of the area of indents is shown in (Fig. 7.22a). The 2D in situ indentation modulus map also reveals more variation in properties across the new paste matrix (Fig. 7.22b). The areas of relatively higher modulus in the sample area correspond to the areas of unhydrated cement, while the lower modulus areas correspond to those with the porosity phases. The variation in indentation modulus of the new paste matrix is similar to that of the old paste matrix, but significantly lower than that in the ITZ regions (Fig. 7.22c). Statistical analysis was applied to obtain the modulus probability of each phase in the new paste matrix (Fig. 7.22d). Comparing the probability density of the new paste matrix and old paste matrix, it is apparent that the volume fractions of the two types of C–S–H are different, which might have contributed to the differences in nanomechanical properties.

Compared to the ITZs, the distributions of indentation modulus of the paste matrices are much less variable (Fig. 7.23a). It reveals that within the old and new paste matrix, the modulus appears to remain constant. Moreover, the variation in properties of paste matrix is similar with the new ITZ with TSMA, but appreciably lower than that of the new ITZ with NMA.

It is observed that the peak of the modulus probability plot for the new paste matrix falls between 15 and 20 GPa, corresponding to low density C–S–H, while the

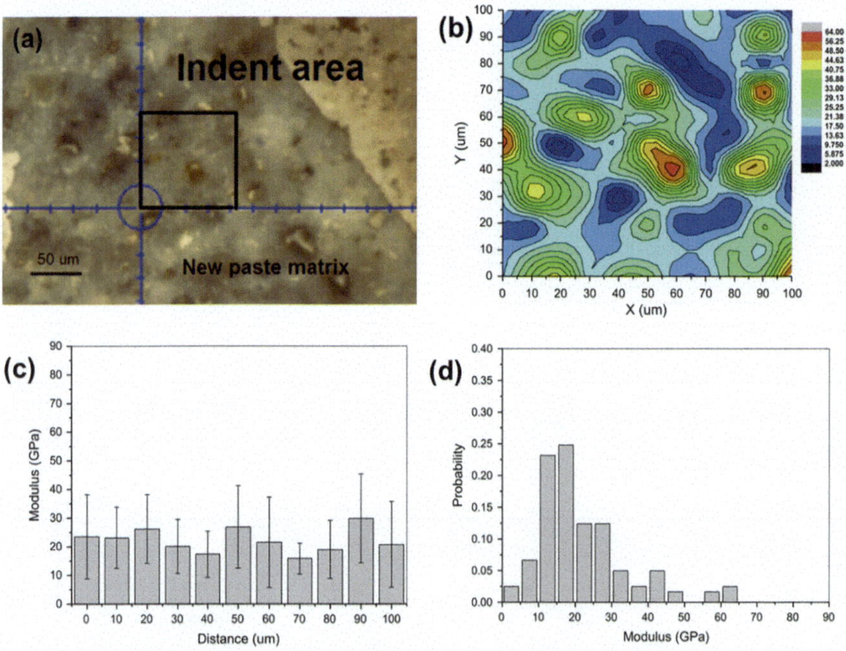

Fig. 7.22 Nanoindentation in new paste matrix: **a** indent area for new paste matrix; **b** contour map of indentation modulus [GPa]; **c** modulus distribution in new paste matrix; **d** statistical analysis of modulus [bin size = 5]

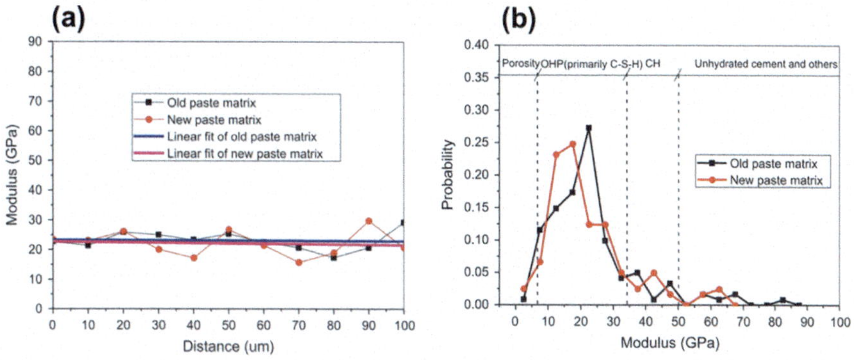

Fig. 7.23 Nanomechanical properties characteristics of old and new paste matrix in RAC: **a** general modulus distribution; **b** general modulus statistic analysis

peak of modulus probability plot for the old paste matrix is between 20 and 25 GPa (Fig. 7.23b). Comparing the new paste matrix and old paste matrix, it is apparent that the amount of low density C–S–H and high density C–S–H phases were changed with the curing age.

7.8.5 Effect of Mix Proportion on New ITZ

Similar nanoindentation test was undertaken on RAC without fly ash also at 90 days old. Figures 7.8b and 7.24a show the indent location of new ITZ in RAC I and RAC II, respectively. It exhibits that the indentation modulus of the old paste matrix close to new ITZ on the old paste aggregate side was higher compared to other regions in both RAC I and RAC II (Figs. 7.8d and 7.24b). Figures 7.8f and 7.24c also show the indentation modulus distribution with the distance across the new ITZs. For the new ITZ in RAC I, the average indentation modulus is found to be around 85% of that of the new paste matrix. Similarly, the average modulus of new ITZ in RAC II is also found to be approximately 85% of that of new paste matrix. However, the average indentation modulus of new ITZ in RAC I is slightly lower than that of new ITZ in RAC II, even though the water-to-binder ratio used in RAC I was slighter lower. This can be attributed to the usage of fly ash in RAC I, as it is widely accepted that although fly ash can introduce pozzonlanic reactions to improve paste properties, it may also delay the hydration process.

To study the effect of mix proportion (fly ash) on the properties of the ITZs in RAC, the cylinder compressive strengths of RAC I and RAC II were also tested. The compressive strength of RAC I at 7, 28 and 90 days is 22.1 MPa, 34.9 MPa and 42.8 MPa, respectively; the value for RAC II is 39.3 MPa, 45 MPa and 50.1 MPa, respectively. The compressive strength of RAC I is lower than that of RAC II, while the strength development rate of RAC I is higher than that of RAC II after 7 days.

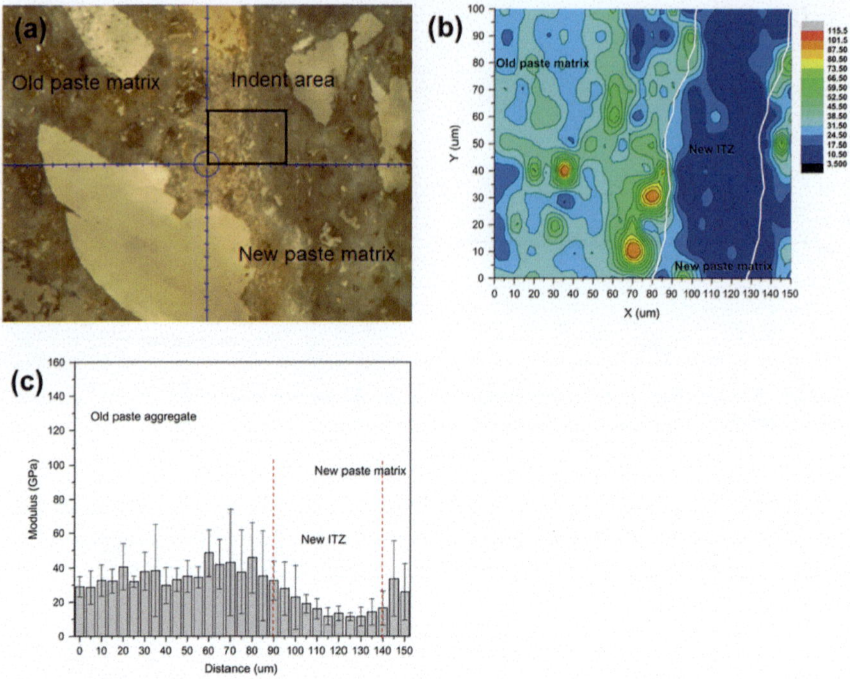

Fig. 7.24 Grid indentation across new ITZ in RAC II at 90 days: **a** new ITZ, indent area: 150 × 100 μm; **b** contour map of modulus in GPa; **c** modulus distribution across new ITZ

This corresponds to the finding that the new ITZ in RAC I is relatively weaker than that of new ITZ in RAC II and the thickness of new ITZ in RAC I is somewhat greater than that of new ITZ in RAC II (Fig. 7.25). Therefore, the nanomechanical properties of new ITZ can be used to explain the difference in strength between RAC I and RAC II. It can be induced that the compressive strength development of RAC are related to the nanomechanical properties of new ITZ.

7.8.6 Old ITZs Around Different Types of Aggregates

Figure 7.26a, b shows the indent location of old ITZs around two types of natural aggregates (limestone and gravel) with similar sizes in RAC I at 90 days. For old ITZ around limestone aggregate, the thickness appeared to be around 50 μm. The average indentation modulus in old ITZ appeared to be around 80% of that of old paste matrix (Fig. 7.26c). As for the old ITZ around gravel, the thickness appeared to be 60 μm, which is more than that of old ITZ around limestone aggregate. This may be due to limestone normally having a rough surface and being more porous compared with

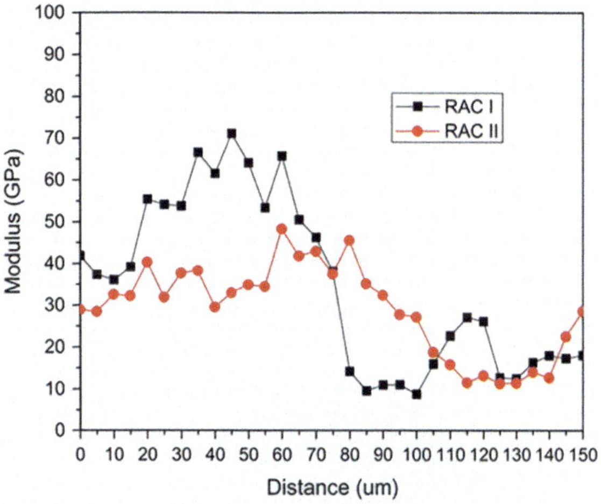

Fig. 7.25 Modulus distribution across new ITZ of RAC I and RAC II

gravel (Fig. 7.26d). The average indentation modulus of old ITZ around the gravel was observed to around 70% of that of the old paste matrix.

From Fig. 7.26e, f, it can be seen that the old ITZ around gravel aggregate achieved a lower indentation modulus than that of old ITZ around limestone, which is consistent with other investigations [50, 51]. These results confirm that the properties of old ITZ in RAC appear to depend more on the original natural aggregate that used.

7.8.7 Age Effect on ITZs

To study the hydration age effect on the ITZs in RAC, the RAC II samples were also studied by nanoindentation at 7, 28 and 90 days, respectively, as shown in Table 7.3. Figure 7.27a, b show the indent locations for old and new ITZs in RAC II at the age of 7 days. The contour maps of indentation modulus of old ITZ and new ITZ are shown in Fig. 7.27c, d, respectively. At 7 days age, both new ITZ and new paste matrix had relatively lower modulus than that of the old ITZ and the old paste matrix, correspondingly. There is no significant difference can be noticed between new ITZ and new paste matrix from Fig. 7.27d. The result suggests that new ITZ and new paste matrix are the weak links in RAC at an early age. At 7 days, the thickness of the old ITZ is found to be around 45 μm, which is consistent with that of old ITZ at later ages (Fig. 7.27e). Moreover, the modulus of the old ITZ was also estimated to be about 75% of old paste matrix, which was roughly the similar value with that of old ITZ at a later age. Therefore, the results can imply that the hydration age has no obvious effects on the nanomechanical properties of old ITZ. On the other hand, the indentation modulus of new ITZ was almost 95% of that of new paste matrix (Fig. 7.27f). The thickness of new ITZ is found to be approximately 85 μm, which indicates that the thickness of new ITZ at an early age is larger than the new ITZ

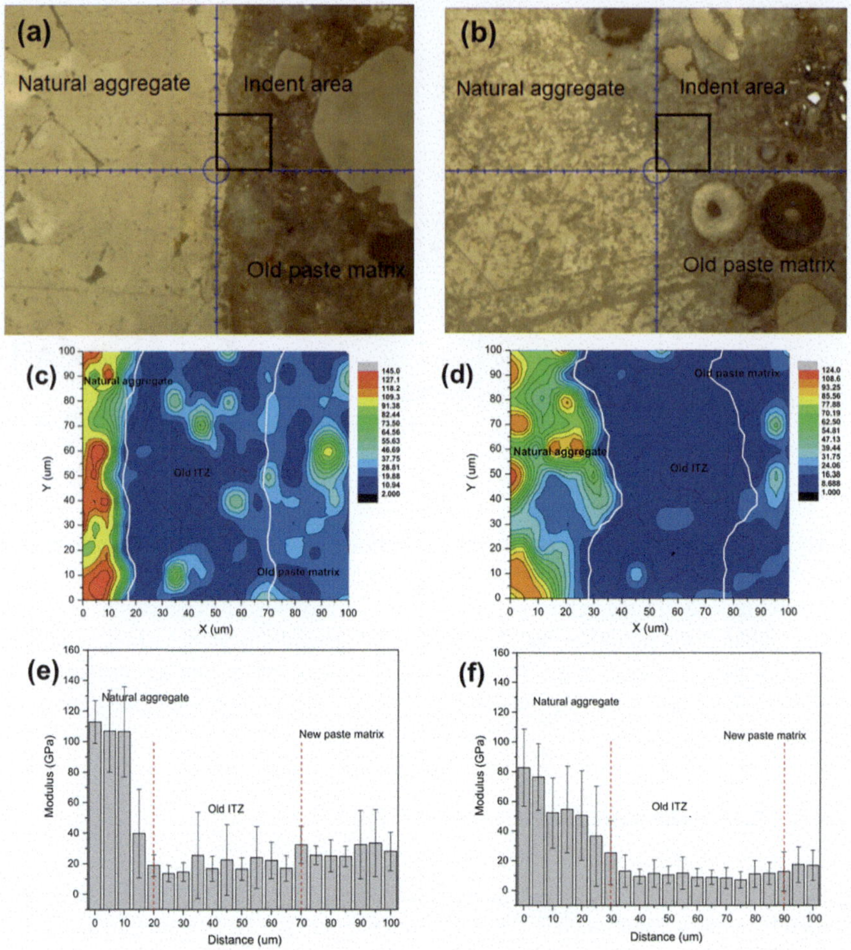

Fig. 7.26 Grid indentation across old ITZ around different aggregates in RAC I at 90 days: **a** old ITZ around limestone: 100 × 100 μm; **b** old ITZ around gravel: 100 × 100 μm; **c** contour map of modulus in GPa (old ITZ); **d** contour map of modulus in GPa (old ITZ); **e** modulus distribution across old ITZ; **f** modulus distribution across old ITZ

at later ages. It can be concluded that the hydration age has a significant influence on the properties and development of new ITZ. This is mainly due to the further hydration in the new ITZs. With the hydration time increasing, the reduction of the porosity in new ITZ is initially because of the packing of anhydrous cement grains. Therefore, the thickness of new ITZ decreases with the hydration age increasing, and the new ITZ region becomes stronger and denser. Furthermore, the results show good agreements with previous studies that the properties of new ITZ are highly changed with the hydration process [11, 52].

Table 7.3 Characteristics of old ITZ and new ITZ in RAC (mean ± standard deviation)

Items		Old ITZ		New ITZ	
		Thickness (μm)	Modulus ratio (M_{OITZ}/M_{OCP})(%)	Thickness (μm)	Modulus ratio (M_{NITZ}/M_{NCP})(%)
Curing age (RAC I)	7 days	43 ± 10	77 ± 8	86 ± 9	97 ± 8
	28 days	45 ± 12	81 ± 12	67 ± 15	91 ± 12
	90 days	49 ± 11	84 ± 11	53 ± 20	84 ± 15
Natural aggregate (RAC II at 90 days)	Limestone	47 ± 18	82 ± 15	–	–
	Gravel	63 ± 5	71 ± 10	–	–

Note M_{OITZ}: indentation modulus of old ITZ; M_{OCP}: indentation modulus of old paste matrix; M_{NITZ}: indentation modulus of new ITZ; M_{NCP}: indentation modulus of new paste matrix

Figure 7.28a, b show the indent locations of ITZs in RAC II at 28 days. There was no significant difference in the indentation modulus of old ITZs between 7 and 28 days of the sample based on the distribution of the indentation modulus (Fig. 7.28c). On the other hand, the indentation modulus of new ITZ and new paste matrix significantly increased at 28 days compared to that of 7 days age. However, the indentation modulus ratio of new ITZ to new paste matrix decreased. The relatively lower modulus in the new ITZ mostly occurred in the vicinity of the old paste aggregate at a distance 20 μm from the old paste matrix surface (Fig. 7.28d). As shown in Fig. 7.28e, the thickness of old ITZ was estimated to be around 50 μm, which was similar to that of old ITZ at 7 days age. The indentation modulus of old ITZ appeared to be around 80% of that of the old paste matrix. As for new ITZ, the average modulus was found to be approximately 90% of that new paste matrix (Fig. 7.28f). The thickness of new ITZ was considered to be around 65 μm, which was less than that of new ITZ at 7 days age. In comparison, the average indentation modulus of new ITZ was slightly lower than that of old ITZ in RAC II at 28 days age.

Figures 7.29a and 7.24a show the indent location of ITZs in RAC II at 90 days. The old ITZ thickness still appeared to be around 50 μm. Indentation modulus in old ITZ appeared to be 85% of that of old paste matrix (Fig. 7.29b). There was no significant difference in the indentation modulus distribution of old ITZ among 7, 28 and 90 days hydration ages. This suggests that there are no further significant hydration progresses took place in the old ITZ during the testing period. As for the new paste matrix, the indentation modulus at 90 days was higher than that at 28 days as expected (Fig. 7.24b).

As shown in Fig. 7.29c, the modulus of old ITZ when RAC is 90 days old was almost similar with that of old ITZ in RAC at 7, 28 days age. As for new ITZ, the average modulus of new ITZ was found to be approximately 85% of that the new paste matrix (Fig. 7.25). The thickness of new ITZ was considered to be around

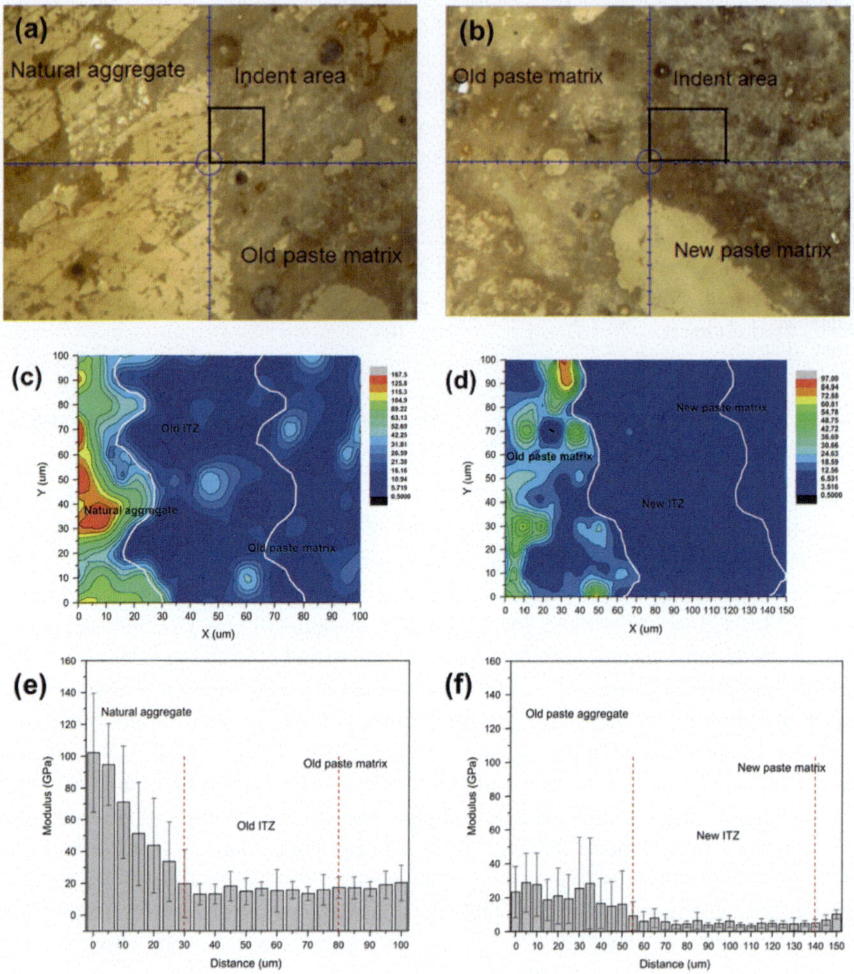

Fig. 7.27 Grid indentation across ITZs of RAC II at 7 days: **a** old ITZ, indent area: 100 × 100 μm; **b** new ITZ, indent area: 150 × 100 μm; **c** contour map of modulus in GPa (old ITZ); **d** contour map of modulus in GPa (new ITZ); **e** modulus distribution across old ITZ; **f** modulus distribution across new ITZ

55 μm, which was less than that of new ITZ in RAC at 28 days age. These results reveal that the thickness of new ITZ decreases with the hydration age.

The indentation modulus distributions of old and new ITZs in RAC II at 7, 28 and 90 days are compared and displayed in Fig. 7.30. It shows that the age effect on the indentation modulus of old ITZ and new ITZ in RAC was different. The properties of old ITZ in RAC do not change with the hydration age, while the indentation modulus of new ITZ and new paste matrix increases with the hydration age (Fig. 7.30a). At a late age, the indentation modulus increases at a lower rate compared to the early age

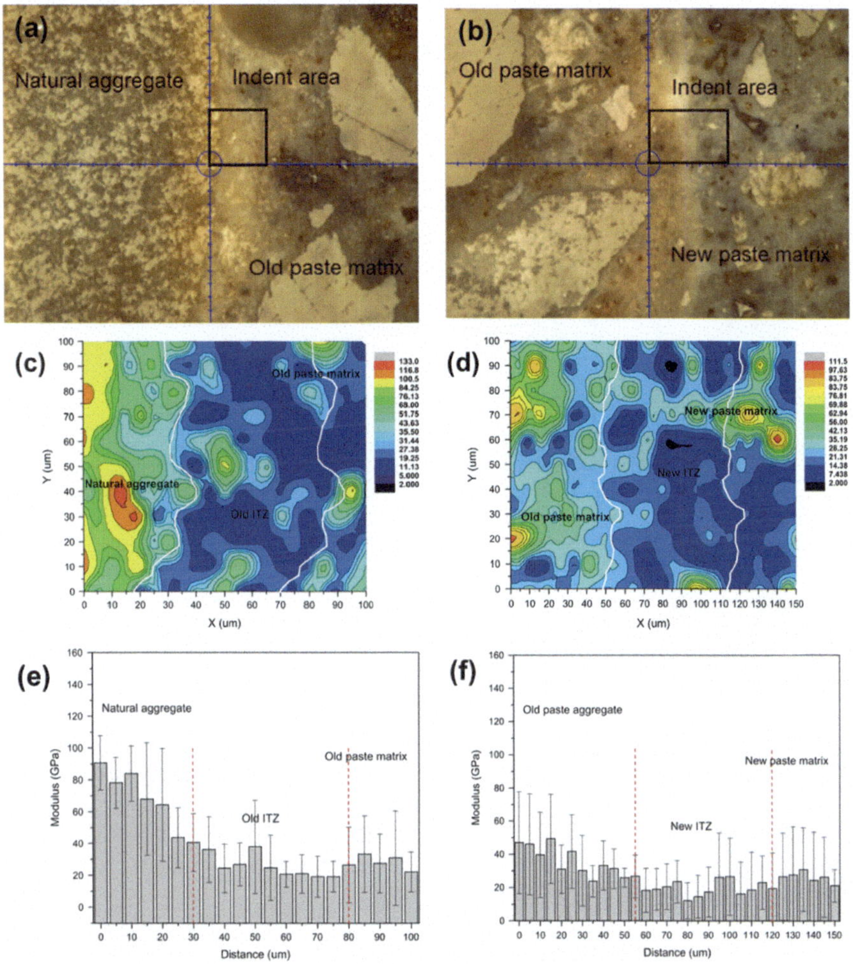

Fig. 7.28 Grid indentation across ITZs in RAC II at 28 days: **a** old ITZ, indent area: 100 × 100 μm; **b** new ITZ, indent area: 150 × 100 μm; **c** contour map of modulus in GPa (old ITZ); **d** contour map of modulus in GPa (new ITZ); **e** modulus distribution across old ITZ; **f** modulus distribution across new ITZ

(Fig. 7.30b). This trend is similar with that of the compressive strength development of RAC. It should be noted that the ratio between indentation modulus of new ITZ and new paste matrix decreases with the age.

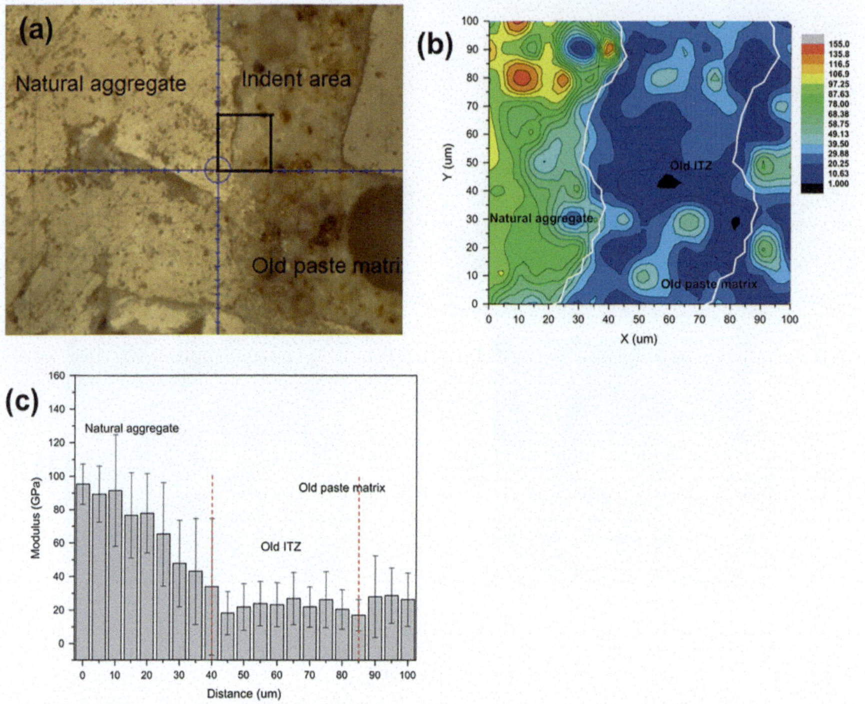

Fig. 7.29 Grid indentation across old ITZs in RAC II at 90 days: **a** old ITZ, indent area: 100 × 100 μm; **b** contour map of modulus in GPa (old ITZ); **c** modulus distribution across old ITZ

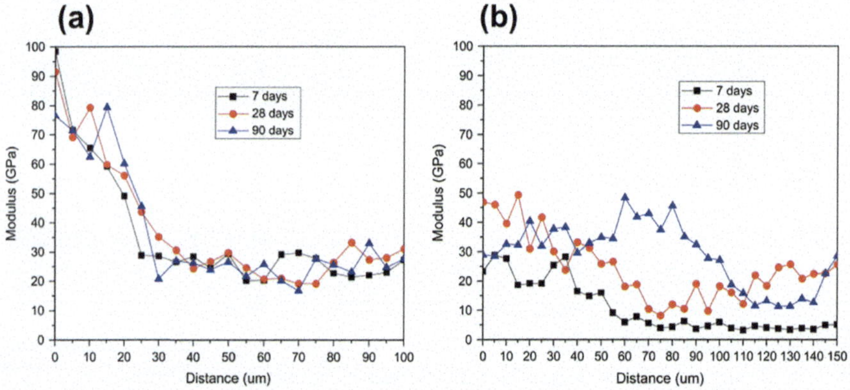

Fig. 7.30 Modulus distribution across ITZs of RAC II different hydration ages: **a** across old ITZ; **b** across new ITZ

7.9 Conclusion

Detailed and quantitative analysis of SEM images and nanoindentation have drawn significant conclusions on microstructural characterization and the phase distributions of porosity, calcium hydroxide (CH), the other hydration production (primarily C–S–H) and unhydrated cement for ITZs in RAC with different mixing approaches. The main conclusions are as follows:

(1) This study developed successful protocols for grinding and polishing specimens and can achieve perfectly acceptable characterization of the surface roughness for nanoindentation, which was verified by AFM. Cement-based samples contain numerous soft C–S–H gels. As a result, the grinding and polishing period, force and speed should be controlled strictly to ensure a eligible flatness and retain main phases. From the SEM images, obvious voids and calcium hydroxide exist in both old and new ITZs in RAC.
(2) Based on the nanoindentation study, the thicknesses of old and new ITZs are found to be around 40–50 μm and 55–65 μm respectively at a late age (90 days). Moreover, the average modulus of old and new ITZs appears to be approximately 70–80% of that of old paste matrix and 80–90% of that of new paste matrix, respectively. From the imaging nanoindentation, for the old ITZ in RAC, the lowest modulus in the vicinity of aggregate is 20–30 μm from natural aggregate surface, while for the new ITZ, the lowest modulus of new ITZ existed in the vicinity of old paste matrix is 10–20 μm from the old paste aggregate surface.
(3) For the old ITZ, the indentation modulus increases as the distance from original aggregate surface increases. For the new ITZ with TSMA, there is no obvious trend indicating that the indentation modulus increases or decreases with the increasing distance from the old paste matrix surface, while in the new ITZ with NMA the modulus distribution appears to decrease significantly from the old paste matrix surface, because NMA creates more voids within new ITZs. However, TSMA can effectively reduce the size and effect of water layers. Therefore, the amount of porosity is reduced, and micromechanical properties of new ITZ with TSMA is improved.
(4) With the different types of natural aggregate, the indentation modulus of old ITZ around limestone is higher than that of old ITZ around gravel, and the thickness of old ITZ around limestone is thinner than that around gravel. However, both the mix proportion and hydration age have no obvious effects on the nanomechanical properties of the old ITZ in RAC.
(5) The average indentation modulus of new ITZ in RAC I was lower than that of new ITZ in RAC II, and the thickness of new ITZ in RAC I was greater than that in RAC II. This is probably attributed to the fly ash which was added in RAC I. With the increase of hydration age, the new ITZ thickness appears to be reduced and the microstructure at new ITZ tends to become denser because of the development of hydrates with the age.

References

1. M. Miller, C. Bobko, M. Vandamme, F.J. Ulm, Surface roughness criteria for cement paste nanoindentation. Cem. Concr. Res. **38**(4), 467–476 (2008)
2. F.J. Ulm, M. Vandamme, C. Bobko, J. Alberto Ortega, K. Tai, C. Ortiz, Statistical indentation techniques for hydrated nanocomposites: concrete, bone, and shale. J. Am. Ceramic Soc. **90**(9), 2677–2692 (2007)
3. J. Xiao, Y. Sun, H. Falkner, Seismic performance of frame structures with recycled aggregate concrete. Eng. Struct. **28**(1), 1–8 (2006)
4. P.E. Stutzman, Scanning electron microscopy in concrete petrography. Mater. Sci. Concr. (2000)
5. C.F. Hendriks, H. Pietersen, Report 22: sustainable raw materials: construction and demolition waste–state-of-the-art report of RILEM technical committee 165-SRM (RILEM Publications, 2000)
6. J.F. Lamond, R.L. Campbell Sr., A. Giraldi, N. Jenkins, T.R. Campbell, W. Halczak, R. Miller, J.A. Cazares, H.C. Hale Jr., P.T. Seabrook, *Removal and reuse of hardened concrete*, vol. 26 (American Concrete Institute, Farmington Hills, MI, USA, 2001)
7. P. Mondal, S.P. Shah, L. Marks, A reliable technique to determine the local mechanical properties at the nanoscale for cementitious materials. Cem. Concr. Res. **37**(10), 1440–1444 (2007)
8. J. Xiao, W. Li, C. Poon, Recent studies on mechanical properties of recycled aggregate concrete in China-a review. SCIENCE CHINA Technol. Sci. **55**(6), 1463–1480 (2012)
9. K.L. Scrivener, A. Bentur, P. Pratt, Quantitative characterization of the transition zone in high strength concretes. Adv. Cem. Res. **1**(4), 230–237 (1988)
10. V.W.Y. Tam, X.F. Gao, C.M. Tam, Microstructural analysis of recycled aggregate concrete produced from two-stage mixing approach. Cem. Concr. Res. **35**(6), 1195–1203 (2005)
11. V.W. Tam, C.M. Tam, Diversifying two-stage mixing approach (TSMA) for recycled aggregate concrete: TSMAs and TSMAsc. Constr. Build. Mater. **22**(10), 2068–2077 (2008)
12. V.W.Y. Tam, X.F. Gao, C.M. Tam, Comparing performance of modified two-stage mixing approach for producing recycled aggregate concrete. Mag. Concr. Res. **58**(7), 477–484 (2006)
13. V.W.Y. Tam, C.M. Tam, Assessment of durability of recycled aggregate concrete produced by two-stage mixing approach. J. Mater. Sci., 3592–3602 (2007)
14. J. Xiao, W. Li, Z. Sun, D.A. Lange, S.P. Shah, Properties of interfacial transition zones in recycled aggregate concrete tested by nanoindentation. Cement Concr. Compos. **37**, 276–292 (2013)
15. I.B. Topçu, S. Şengel, Properties of concretes produced with waste concrete aggregate. Cem. Concr. Res. **34**(8), 1307–1312 (2004)
16. H. Jeong, Processing and properties of recycled aggregate concrete (University of Illinois at Urbana-Champaign, 2011)
17. K.O. Kjellsen, A. Monsøy, K. Isachsen, R.J. Detwiler, Preparation of flat-polished specimens for SEM-backscattered electron imaging and X-ray microanalysis—importance of epoxy impregnation. Cem. Concr. Res. **33**(4), 611–616 (2003)
18. A. Shayan, A. Xu, Performance and properties of structural concrete made with recycled concrete aggregate. Materials Journal **100**(5), 371–380 (2003)
19. J. Xiao, W. Li, Y. Fan, X. Huang, An overview of study on recycled aggregate concrete in China (1996–2011). Constr. Build. Mater. **31**, 364–383 (2012)
20. P. Trtik, J. Kaufmann, U. Volz, On the use of peak-force tapping atomic force microscopy for quantification of the local elastic modulus in hardened cement paste. Cem. Concr. Res. **42**(1), 215–221 (2012)
21. A.R. Sakulich, V.C. Li, Nanoscale characterization of engineered cementitious composites (ECC). Cem. Concr. Res. **41**(2), 169–175 (2011)
22. P. Mondal, S.P. Shah, L.D. Marks, Nanoscale characterization of cementitious materials. ACI Mater. J. **105**(2), 174 (2008)

23. L. Sorelli, G. Constantinides, F.J. Ulm, F. Toutlemonde, The nano-mechanical signature of ultra high performance concrete by statistical nanoindentation techniques. Cem. Concr. Res. **38**(12), 1447–1456 (2008)
24. P. Mondal, S. Shah, L. Marks, Nanomechanical properties of interfacial transition zone in concrete, in *Nanotechnology in Construction 3: Proceedings of the NICOM3* (Springer, 2009), pp. 315–320
25. A.C. Fischer-Cripps, D. Nicholson, Nanoindentation. Mechanical engineering series. Appl. Mech. Rev. **57**(2), B12-B12 (2004)
26. G. Constantinides, F.J. Ulm, The nanogranular nature of C-S-H. J. Mech. Phys. Solids **55**(1), 64–90 (2007)
27. R. Chamrova, Modelling and measurement of elastic properties of hydrating cement paste (EPFL, 2010)
28. S. Diamond, The microstructure of cement paste and concrete—a visual primer. Cement Concr. Compos. **26**(8), 919–933 (2004)
29. M. Etxeberria, E. Vázquez, A. Marí, Microstructure analysis of hardened recycled aggregate concrete. Mag. Concr. Res. **58**(10), 683–690 (2006)
30. S. Igarashi, A. Bentur, S. Mindess, Microhardness testing of cementitious materials. Adv. Cem. Based Mater. **4**(2), 48–57 (1996)
31. A. Domingo, C. Lázaro, F.L. Gayarre, M.A. Serrano, C. López-Colina, Long term deformations by creep and shrinkage in recycled aggregate concrete. Mater. Struct./Materiaux et Constructions **43**(8), 1147–1160 (2010)
32. W. Li, J. Xiao, Z. Sun, S. Kawashima, S.P. Shah, Interfacial transition zones in recycled aggregate concrete with different mixing approaches. Constr. Build. Mater. **35**, 1045–1055 (2012)
33. J. Maso, in *Interfacial Transition Zone in Concrete* (CRC Press, 1996)
34. K.L. Scrivener, A.K. Crumbie, P. Laugesen, The interfacial transition zone (ITZ) between cement paste and aggregate in concrete. Interface Sci. **12**, 411–421 (2004)
35. S. Erdem, A.R. Dawson, N.H. Thom, Influence of the micro- and nanoscale local mechanical properties of the interfacial transition zone on impact behavior of concrete made with different aggregates. Cem. Concr. Res. **42**(2), 447–458 (2012)
36. N. Otsuki, S.-I. Miyazato, W. Yodsudjai, Influence of recycled aggregate on interfacial transition zone, strength, chloride penetration and carbonation of concrete. J. Mater. Civ. Eng. **15**(5), 443–451 (2003)
37. A. Padmini, K. Ramamurthy, M. Mathews, Influence of parent concrete on the properties of recycled aggregate concrete. Constr. Build. Mater. **23**(2), 829–836 (2009)
38. C.S. Poon, Z.H. Shui, L. Lam, H. Fok, S.C. Kou, Influence of moisture states of natural and recycled aggregates on the slump and compressive strength of concrete. Cem. Concr. Res. **34**(1), 31–36 (2004)
39. A.W. Pope, H.M. Jennings, The influence of mixing on the microstructure of the cement paste/aggregate interfacial zone and on the strength of mortar. J. Mater. Sci. **27**, 6452–6462 (1992)
40. A. Elsharief, M.D. Cohen, J. Olek, Influence of aggregate size, water cement ratio and age on the microstructure of the interfacial transition zone. Cem. Concr. Res. **33**(11), 1837–1849 (2003)
41. W.C. Oliver, G.M. Pharr, An improved technique for determining hardness and elastic modulus using load and displacement sensing indentation experiments. J. Mater. Res. **7**(6), 1564–1583 (1992)
42. A. Bordelon, V. Cervantes, J.R. Roesler, Fracture properties of concrete containing recycled concrete aggregates. Mag. Concr. Res. **61**(9), 665–670 (2009)
43. D. Davydov, M. Jirásek, L. Kopecký, Critical aspects of nano-indentation technique in application to hardened cement paste. Cem. Concr. Res. **41**(1), 20–29 (2011)
44. G. Fathifazl, A.G. Razaqpur, O.B. Isgor, A. Abbas, B. Fournier, S. Foo, Flexural performance of steel-reinforced recycled concrete beams. ACI Struct. J. **106**(6) (2009)
45. W. Li, J. Xiao, Z. Sun, S.P. Shah, Failure processes of modeled recycled aggregate concrete under uniaxial compression. Cement Concr. Compos. **34**(10), 1149–1158 (2012)

46. Q. Liu, J. Xiao, Z. Sun, Experimental study on the failure mechanism of recycled concrete. Cem. Concr. Res. **41**(10), 1050–1057 (2011)
47. M. Zhang, J. Chen, J. Zhu, J. Chen, Experimental evaluation on modulus of equivalent homogeneous Ettringite. Acta Mech. Solida Sin. **22**(4), 320–327 (2009)
48. G. Constantinides, F.J. Ulm, The effect of two types of C-S-H on the elasticity of cement-based materials: results from nanoindentation and micromechanical modeling. Cem. Concr. Res. **34**(1), 67–80 (2004)
49. X.H. Wang, S. Jacobsen, S.F. Lee, J.Y. He, Z.L. Zhang, Effect of silica fume, steel fiber and ITZ on the strength and fracture behavior of mortar. Mater. Struct./Materiaux et Constructions **43**(1–2), 125–139 (2010)
50. C.S. Poon, Z.H. Shui, L. Lam, Effect of microstructure of ITZ on compressive strength of concrete prepared with recycled aggregates. Constr. Build. Mater. **18**(6), 461–468 (2004)
51. D. Zampini, S.P. Shah, H.M. Jennings, Early age microstructure of the paste-aggregate interface and its evolution. J. Mater. Res. **13**(7), 1888–1898 (1998)
52. R.A. Tarefder, A.M. Zaman, W. Uddin, Determining hardness and elastic modulus of asphalt by nanoindentation. Int. J. Geomech. **10**(3), 106–116 (2010)

Chapter 8
Nano/Microscopic Investigation on the Interfacial Transition Zones in Modelled Aggregate Concrete

8.1 Introduction

Most existing studies on the microscopic properties of RAC are based on irregular aggregate [1–4]. However, rough and angular aggregate surface brings severe challenges to the research of ITZ [5, 6]. In nanoindentation tests, for example, the indenter may touch the aggregate hidden under the surface cement paste [7, 8]. The number of test points available for ITZ analysis in each column of grid nanoindentation is different [1]. It is also difficult to determine the distance of each row of test points from the aggregate boundary [9]. Testers have to reduce indentation depth and increase the density of the grid. This greatly increases the workload and makes it difficult to achieve accurate statistical results [10]. SEM-based analysis faces a similar problem. It is well-known that the formation of ITZ is attributed to the 'wall effect' of aggregate [7]. The uneven surface of aggregate is destined to the heterogeneity and complexity of ITZ [1, 7]. In addition, the bleeding around the aggregate seriously affects the properties of ITZ at different locations [9]. This requires researchers to extract a large number of ITZ images at different positions to obtain a passable average result.

To solve the above problems, Shah and Winter [11] are the first to propose a novel modelled concrete with cylindrical aggregate for analysing the inelastic mechanical properties of concrete. Subsequently, Xiao et al. [12] extended the study of this concrete to the microscopic level. In recent years, Luo et al. [9, 11], Fu et al. [13], and Zhan et al. [14] used a similar concept to simulate ITZ by aggregate blocks and obtained satisfactory microstructure, chemical composition and micromechanical performance test results. It has to be admitted that the ITZ simulated by modelled concrete idealises the actual situation and ignores the randomness and complexity of the real ITZ [15, 16]. However, with the help of characteristics of modelled concrete, the bonding mechanism of aggregate and mortar can be more effectively explored.

This chapter discusses the nano/microscopic characterizations of the ITZs in modelled aggregate concrete (MAC) and modelled recycled aggregate concrete

W. Li et al., *Nanocharacterization of Cementitious Composites*,
https://doi.org/10.1007/978-981-96-2917-6_8

(MRAC) samples. The ITZ width was identified by Backscattered electron (BSE) combined with image analysis. Furthermore, the influence of waste glass powder on ITZs was evaluated through multi-technique holistic investigation approach. The differences in chemical composition and micromechanical properties were systematically tested by EDS, TGA, NMR, XRD, Fourier transform infrared spectroscopy (FTIR), nanoindentation and nanoscratch techniques to evaluate the improving performance of WGP within the new ITZs.

8.2 Nano/Microscopic Investigation on the Interfacial Transition Zones in MAC

8.2.1 Raw Materials

General purpose Portland cement (GP cement) conforming to AS 3972 was the main cementitious material in this study. Waste glass (soda-lime glass), preliminary composed of recycled municipal kerbside council collection, was milled to fine powder by a mill machine as shown in Fig. 8.1 in the laboratory to replace 20 wt% of cement as SCM. The particle size distribution (PSD) curves of cement and WGP measured by Malvern Mastersizer 3000 are shown in Fig. 8.2. The WGP was ground to achieve a similar PSD to cement with a median particle size of ~20 μm. The detailed chemical compositions determined using X-ray fluorescence (XRF) are shown in Table 8.1. The loss on ignition (LOI) of WGP in Table 8.1 suggests that before supply, the glass fragments went through a furnace at a high temperature to reduce the organic content and residuals (liquid and paper tags) to a minimum. The main oxide components of cement and WGP were calcium oxide and silica, respectively. Each MAC contained a granite block as natural coarse aggregate, ground from the single aggregate particle. Naturally occurring silica sand with a fineness modulus of 1.9 was selected as the fine aggregate.

8.2.2 Sample Preparation

Modelled concrete samples were made by placing a piece of aggregate into the mortar. The mortar had a water-cement ratio of 0.45 and a sand-cement ratio (by mass) of 1.5. In one group of samples, 100% cement was used (MAC-C), while in the other, 20 wt% of cement was replaced by WGP (MAC-G). Following previous research on modelled samples [1, 11], the MAC processing steps were designed and are shown in Fig. 8.3. Firstly, the granite gravel with a suitable size was selected and ground to a 13 mm × 13 mm × 8 mm block. The granite block was further ground using a series of abrasive paper with different grits for 30 min and polished by 0.3 μm and 0.05 μm alumina slurries for 20 min. The alumina particles attached to the surface

Fig. 8.1 Mill machine

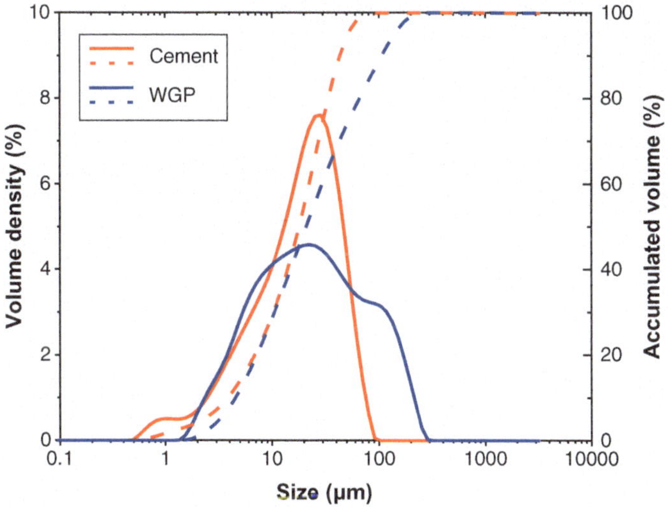

Fig. 8.2 Particle size distribution of cement and WGP

Table 8.1 Chemical composition of cement and WGP

Composition (%)	CaO	SiO_2	Al_2O_3	SO_3	Fe_2O_3	MgO	K_2O	Na_2O	LOI
Cement	64.9	18.6	4.1	3.0	3.1	1.6	0.5	–	4.2
WGP	10.8	68.4	0.9	0.1	0.4	1.8	0.3	13.5	3.8

Note LOI means the loss on ignition

of the granite were removed in an ultrasonic bath. Subsequently, the surfaced-dried granite block was put in a Φ35 mm cylindrical mould and poured with fresh mortar. Large air bubbles were discharged by tapping. After curing in the mould for one day and being transferred and stored in a water tank at 23 ± 3 °C for 3 months, the semi-finished sample was briefly ground and covered with a layer of epoxy resin to avoid damage to the ITZ during the following cutting. The sample that was cut and exposed to the ITZ was re-covered with a layer of epoxy resin in a Φ35 mm cylindrical mould. After the observation side was ground for 30 min and polished for 40 min, the sample was put in an ultrasonic bath for 2 min to remove alumina particles on the surface. Finally, the samples were put in a vacuum drying oven at 40 °C for 2 days. The MAC covered with epoxy resin was subjected to nanoindentation, nanoscratch and Backscattered Electron-Energy dispersive spectrometer (BSE-EDS) tests. Then, the mortar on the MAC was cut out and ground into powder for XRD and FTIR tests. Although it is impossible to accurately measure the change of the mineral phase and chemical composition within ITZs, the pozzolanic reaction and its influence on the properties of ITZ with WGP was reasonably inferred through the analysis of the chemical composition differences between mortar matrix in MAC with and without WGP.

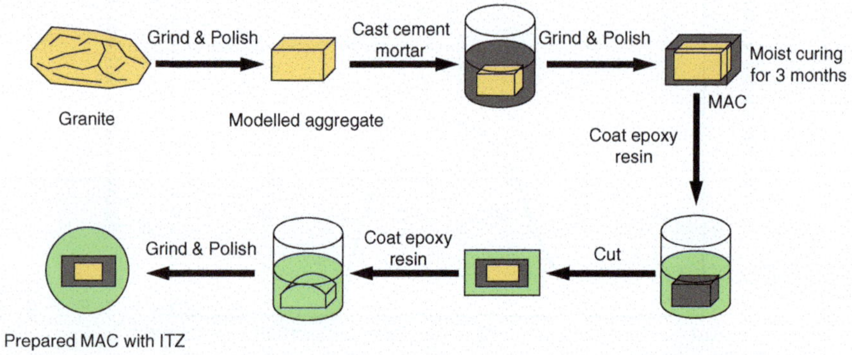

Fig. 8.3 Preparation process of ITZ simulated by MAC

8.2.3 *Constituent Distribution Within ITZs Through BSE-Image Analysis*

To improve conductivity of concrete samples, the surface was coated with 5 nm of gold. Thirty BSE images with a magnification of 500× were randomly extracted from the ITZs with and without WGP along the lateral boundary of the granite block. The probe current and accelerating voltage were set to 875 pA and 15 kV, respectively. The working distance was around 10 mm. To reduce unnecessary workload in subsequent analysis, the brightness and contrast during the observation process were kept consistent.

For statistical analysis of BSE images, modelled coarse aggregate can be easily identified and removed. Due to the significant difference in hardness between granite and cement paste, the grinding and polishing process may cause a height difference between the two phases [17, 18]. Along the aggregate and mortar interface, this appeared as a black narrow strip, which would be mistaken for pores or cracks by image analysis software according to the grey value, but it is the shadow produced by the height difference under the electron microscope [17, 18]. Therefore, the shadow portion was removed together with the aggregate. The grey threshold recognition technique proposed by Wong et al. [19, 20] was adopted to distinguish pores and cracks, unreacted clinkers and hydration products in ITZ and adjacent mortar matrix. As shown in Figs. 8.4b and 8.5b, there are two inflections in the cumulative grey curves of cement paste. Two intersections obtained by linear fitting on both sides can be used as the grey threshold for identifying constituents. Taking Fig. 8.4b as an example, grey values 0–129; 129–139; 139–254 represent pores and cracks; hydration products; unreacted clinkers, respectively. The 20 continuous strips with a width of 5 μm were taken out from the aggregate boundary to the mortar matrix. For ITZs with and without WGP, 30 BSE images were used to statistically analyse the relationship between the volume fraction of each constituent and the distance from the aggregate boundary. The width of ITZs can be easily determined by image analysis results. The processing procedure of the BSE image is shown in Figs. 8.4 and 8.5.

The statistical results of image analysis are shown in Fig. 8.6. The proportion of hydration products was directly related to unreacted clinkers [21, 22]. That is, when the content of hydration products increased with the distance from the aggregate boundary, the proportion of unreacted particles decreased, and vice versa. The proportion of pores and cracks was the highest near the aggregate boundary and decreased rapidly to a stable value with distance from the modelled aggregate. Due to the 'wall effect' on the distribution of clinkers and free water in concrete, the width of ITZ can be determined by the region where the proportion of hydration products and unreacted clinkers changed most significantly [23]. Therefore, the ITZ widths of the modelled samples with and without WGP were approximately 10 μm and 30 μm, respectively. The ITZ and matrix in Fig. 8.6 were separated by brown dash lines. The average proportions and corresponding standard deviations of phases in ITZs and the adjacent matrix are listed in Table 8.2. It can be roughly seen from the

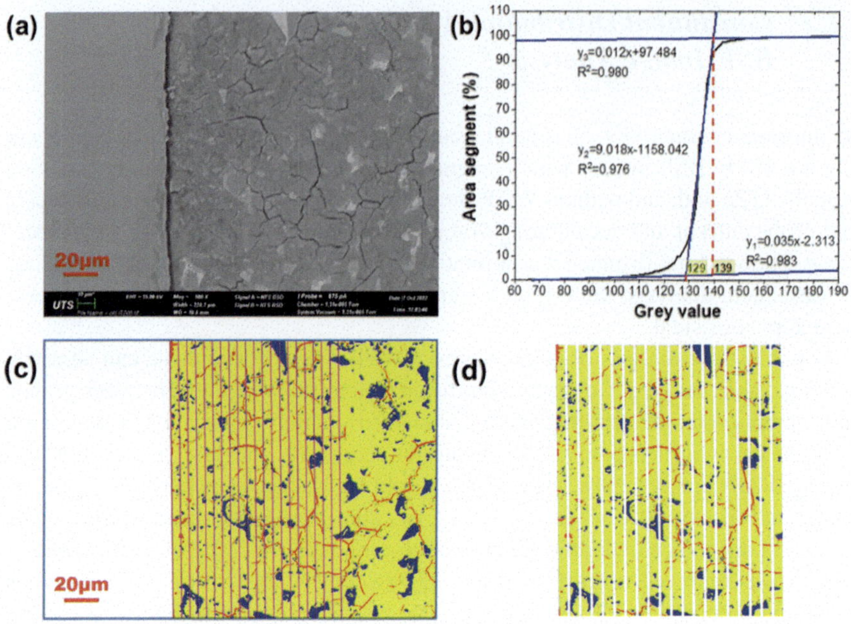

Fig. 8.4 BSE image processing procedure of ITZ without WGP: **a** BSE image of ITZ without WGP; **b** grey value threshold determination; **c** constituent segmentation; **d** strips division. *Note* Yellow area means hydration products; blue area means unreacted clinkers; red area means pores and cracks

phase segmentation image in Figs. 8.4 and 8.5 that the content of unreacted clinkers in the sample with WGP was higher than that in the counterpart, and the particle size gradually increased with the distance from the boundary of the granite block. In fact, the statistical results in Table 8.2 show that the ITZ containing WGP possessed a lower proportion of unreacted clinkers and a higher proportion of hydration products compared with the ITZ without WGP, while the trends were opposite for the matrix, and the data dispersion of samples with WGP was larger. This is attributed to the nucleation effect of WGP, which offers additional space and nucleation sites for cement particles to hydrate [24–26]. A similar phenomenon was also found in the analysis of the distribution of fly ash and slag clinkers in geopolymer concrete [9, 23]. For example, Fang and Zhang [23] found that the diameter of grains distributed in ITZ was much smaller than that in the mortar matrix. For pores and cracks, the proportion in the ITZ with WGP was slightly higher than that in the sample without WGP, while the proportion in the matrix with WGP was significantly reduced. Part of WGP participated in the pozzolanic reaction to generate more C–S–H gels and filled gaps [27]. In addition, the rough and angular surface of WGP provided a site for C–S–H nucleation reaction, which promoted cement hydration, and the rest unreacted WGP can also play the role of filler to improve the pore structure [28, 29].

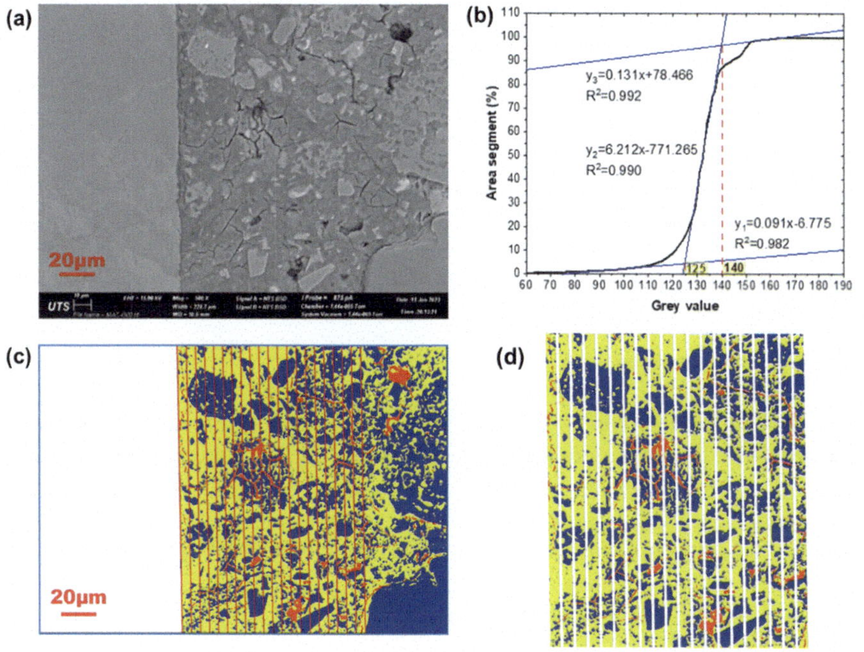

Fig. 8.5 BSE image processing procedure of ITZ with WGP: **a** BSE image of ITZ with WGP; **b** grey value threshold determination; **c** constituent segmentation; **d** strips division. *Note* Yellow area means hydration products; blue area means unreacted clinkers; red area means pores and cracks

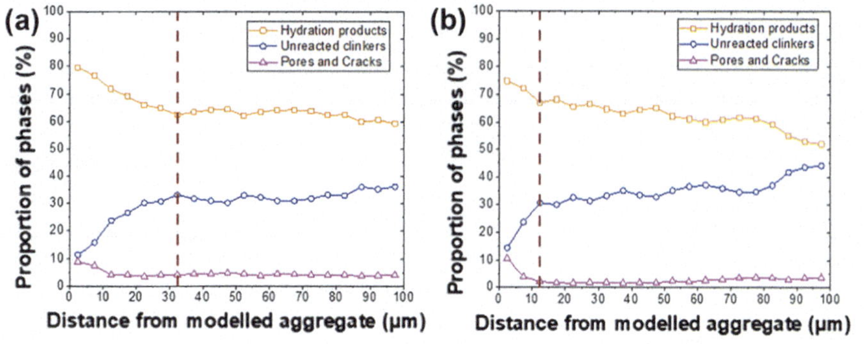

Fig. 8.6 Phase distributions in ITZs and adjacent mortar matrix: **a** modelled sample without WGP; **b** modelled sample with WGP

Table 8.2 Average proportion of different phases in image analysis

	Hydration products (%)		Unreacted clinkers (%)		Pores and cracks (%)	
	ITZ	Mortar matrix	ITZ	Mortar matrix	ITZ	Mortar matrix
MAC-C	70.16 ± 5.84	62.80 ± 1.68	24.61 ± 7.58	32.89 ± 1.89	5.23 ± 1.92	4.31 ± 0.32
MAC-G	71.45 ± 3.23	61.52 ± 4.44	22.99 ± 6.68	35.94 ± 3.87	5.56 ± 3.68	2.54 ± 0.79

8.2.4 Elements and Mineralogical Composition of ITZs

8.2.4.1 EDS Analysis

EDS mapping analysis was used to observe the chemical elements distribution within ITZs and the adjacent regions under the background of BSE images with a magnification of 1500×. During the EDS test, the accelerating voltage and working distance remained at 15 kV and 10 mm, and the probe current was lower than that of the BSE to reduce dead time to a reasonable level. The distributions of Al, Si, Ca, and Na were observed. In addition, the atomic ratio of Si/Ca versus Si/Na was stoichiometrically quantified by a peak to the background (P/B)-ZAF correction and deconvolution of carbon.

GP cement and soda-lime glass powder were the main cementitious materials in this study. A BSE image was used as background, and the distributions of Al, Si, Ca, and Na are shown in Fig. 8.7 by EDS mapping. The region with a relatively uniform distribution of Si and Ca was C–S–H gel. It can be understood that cement hydration in MAC with and without WGP has reached a high level. There were more Al-rich regions in the ITZ without WGP, whereas more Si- and Na-rich regions appeared in the ITZ with WGP. This is because, in the sample containing WGP, there was less cement, but a good amount of partially reacted WGP containing higher concentrations of SiO_2 and Na_2O remained within the ITZ.

Figure 8.8 depicts the quantitative analysis results based on EDS mapping in random regions of ITZ and matrix. The data cluster area represented C–S–H gel, except some data with Si/Ca and Si/Na close to 0 may refer to pores and cracks. The Si/Ca of C–S–H in the ITZ and matrix without WGP was mainly concentrated around 0.5, while WGP generated more C–S–H gels with Si/Ca in the range of 1–2. It has been known that C–S–H with high Si/Ca had higher cohesion strength [30, 31]. Combining the proportion of hydration products in Table 8.2, it can be reasonably speculated that ITZ exhibited higher performance in samples containing WGP. In addition, the dot where the Si/Ca ratio was close to 7 represented WGP according to the oxide components listed in Table 8.1. As expected, after adding 20% WGP, the number of these dots significantly increased in the ITZ and matrix. However, the existing data with a high Si/Ca ratio in ITZ and matrix without WGP may be due to the interference of silica sand in the samples. It is to be noted that although locations for EDS point mapping were carefully chosen to avoid sand grains in the microstructure, a few might have interfered resulting in outliers with high Si/Ca

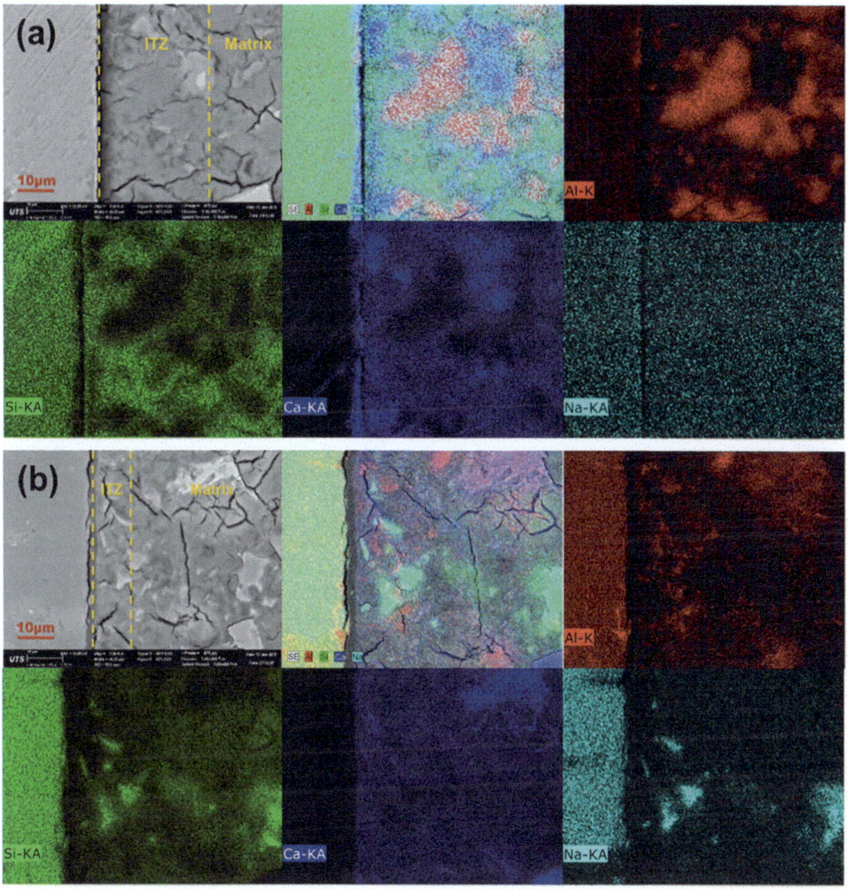

Fig. 8.7 EDS mapping analysis of ITZs and adjacent matrix: **a** chemical elements distribution of ITZ without WGP; **b** chemical elements distribution of ITZ with WGP

ratios. In addition, the Si/Na ratio of the ITZ and matrix were lower in the sample with WGP because the cement was replaced by high-Na glass powder.

8.2.4.2 XRD Analysis

XRD patterns of cement mortar with and without WGP were obtained by a Bruker D8 Discover diffractometer. The mineralogical characterisations of the two groups of samples were analysed at a range of 5°–70° 2θ, with a 0.02° step. CuKα radiation of wavelength 1.54 Å was applied to measure diffraction patterns.

Specific crystals exposure to a beam of incident X-rays can cause diffraction in specific directions. Comparing the diffraction intensities in different directions with the standard powder diffraction data can determine the crystal phases in the material

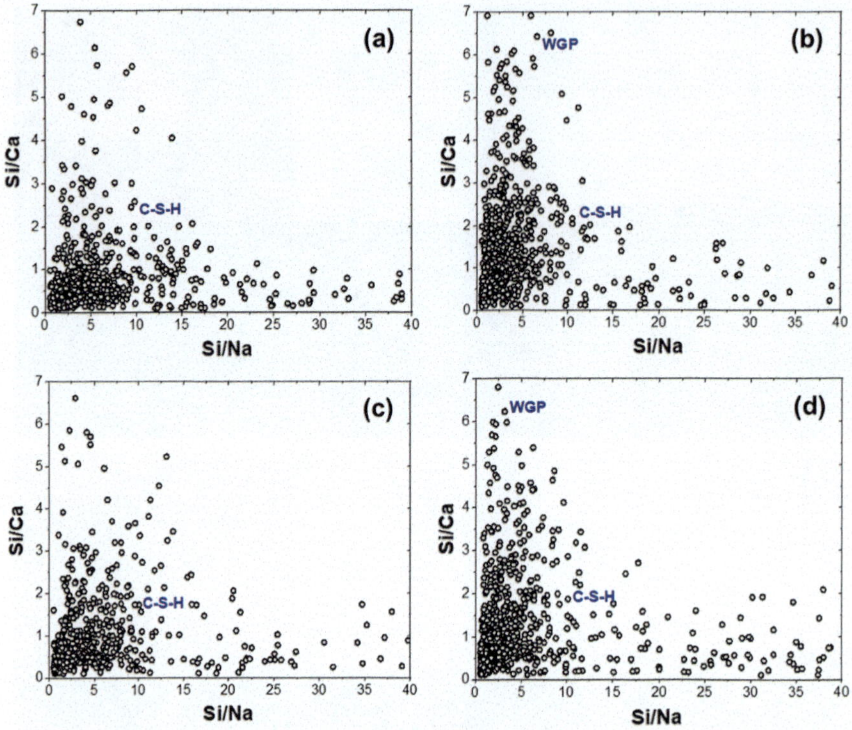

Fig. 8.8 Statistical analysis results of EDS mapping: **a** ITZ without WGP; **b** ITZ with WGP; **c** mortar matrix without WGP; **d** mortar matrix with WGP

[32]. The mortar part of MAC with and without WGP was cut and ground into powder for XRD testing, and the corresponding XRD patterns are shown in Fig. 8.9. Because the powder sample contained silica sand, there were much more quartz diffraction peaks in the XRD patterns than previous research on paste samples [4, 29]. The diffraction peaks corresponding to Portlandite appeared at around 18.09° 2θ, 34.14° 2θ, 47.16° 2θ and 50.82° 2θ. Comparing the XRD spectra, the content of portlandite in the sample decreased when 20% of cement was substituted by WGP. It should be admitted that the cement dilution reduced the peak density of portlandite. According to the change of Si/Ca ratio in Fig. 8.8 and previous research results [26, 33], the reaction of amorphous silica in WGP with $Ca(OH)_2$ to generate C–S–H was another important reason leading to the decrease of portlandite peaks.

8.2.4.3 FTIR Analysis

FTIR spectra of samples with and without WGP were obtained by a Nicolet 6700 FTIR spectrometer. The wavenumber ranged from 4000 to 500 cm^{-1} and a resolution

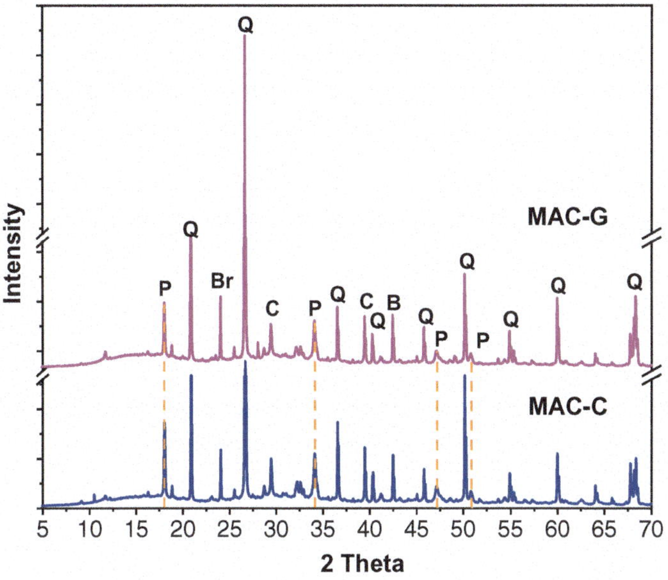

Fig. 8.9 XRD patterns of cement mortar with (MAC-G) and without (MAC-C) WGP at age of 3 months. *Note* Q-Quartz; P-Portlandite; B-Belite; Br-Brownmillerite; C-Calcite

of 8 cm^{-1} was adopted to analyse the contents of portlandite, ettringite, C–S–H gel etc. through chemical bonds.

Figure 8.10 shows the FTIR spectrum of samples with and without WGP in this study. Band 1 at around 3642 cm^{-1}, typical of the O–H stretching vibrations generated by portlandite ($Ca(OH)_2$) [34, 35], indicates that WGP consumed a certain amount of portlandite to generate C–S–H gel. This is consistent with the above XRD analysis results and the conclusions observed by other researchers through SEM and TGA [28, 36, 37]. The rest bands were mainly concentrated between 1500–500 cm^{-1} and the corresponding band assignments are listed in Table 8.3. To compare the content of components in samples, the FTIR spectra between 1300 and 700 cm^{-1} were separated into four independent peaks at around 785 cm^{-1}, 869 cm^{-1}, 975 cm^{-1}, 1100 cm^{-1}, with a shoulder at 1180 cm^{-1} by Gaussian mixture models (see Fig. 8.11). The frequency values of the peaks and respective area fractions obtained from statistical deconvolution are shown in Table 8.4. The peaks at around 785 cm^{-1} and 975 cm^{-1} were characteristic of the stretching vibrations generated by Si–O bonds in Q^1 and Q^2 (C–S–H), respectively [34]. The area fractions of Q^1 and Q^2 in the cement mortar with WGP increased by 2.06% and 4.19%, respectively, compared with the sample without WGP, indicating that more C–S–H gel was generated after adding WGP. In addition, the Q^2 peak tends to shift toward higher frequencies with the improvement of the Si/Ca ratio due to the progressive depolymerisation of the silicate chain [34]. The frequency shifted from 974 cm^{-1} to 980 cm^{-1} when 20% WGP was added to the sample representing more C–S–H with a high Si/Ca ratio was generated, which is corresponding to the results in Fig. 8.8. The appearance of the peak at frequencies around 1180 cm^{-1} is attributed to silica-rich gel. The signal of this peak was stronger in the sample with WGP, suggesting a more condensed microstructure. The peaks at around 869 cm^{-1} and 1100 cm^{-1} corresponded to C–O (CO_3^{2-}) and S–O (mainly for ettringite), respectively. The data in Table 8.4 shows that the area fraction of the peak corresponding to C–O increased by 0.29% in the sample containing WGP, while the figure corresponding to S–O decreased by 7.3% compared to the counterpart. The generation of ettringite and carbonation of cement paste were largely controlled by the content of soluble alkali metal oxides (mainly CaO and Na_2O) [38]. A high concentration of Na_2O could increase the pH value of the pore solution. The alkali environment accelerated the carbonisation speed of cement paste and enhance the solubility of ettringite. Moreover, ettringite was formed by combining CaO with Al^{2+} and SO_4^{2-} [39]. WGP contained more Na_2O than CaO in Table 8.1. As a result, compared with the sample without WGP, the peak corresponding to C–O was strengthened after adding WGP, while the peak corresponding to S–O weakened.

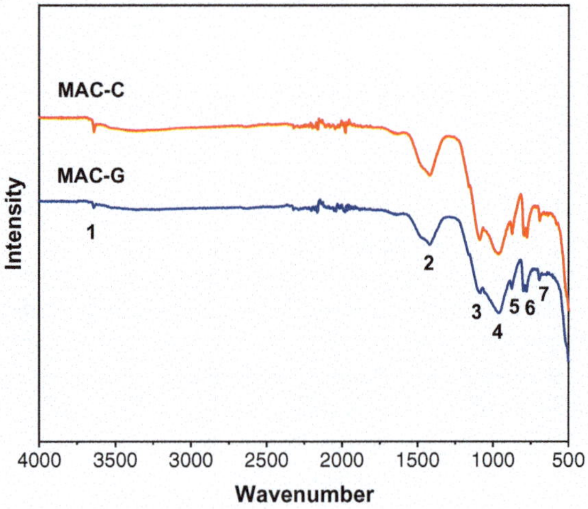

Fig. 8.10 FTIR spectra of cement mortar with (MAC-G) and without (MAC-C) WGP at the age of 3 months

Table 8.3 Sample powder with and without WGP assignments

Band	Frequency (cm^{-1})	Assignment
1	3642	v OH ($Ca(OH)_2$)
2	1420	v_3 CO (CO_3^{2-})
3	1090	v Si–O
4	965	v Si–O (C–S–H) Q^2
5	874	v_2 CO (CO_3^{2-})
6	797	v Si–O (C–S–H) Q^1
7	693	v_2 CO (CO_3^{2-})

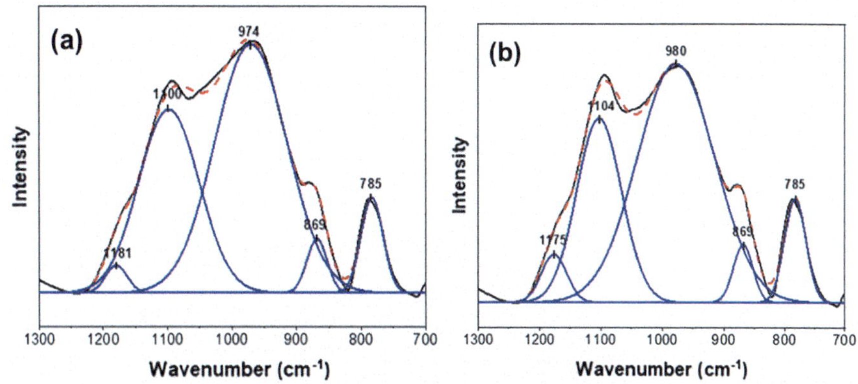

Fig. 8.11 Deconvolution spectra of mixtures from 700 to 1300 cm^{-1}: **a** cement mortar without WGP; **b** cement mortar with WGP

Table 8.4 Results obtained from the statistical deconvolution of FTIR spectra from 1300 to 700 cm^{-1}

MAC-C		MAC-G	
Frequency (cm^{-1})	Area (%)	Frequency (cm^{-1})	Area (%)
1181	1.82	1175	3.88
1100	33.79	1104	26.49
974	54.68	980	58.87
869	3.30	869	3.59
785	6.40	785	7.16

8.2.5 Nano/Micromechanical Properties of ITZs

The Agilent G200 Nano Indenter can simultaneously apply normal and lateral forces to a sample. This enables the Indenter to perform nanoindentation and nanoscratch tests, simply by choosing the appropriate tip and testing method. The schematic diagram of nanoindentation and nanoscratch tests is shown in Fig. 8.12.

8.2.5.1 Nanoindentation Test Results

When conducting nanoindentation tests, a Berkovich tip with a radius of 20 nm was installed. Grid nanoindentation combined with statistical analysis was used to reveal the micromechanical properties of the ITZ and its adjacent regions, which contributed to the understanding of the bond strength between aggregate and cement paste in concrete. Concrete is a heterogeneous and multi-phase structure. Although the sample surface was ground and polished during processing, it was still difficult to ensure that the region to be tested had perfect flatness. Therefore, the nanoindentation depth was made enough to avoid the negative influence of sample surface roughness.

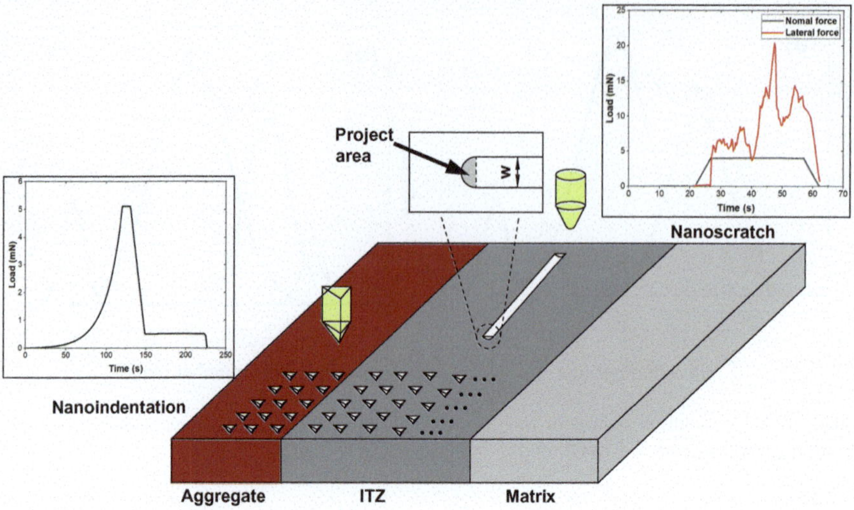

Fig. 8.12 Schematic diagram of nanoindentation and nanoscratch

According to previous experience [21], the average depth was set to 800 nm to satisfy the test accuracy as well as the number of effective testing points within the ITZs. The loading force was controlled by indentation depth. When the indentation depth was close to 800 nm, the load was maintained for 10 s to eliminate the influence of residual strain. After unloading to 10% of the peak load and maintaining for another 75 s, the indenter was completely detached from the sample surface. Five regions avoiding large cracks and fine aggregate were selected for nanoindentation tests of ITZ of both with and without WGP specimens. For each nanoindentation region, as shown in Fig. 8.12, 3 × 5 and 11 × 5 measuring points were conducted on the granite block and cement paste, respectively. The vertical spacing of the measuring points was 15 μm. The distance between the aggregate boundary and adjacent measuring points was 4 μm. No measuring points were arranged on the aggregate boundary. The lateral spacing of the remaining measuring points was 7 μm. According to previous research, the ITZ width in concrete is about 15–50 μm [16, 22]. Therefore, the designed grid nanoindentation area can cover the ITZ and a part of the adjacent matrix. The following image analysis technique will be used to identify the ITZ width of the samples in this study and confirm this inference. Finally, the elastic modulus and hardness within ITZ at different distances from the aggregate boundary can be obtained by load-indentation depth curve (P—indentation load; h—indentation depth; E_r—reduced elastic modulus), characteristics of the tip (E_i—elastic modulus; v_i—Poisson's ratio), characteristics of the sample (E—elastic modulus; v—Poisson's ratio), and projected contact area (A) according to Eqs. 8.1–8.3.

$$H = \frac{P_{\max}}{A} \tag{8.1}$$

$$S = \frac{dP}{dh}\bigg|h = h_{\max} = \frac{2}{\sqrt{\pi}} E_r \sqrt{A} \tag{8.2}$$

$$\frac{1}{E_r} = \frac{1 - v^2}{E} + \frac{1 - v_i^2}{E_i} \tag{8.3}$$

The elastic modulus of materials can be determined from the slope of the Load versus Indentation depth unloading curve. The frequency density distribution histograms (FDH) of elastic modulus for ITZ and the adjacent matrix for specimens with and without WGP are shown in Fig. 8.13. The results show that the elastic modulus of ITZ and matrix of MAC without WGP followed an approximately normal distribution. For ITZ and adjacent matrix without WGP (see Fig. 8.13a, b), the elastic modulus with a relative frequency higher than 0.04 were distributed in the range of 5–20 GPa and 5–23 GPa, respectively. The weighted mean elastic modulus was 13.29 GPa and 13.64 GPa, respectively, which is in line with the traditional concept of the micromechanical difference between ITZ and matrix in cement-based materials [8, 40]. Some recent studies reported an opposite conclusion [11, 16]. There was more CH in the ITZ, resulting in a slightly higher measured elastic modulus of the ITZ than that of the mortar matrix [11]. However, the test results still approved the conclusion that the ITZ was the weakest part of concrete because the cohesiveness of CH can be neglected. The curing age of the MAC samples in this study was more than 3 months. Combined with the EDS mapping analysis in Fig. 8.7a, there was almost no obvious Ca^{2+}-rich area within ITZ, indicating that the cement hydration has reached a relatively high level [30, 31]. Most of the nanoindentation data in Fig. 8.13a, b can be regarded as the mechanical properties of gel products. As a result, it can be inferred that the cohesion of C–S–H in the mortar matrix was greater than that in ITZ without WGP which explains why cement composites are prone to fail along the ITZ; better cohesion imparts better mechanical properties (compressive strength, flexural strength, tensile strength and elastic modulus etc.)

It is worth noting that the elastic modulus of ITZ and matrix with WGP were distributed over a wider range and two remarkable peaks appeared. The first peak ranged from 10 to 15 GPa, mainly reflecting the mechanical properties of the C–S–H gel. According to the literature, the elastic modulus of cement paste was generally between 10 and 20 GPa, while the elastic modulus of glass particles can be as high as 80 GPa [7, 41, 42]. The second peaks at 19–27 GPa in Fig. 8.13c, d could represent the mixed phase of gel and partially reacted glass particles. It can also be concluded that the heterogeneity of ITZ and matrix with WGP was more significant.

8.2.5.2 Nanoscratch Results

For nanoscratch tests, a hemispherical tip with a radius (r) of 5 μm and apex angle (α) of 60° was adopted to avoid the errors caused by the asymmetry of the Berkovich tip on the test results. Different from nanoindentation, nanoscratch depth (d) was

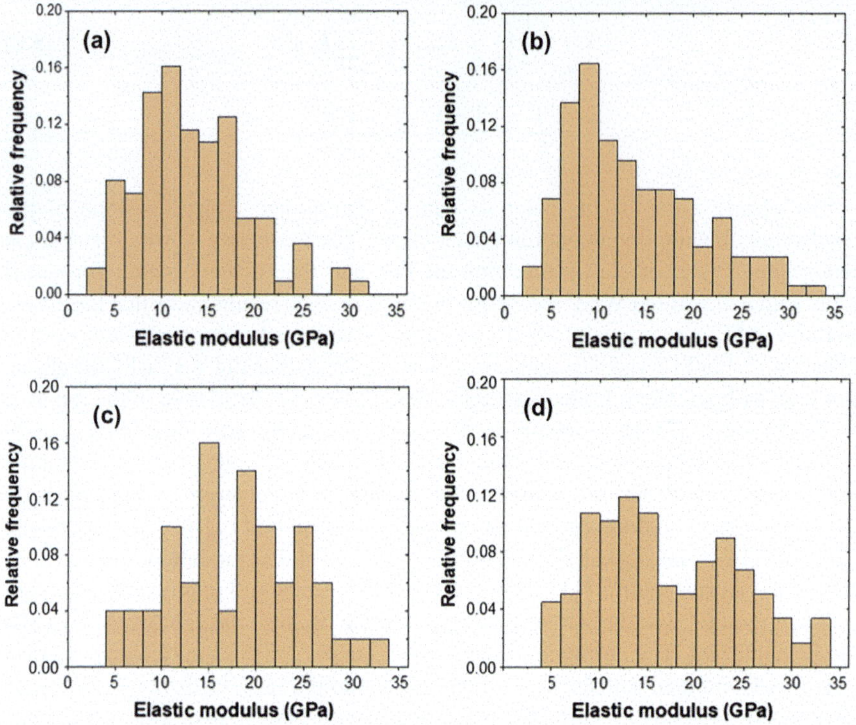

Fig. 8.13 Elastic modulus statistical results obtained from nanoindentation tests (Bin size = 2 GPa): **a** ITZ without WGP; **b** mortar matrix without WGP; **c** ITZ with WGP; **d** mortar matrix with WGP

controlled by the normal force. The maximum normal force ($F_{N\max}$), scratch speed (v_s) and scratch distance (L) can be defined by the test method according to standards (ASTM G171-03) and previous research results [1, 9]. A moderate maximum normal force can ensure relatively accurate test results while avoiding the interference of aggregate constraints [1, 9]. In this study, the normal force was uniformly increased to 4 mN during the pre-scratch process, and then, the normal force was kept at peak value and 4 μm/s of scratch speed until the scratch distance obtained was 120 μm. Finally, the loading procedure entered the post-scratch stage until the tip left the sample surface. The data acquisition rate was set to 1 point/μm. A total of 20 scratches were randomly conducted along the centre of the ITZ and the matrix 30 μm away from the ITZ, respectively. Regions close to fine aggregate and large cracks were artificially avoided. The scratch hardness (HS_p) and fracture toughness (K_c) of samples were calculated based on Eqs. 8.4–8.7. The geometric depth (d_{gt}) and the corresponding scratch width (w_{gt}) were 2500 nm and 5 μm, respectively. The transverse force (F_T) was obtained directly from the test results.

$$HS_p = \frac{8F_{N\max}}{\pi w^2} \tag{8.4}$$

$$K_c = \frac{2\sqrt{2}F_T}{\pi\sqrt{w^3}} \tag{8.5}$$

$$w = \begin{cases} 2\sqrt{r^2 - (r-d)^2} d \le d_{gt} \\ w_{gt} + 2\tan\left(\frac{\alpha}{2}\right) \times \left(d - d_{gt}\right) d > d_{gt} \end{cases} \tag{8.6}$$

$$d_{gt} = r\left(1 - \sin\left(\frac{\alpha}{2}\right)\right) \tag{8.7}$$

Nanoscratch can provide far more test data than nanoindentation in a short operation time. The huge amount of data made it possible to use a very small bin size for frequency density statistical analysis, revealing more phases but not disordering the shape of the histogram [1]. After trial and error, the bin size was finally determined to be 0.05 GPa. However, the tip radius and the displacement into the surface of the nanoscratch were usually higher than those of nanoindentation, which led to a higher project area than nanoindentation tests. As a result, the nanoscratch results often reflect the mixed results of multiple phases. Nevertheless, most of the test values can still represent the micromechanical properties of C–S–H due to the deliberate avoidance of large pores and aggregate in the test.

Figure 8.14 shows the FDH of fracture toughness of ITZ and matrix with and without WGP obtained from nanoscratch tests. Fracture toughness quantitatively evaluates the ability of a material with natural defects to resist rapid and unlimited crack propagation [43]. The FDH of fracture toughness of ITZ without WGP can be separated into two components by the Gaussian mixture model. The expected values of fracture toughness for these two components were 0.18 MPa·m$^{1/2}$ and 0.39 MPa·m$^{1/2}$, which could be regarded as low-density C–S–H and high-density C–S–H, respectively. In contrast, there was only one independent peak in the mortar matrix without WGP. The expected value of fracture toughness of this component was 0.40 MPa·m$^{1/2}$, indicating that the gel phase in the matrix without WGP was mainly composed of high-density C–S–H with better homogeneity and cohesion strength, compared with weaker C–S–H gel structure in the ITZ. For ITZ and mortar matrix with WGP, the FDH of fracture toughness can be segmented into an independent peak with a shoulder. It can be seen from deconvolution analysis that the standard deviation of fracture toughness of ITZ and matrix with WGP was higher than their counterparts, indicated by shorter and wider distribution curves. The expected value of the peak was around 0.37 MPa·m$^{1/2}$. It can be explained that the WGP promoted the formation of C–S–H with a high Si/Ca ratio, improving the density of the ITZ and matrix, that is, the sample containing 20% WGP had more high-density C–S–H. The explanation of the shoulder with the expected value of 0.86 MPa·m$^{1/2}$ can refer to the second peak in Fig. 8.13c, d, which is the mixed phase of C–S–H and incompletely reacted WGP.

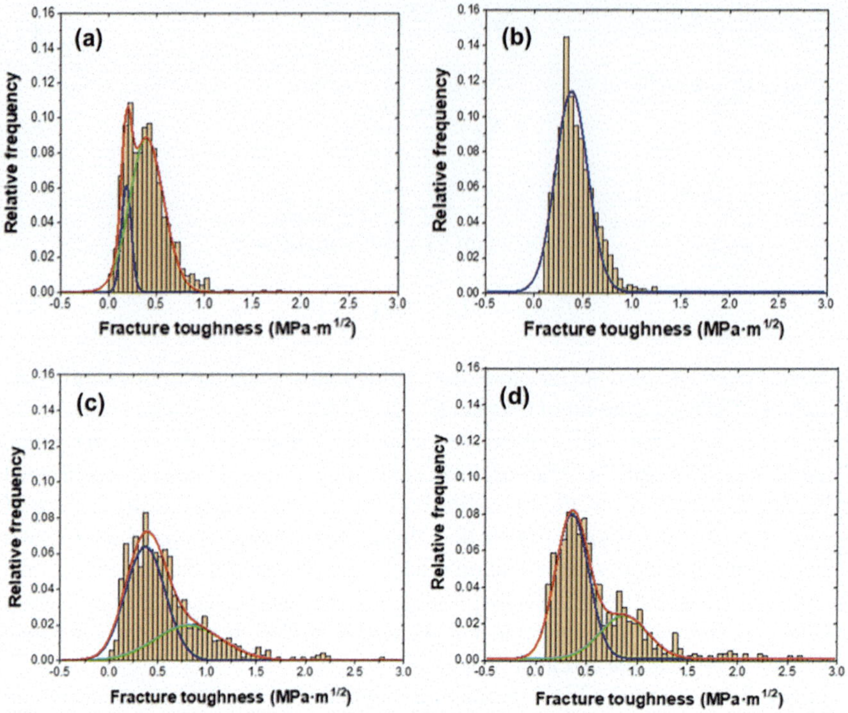

Fig. 8.14 Fracture toughness statistical results and deconvolution analysis obtained from nano-scratch (Bin size = 0.05 GPa): **a** ITZ without WGP; **b** mortar matrix without WGP; **c** ITZ with WGP; **d** mortar matrix with WGP

8.2.5.3 Comparison of Nanoindentation and Nanoscratch Results

Figure 8.15 compared the nanoscratch hardness and nanoindentation hardness of samples with and without WGP. Because of the huge amount of data, the nano-scratch hardness of samples still followed an approximate normal distribution even if a very small bin size was adopted in histograms. In contrast, the nanoindentation hardness data was more discrete, and there were gaps between some values. Nevertheless, for samples without WGP, the main data of nanoindentation hardness and nanoscratch hardness can still achieve good concordance. By observing Fig. 8.15a, b, it can be reasonably identified that the hardness of ITZ and matrix without WGP ranged from 0.125 GPa to 0.375 GPa. In comparison, when 20% WGP was added to the sample, the hardness of ITZ and matrix was distributed in a wider range, which is similar to elastic modulus in Fig. 8.13 and fracture toughness in Fig. 8.14. In the test results of nanoscratch, the value of the first two bars of the sample with WGP was higher than that of the sample without WGP. This is due to the presence of softer C–S–H gel within the sample containing WGP although the microstructure was denser, and the unreasonable data corresponding to the large pores and cracks in the ITZ and matrix without WGP were removed in the statistical analysis. Therefore, there were more low-hardness constituents (≤ 0.1 GPa) in the nanoscratch test results of ITZ and matrix with WGP. It is worth noting that the nanoindentation hardness of ITZ and matrix with WGP was slightly higher than the nanoscratch hardness. As mentioned before, the project area of nanoindentation was much smaller than that of nanoscratch, which means nanoindentation could measure the micromechanical properties of some individual phases, such as CH or glass particles. The nanoscratch hardness was determined according to the scratch depth of the tip under a certain normal load [44]. The difference in test methods made it difficult for the nanoscratch to accurately reflect the hardness of glass particles. For example, when a hemispherical tip touched a small glass particle during scratching, the tip may push it aside rather than swipe directly across the glass. In this condition, the hardness of the combined phase of glass and the surrounding C–S–H would be mistaken for the hardness of the glass particles. As a result, the concordance between nanoscratch hardness and nanoindentation hardness was poor in samples containing WGP. In general, nanoindentation is likely to reflect the real micromechanical properties of materials, but this is based on a large amount of test time to produce massive test data, otherwise, it is difficult to observe the right frequency density distribution of mechanical properties [8]. For materials with high homogeneity, nanoscratch is an ideal technique which can not only ensure the accuracy of test results but also greatly shorten test time [45].

8.3 Nano/Microscopic Investigation on the Interfacial Transition Zones in MRAC

8.3.1 Materials and Mixture Proportion

A MRAC sample consists of three parts: a rectangular granite block, old mortar on the surface of the stone block, and new mortar on the outermost layer of the sample. The schematic diagram of MRAC is shown in Fig. 8.16. Rectangular granite blocks were seen as virgin coarse aggregate. Fine aggregate in the old and new mortar was silica sand with a fineness modulus of 1.9. General purpose Portland cement was used as the main cementitious material. In the MRAC sample with WGP, 20% of cement in the new mortar was replaced by milled waste soda-lime glass powder with average particle size of 50 μm. The particle size distribution and oxide composition of cement and WGP is shown in Fig. 8.17 and Table 8.1, respectively. Considering that the interface between old and new mortar may be difficult to distinguish under an electron microscope, different water-cement ratios were designed in the preparation of new and old mortar. According to previous studies, the strength of RAC is normally lower than that of NAC under the same conditions [14]. Civil engineers usually reduce water-cement ratio of new mortar or choose cement with higher strength grade to ensure that the RAC structures meet the strength requirements [46]. Therefore, the

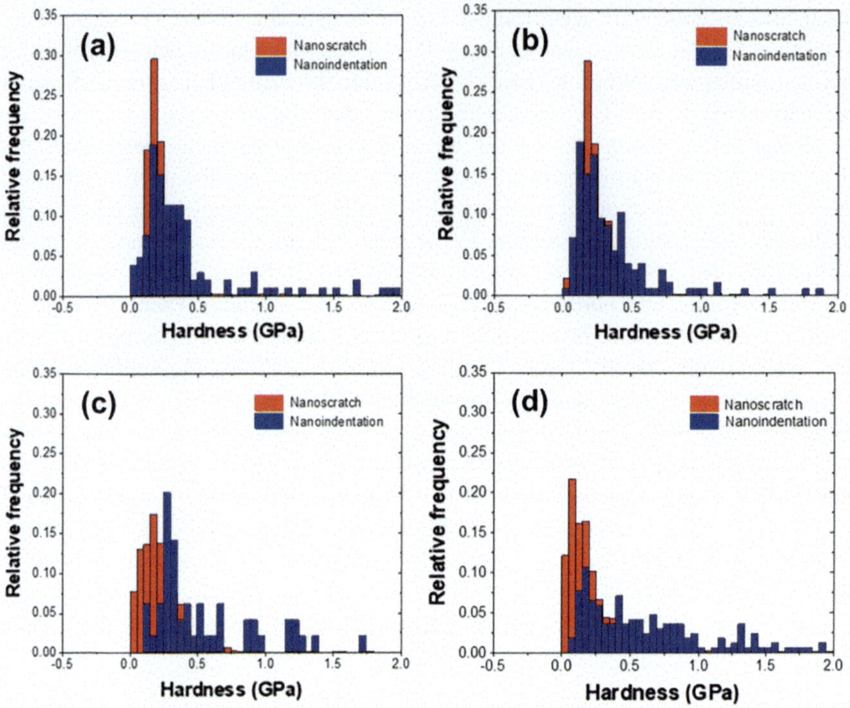

Fig. 8.15 Comparison of nanoindentation and nanoscratch results (Bin size = 0.05 GPa): **a** ITZ without WGP; **b** mortar matrix without WGP; **c** ITZ with WGP; **d** mortar matrix with WGP

water-cement ratios of old and new mortars in this study were designed to be 0.45 and 0.4, respectively. The detailed mix proportion of new and old mortar is shown in Table 8.5.

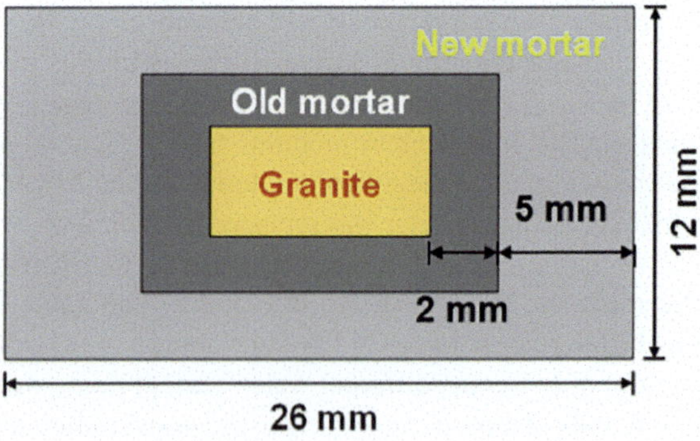

Fig. 8.16 Schematic diagram of modelled recycled aggregate concrete

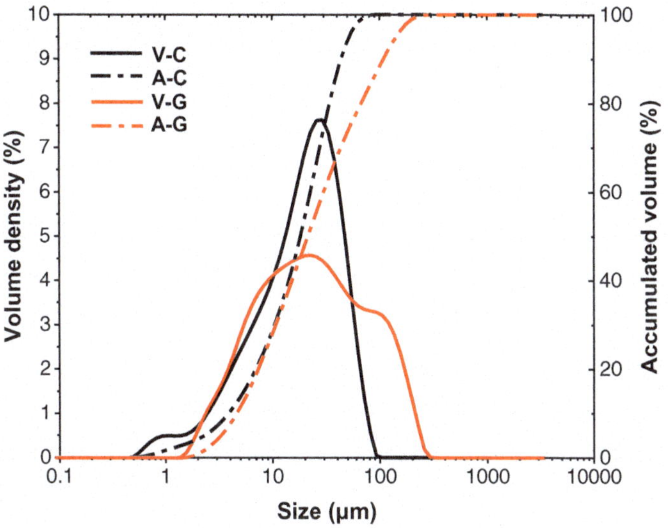

Fig. 8.17 Particle size distribution of cementitious materials tested by Mastersizer 3000. *Note* 'V-C' and 'V-G' represent volume density of cement and WGP respectively; 'A-C' and 'A-G' represent accumulated volume of cement and WGP respectively

Table 8.5 Mix proportions of MRAC

Mix ID	Phases	Cement (g)	Glass (g)	Sand (g)	Water (g)	Granite
MRAC-C	Old mortar	100	0	150	45	1
	New mortar	100	0	150	40	
MRAC-G	Old mortar	100	0	150	45	1
	New mortar	80	20	150	40	

8.3.2 Sample Preparation

According to previous experience [9, 11], the main steps of MRAC sample preparation were designed and shown in Fig. 8.18. Granite gravels with appropriate size were selected and ground on 320 grits abrasive papers into 13 mm × 13 mm × 8 mm rectangular blocks. Then, the surface of the granite blocks was smoothed by 600, 800 and 1200 grits abrasive papers with each grade lasting for 15 min. 0.3 μm and 0.05 μm alumina slurry were used to polish granite blocks under a small force, and the polish process lasted for 1 h. After grinding and polishing, granite blocks were cleaned in an ultrasonic bath for 3 min to remove alumina particles on the surface of samples. The processed granite blocks are called modelled NCA. According to the mix proportion in Table 8.5, the saturated surface-dried modelled NCA were covered with old mortar in Φ25 mm cylindrical moulds, and discharged air bubbles in the mortar by tapping. After curing in the mould for one day, samples were put in

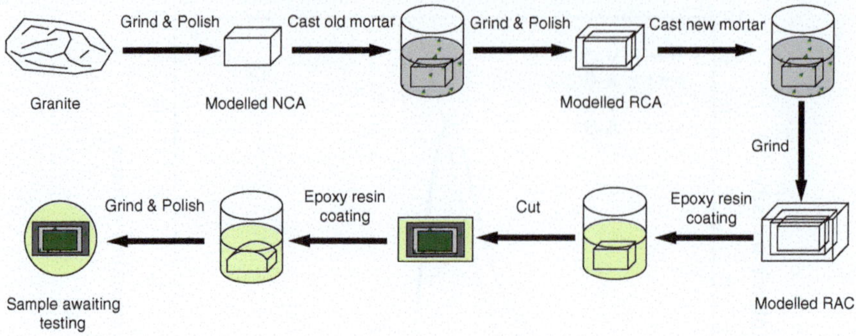

Fig. 8.18 Preparation schematic of modelled recycled aggregate concrete

a moist environment at 23 ± 3 °C for 28 days. Grinding and polishing were consistent with the above-mentioned processes of modelled NCA. Then, new mortar with/without WGP was prepared and covered around the modelled RCA. After 28 days, the MRAC with old and new mortar was ground and put in Φ35 mm cylindrical moulds and poured with epoxy resin. After one day, samples were wet-cut by a low-speed diamond saw under the protection of hardened epoxy resin to expose old and new ITZs into the air. Finally, samples were coated with an additional layer of epoxy resin. The surface waiting for observation was ground and polished. It should be noted that to avoid the height difference between the aggregate and mortar caused by excessive polishing, the entire polishing time was reduced to 30 min. Photos of MRAC manufacturing process are presented in Fig. 8.19. The modelled samples were used for BSE-EDS tests and nanoindentation. After that, the new mortar was cut out and ground into powder for Thermogravimetric-derivative thermogravimetry (TG-DTG) and ^{29}Si NMR tests.

8.3.3 Phases Distribution of ITZs According to BSE-Image Analysis

Figure 8.20 shows typical BSE images of three ITZs in this study. Compared with common SEM images, BSE is easier to distinguish between phases in cement-based materials according to grey value, even though the resolution of BSE image may be lower than SEM images [20]. Many researchers have adopted BSE images combined with image analysis techniques to investigate the constituent distribution in ITZs [15, 47].

Taking the old ITZ as an example, Fig. 8.21 shows the schematic view of BSE image processing for statistical analysis. The boundary between the aggregate and the matrix can be easily captured manually and the aggregate part can be removed by Photoshop. For the remaining mortar matrix region, the image analysis software, Image Pro was used to acquire its grey value-area segmentation curve. Wong's grey

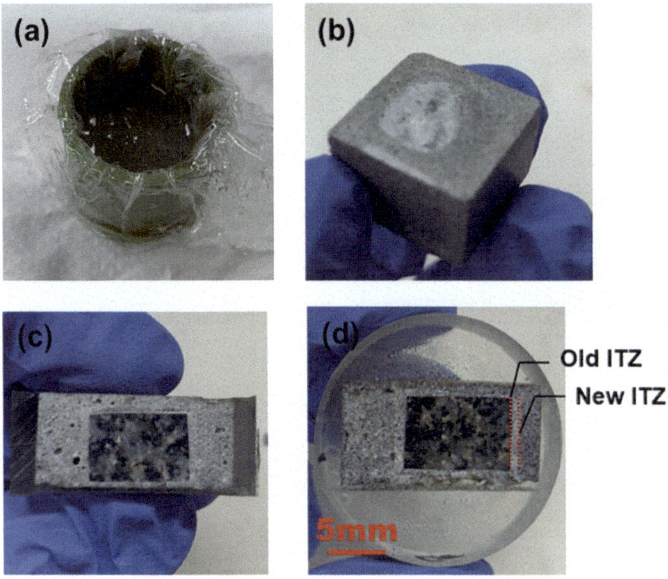

Fig. 8.19 Preparation of modelled recycled aggregate concrete: **a** casting mortar; **b** modelled RAC after surface polishing; **c** cutting modelled sample with epoxy resin; **d** modelled sample after grinding and polishing

value segmentation method was adopted [19, 48]. The grey value threshold and different phases can be identified by the intersection between the two liner fittings at the inflection of the cumulative area curve with grey values, as shown in Fig. 8.21b. However, it must be noted that Wong et al. declared that this method may slightly overestimate the actual grey threshold. Therefore, in this study, Wong's method was used to roughly identify the grey threshold. The grey range and segmentation of each phase were finally determined by manual adjustment [19, 20]. The image after phase segmentation is shown in Fig. 8.21c (red represents pores and cracks; green represents reaction products; blue represents unreacted products). The matrix within 100 μm from the aggregate boundary was continuously divided into 20 strips with a width of 5 μm. For each ITZ, 30 BSE images were selected for statistical analysis of phase volume fraction within ITZ and a part of the mortar matrix.

The volume fractions of unreacted products, reaction products and pores/cracks are shown in Fig. 8.22 as stacked area graphs. The 'wall effect' can be clearly detected by observing the volume fraction change of unreacted products from the aggregate surface to mortar matrix. The content of unreacted products in the region close to the aggregate surface was lower and the percentage gradually improved to a stable value with the increase of distance to the aggregate surface. Compared to pores and cracks, reaction and unreacted product contents have a lower variability degree at different positions [49, 50]. As a result, the ITZ width can be determined according to the volume fraction of unreacted products [15, 47]. Taking the middle of the

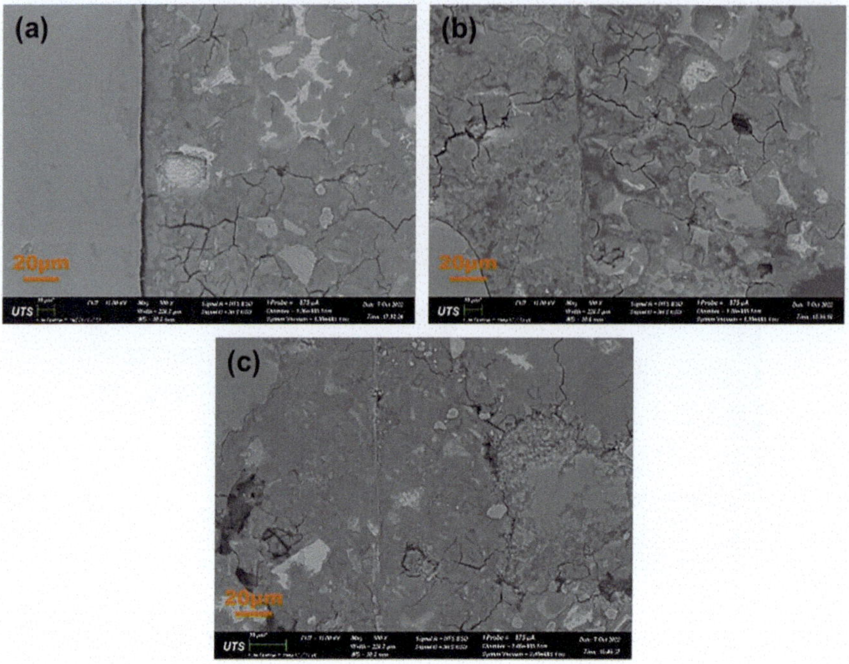

Fig. 8.20 BSE images of old and new modelled ITZs: **a** old ITZ; **b** new ITZ without WGP; **c** new ITZ with WGP

strips as the coordinate points, the region where the content of unreacted products increased significantly was identified as ITZ. Therefore, the widths of old ITZ, new ITZ with and without WGP were 52.5, 37.5 and 22.5 μm, respectively. The width of the new ITZ without WGP was narrower than the old ITZ, probably because the lower water-cement ratio designed in the new mortar improved the properties of the matrix [51]. Comparing the volume fractions of reaction and unreacted products in ITZ, it is seen that the content of unreacted products in the new ITZ without WGP was 11.98%, lower than 14.60% of the old ITZ. However, the volume fractions of reaction products were similar, both around 80.5%, because of the higher proportion of pores and cracks (7.29%) within the new ITZ without WGP than about 4.33% of the old ITZ. The surface of old mortar was looser and more porous than natural granite aggregate, and it was difficult to form a dense bonding between old and new mortar [52]. In comparison, the width, unreacted products, and void volume fractions in new ITZ containing WGP were significantly reduced to 22.5 μm, 10.27% and 4.30% respectively. Correspondingly, the proportion of reaction products increased to 85.43%, indicating that WGP can effectively reduce the volume of ITZ in RAC and improve the bonding performance between new mortar and RCA surface. By observing the volume fraction of pores and cracks in three ITZs, the volume fraction close to the aggregate surface was obviously higher than in other regions and decreased rapidly within 10 μm from the aggregate surface. This is not only due to

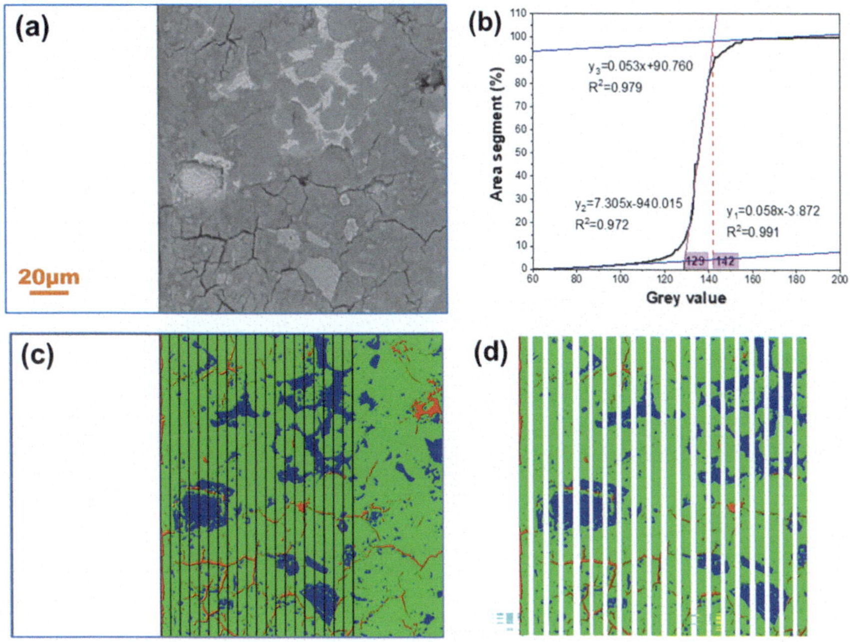

Fig. 8.21 Schematic view of BSE image processing: **a** aggregate region removal; **b** grey threshold determination; **c** phases segmentation; **d** strips division

the 'wall effect' improving local water-cement ratio, but the self-shrinkage of cement paste was more obvious than aggregate because of lower elastic modulus [53]. Some researchers have even discovered the debonding phenomenon between mortar and aggregate [54, 55]. This proves the feasibility of the image analysis technique in this work from another aspect.

To conduct a more detailed statistical analysis of the phases within the ITZ, the analysis of variance (ANOVA) proposed by R. A. Fisher was used to evaluate the difference significance between ITZ-mortar matrix and ITZ-ITZ. The F value in ANOVA represents the ratio of the between-group mean variance to the within-group mean variance [23]. The larger F value means a significant difference between the two groups of data. However, the *P* value can more intuitively reveal the difference between two groups of data. Each F value corresponds to a *P* value. A *P* value higher than 0.05 indicates that the two groups of data are similar. Otherwise, it is believed that there is a significant difference between the two groups of data, and when the *P* value is less than 0.01, it can be considered that the difference between the two groups of data is extremely significant. The statistical results of the volume fraction of phases between ITZ-mortar matrix and ITZ-ITZ are shown in Table 8.6.

The data shows that the content of reaction/unreacted products between old ITZ and old mortar matrix was significantly different. This was equally evident between the new mortar matrix and the new ITZ without WGP. The volume fractions of

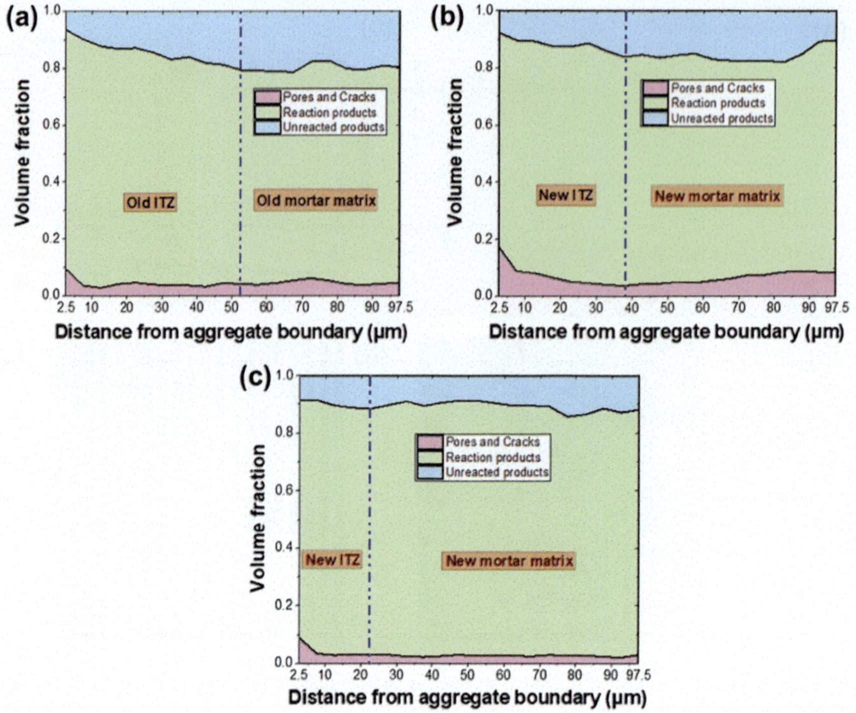

Fig. 8.22 Phases volume fraction from aggregate boundary to mortar matrix: **a** old ITZ; **b** new ITZ without WGP; **c** new ITZ with WGP

Table 8.6 Difference comparison of phases between ITZ and mortar matrix

	Pores and cracks		Reaction products		Unreacted products	
	P value	F value	*P* value	F value	*P* value	F value
Old ITZ Old mortar matrix	0.68204	0.17339	2.44×10^{-4}**	20.76867	0.00202**	13.00413
New ITZ without WGP New mortar matrix without WGP	0.66176	0.19785	0.02505*	5.97313	0.00543**	9.98152
New ITZ with WGP New mortar matrix with WGP	0.01994*	6.52166	0.2609	1.3474	0.53176	0.40653

Note The symbol '*' and '**' represent significant difference and extremely significant difference between two groups of data, respectively

Table 8.7 Difference comparison of phases between ITZs

	Pores and cracks		Reaction products		Unreacted products	
	P value	F value	*P* value	F value	*P* value	F value
Old ITZ New ITZ without WGP	0.065	3.8921	0.81863	0.05424	0.13371	2.48027
New ITZ without WGP New ITZ with WGP	0.2115	1.76007	0.00662**	11.13918	0.19851	1.87228

Note The symbol '*' and '**' represent significant difference and extremely significant difference between two groups of data, respectively

reaction and unreacted products directly reflect the performance of ITZs [56]. It can be inferred from the statistical results that the characteristics of pure cement-based ITZ were obviously weaker than the mortar matrix. In comparison, the content of reaction/unreacted products in the new mortar matrix and new ITZ with 20% WGP were similar, while the proportions of pores and cracks were obviously different, indicating that WGP can effectively refine the percentage of reaction and unreacted products, but the ability to reduce porosity in ITZ was insufficient. Nevertheless, it still can be seen from Fig. 8.22 that the volume fraction of pores and cracks in the ITZ with WGP was lower than that of old mortar and new mortar without WGP.

Table 8.7 presents the statistical results of phase differences among ITZs. The difference of phases between old ITZ and new ITZ without WGP was not obvious. However, the volume fraction of reaction products was significantly different in the new ITZs with and without WGP. Combining with Fig. 8.22, it can be seen that the increase in the volume fraction of reaction products in the new ITZ with WGP was the result of the decrease in both the proportion of pores/cracks and unreacted products. Therefore, the difference in the content of the reaction products within the two groups of new ITZ was the most obvious.

8.3.4 Chemical and Mineralogical Composition of ITZs

8.3.4.1 EDS Element Mapping Analysis

A scanning electron microscope (SEM, Zeiss EVO LS15) equipped with a BSE image detector was used to randomly extract images of ITZs. The accelerating voltage and working distance were set as 15 kV and 10 mm respectively. For each ITZ, 30 BSE images at 500× magnification were used to statistically analyse the volume fraction of constituents [15, 56]. The brightness and contrast of BSE images of each ITZ were kept consistent to reduce the workload of image analysis. EDS data were acquired at accelerating voltage consistent with BSE and lower probe current. EDS mapping was used to show phase distribution.

Considering that soda-lime WGP was used as an additive in this study, in addition to the main elements (Al, Si, Ca) belonging to cement-based materials, the distribution of Na is also presented in Fig. 8.23 by EDS mapping. The natural aggregate used in this research was granite, whose main component was silica and feldspar. The boundary between natural aggregate and old mortar can be easily distinguished by the observation of Al, Si, Na distribution in Fig. 8.23a. Granite possesses an extremely low water absorption ratio. Free water attached below the surface of granite is easy to bleed into ITZs and form a local high water-cement ratio region with less Al, Si and Ca content near the boundary of aggregate, as shown in Fig. 8.23a. This is an important reason why the cohesion of C–S–H in ITZ is weaker than that in the mortar matrix [41]. Al was mainly distributed in some phases, such as C–A–S–H, which is consistent with the distribution of Al in previous cement paste samples [39, 55]. In addition, Si and Ca were dispersed in ITZs uniformly in old ITZ and new ITZ without WGP, indicating that cement in these two regions had high degree of hydration. In comparison, there were obvious Si and Na rich zones in the new ITZ with WGP, which were incompletely reacted glass clinkers, and the proportion increased with the distance to the boundary of old mortar. This is consistent with the unreacted products distribution in Fig. 8.22c and previous studies on the fly ash clinkers distribution in ITZs of geopolymer concrete [3, 57]. According to Chen et al. [58] and Du et al. [29], the pozzolanic reaction involved in WGP is a long process, which may last for one year or more. However, this did not prevent the amorphous silica in small-size and parts of large-size glass clinkers from dissolving and reacting with CH to form more C–S–H gel, which improved the performance of ITZ [59].

8.3.4.2 EDS Element Quantitative Analysis

Stoichiometric quantification with P/B-ZAF on Si, Ca and Na was conducted to deeply analyse the element ratio within ITZs and adjacent mortar matrix. Numerous areas were randomly selected and more than 10,000 points were extracted for quantitative analysis of EDS results. The Si/Ca versus Si/Na atomic ratios of effective points in ITZs and the adjacent matrix were collected and shown in Fig. 8.24. Except several data points close to 0 may represent pores or cracks, the densest point cluster within the blue ellipse represents C–S–H gel. The centre of the ellipse was marked with a red dot. It can be clearly seen from Fig. 8.24a that the number of effective points in the old ITZ was lower than that in other regions, indicating that the constituents in the old ITZ was relatively loose. The number of effective points in the corresponding old mortar matrix increased significantly, and the average Si/Ca ratio of C–S–H was about 1.2, higher than that of the old ITZ (0.6 on average). It has been known that C–S–H exhibits a better performance with a higher Si/Ca ratio, which is consistent with the common knowledge that ITZ is the weakest region in concrete [15]. Since a lower water-cement ratio was designed to prepare new mortar, the distance between cement clinkers in the new ITZ and the matrix reduced. As a result, the

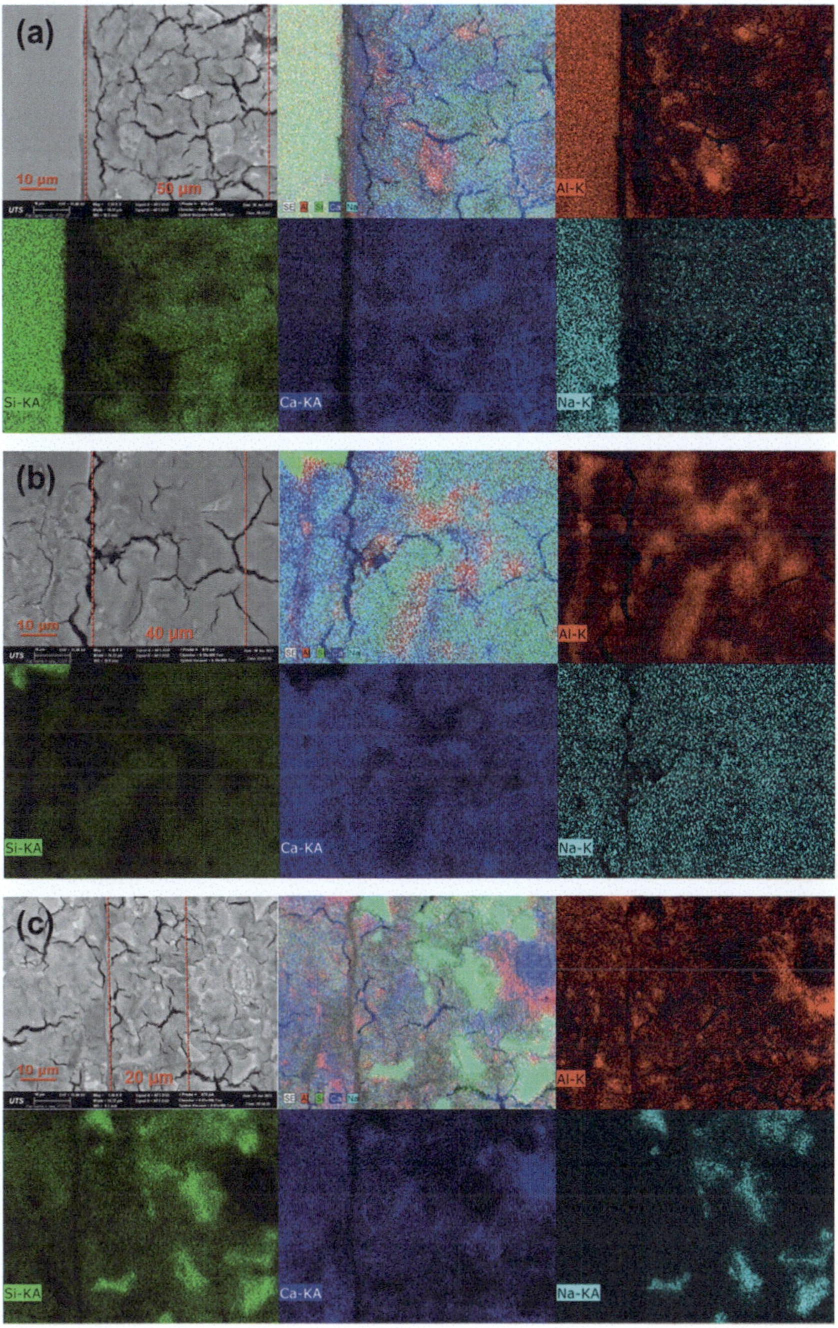

Fig. 8.23 EDS mapping analysis of ITZs: **a** old ITZ; **b** new ITZ without WGP; **c** new ITZ with WGP

number of effective points in the new ITZ and the matrix without WGP were both higher than old ITZ. In addition, the average Si/Ca ratio in the two regions reached around 1.0 in Fig. 8.24c, d. When 20% of the cement in the sample was replaced by WGP, the Si/Na ratio reduced, and the Si/Ca ratio increased significantly to around 1.5 as shown in Fig. 8.24e, f. Cement dilution as well as unreacted glass clinkers were important factors for the above element ratio changes [60]. However, it cannot be ignored that part of the WGP participated in the pozzolanic reaction, generating more C–S–H gel with high Si/Ca ratio, which enhanced the bonding strength of ITZ [35]. In addition, from the physical level, the rough and angular surface of the glass clinkers provided a natural place for the C–S–H nucleation reaction, which promotes the cement hydration, thereby increasing the overall Si/Ca ratio [61].

8.3.4.3 TG-DTG Analysis

TG tests were conducted by NETZSCH STA 449 F3 Jupiter. 20–30 mg sample powder was weighed in an alumina crucible, and then kept in a 40 °C/N_2 environment. After maintaining this environment for 30 min, the temperature in the furnace was heated to 1000 °C at a speed of 10 °C/min, and the mass loss of the sample powder between 50 and 1000 °C was obtained. The TG curves were derived to highlight the relative content differences of C–S–H, Portlandite and Calcite in different groups of samples.

The new mortar part of the sample was cut out and ground into powder for TG tests in this section and ^{29}Si NMR tests in Sect. 8.3.4.4. Although the chemical composition in the new ITZs cannot be understood exactly, the influence of WGP on the performance of ITZs can still be deduced indirectly through the analysis of the new mortar matrix.

Figure 8.25 shows the TG-DTG results for new mortar with and without WGP. The apparent mass loss at 100–150 °C, 385–440 °C and 600–690 °C represent the decomposition of C–S–H/AFt, Portlandite and Calcite, respectively [59]. From the results of DTG analysis, it can be clearly found that the content of CH in the new mortar containing WGP was lower than that of the new mortar without WGP. Moreover, the C–S–H/AFt content of new mortar containing WGP was slightly higher than that of the counterpart. These results were consistent with the EDS chemical elemental analysis in Sect. 8.3.4.2. WGP consumed CH in the mortar and generated C–S–H/AFt, which helped to improve the cohesion and density of ITZs. In addition, WGP increased the mass ratio of $CaCO_3$ in the new mortar.

To further confirm the hydration degree of the two groups of mortar, the total amount of CH (including the existing portion and the portion transformed into $CaCO_3$) and nonevaporable water were calculated by Eqs. 8.8 and 8.9 [13, 62]. The results were shown in Fig. 8.25b.

$$CH = \frac{74}{18} \cdot \frac{M_{440} - M_{385}}{M_0} + \frac{74}{44} \cdot \frac{M_{690} - M_{600}}{M_0} \tag{8.8}$$

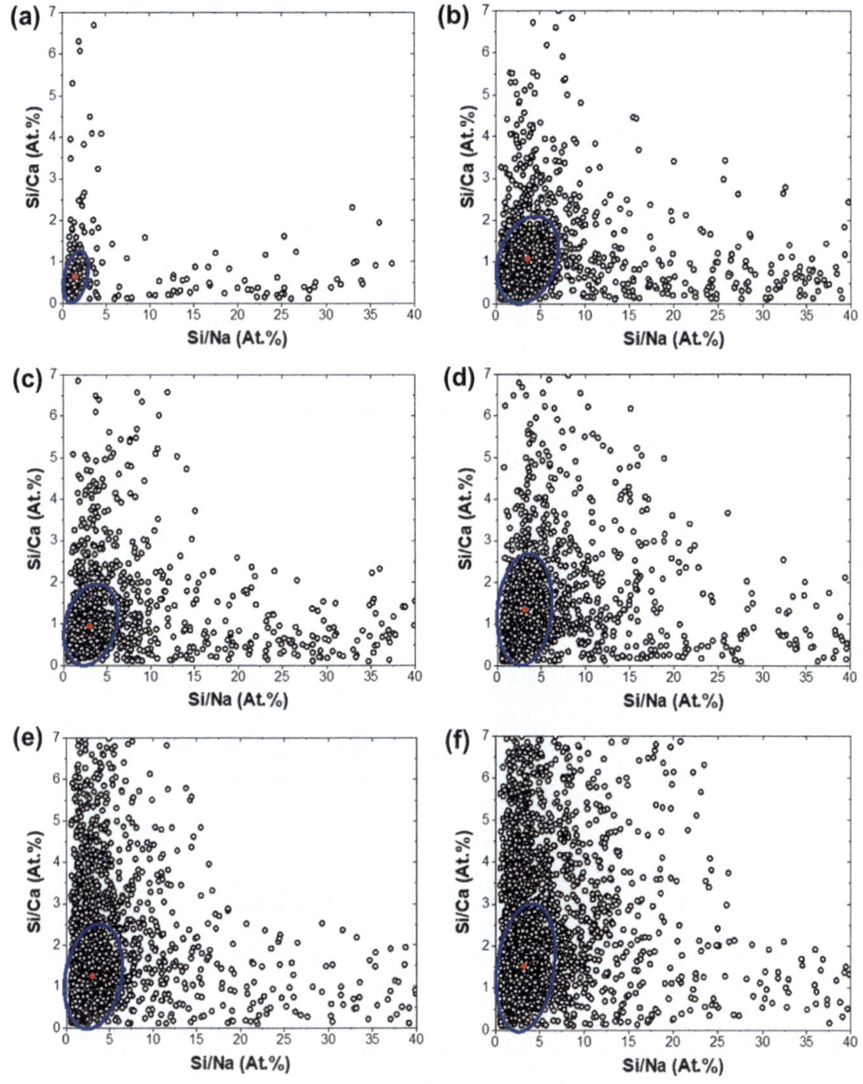

Fig. 8.24 Atomic proportions of element statistical results in ITZs and mortar matrix: **a** old ITZ; **b** old mortar matrix; **c** new ITZ without WGP; **d** new mortar matrix without WGP; **e** new ITZ with WGP; **f** new mortar matrix with WGP

$$NEW = \frac{M_{1000} - M_{105}}{M_0} \tag{8.9}$$

Among the equations, M_0 represents the initial mass of the sample; M_x represents the residual mass of the sample at x °C; The numbers 74, 18 and 44 represent the relative molecular weight of CH, water, and CO_2 respectively.

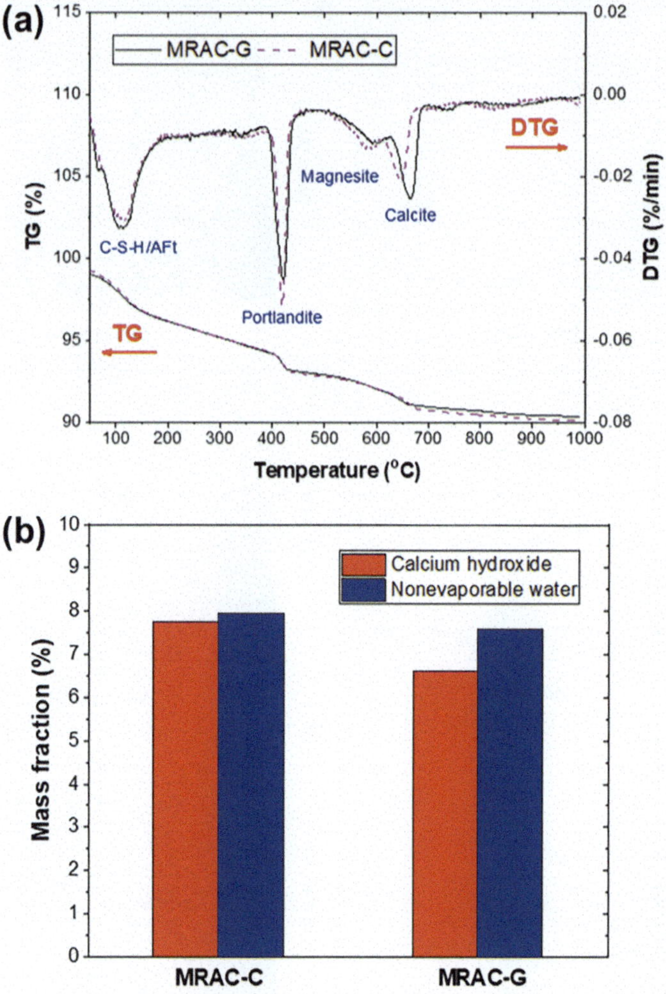

Fig. 8.25 TG and DTG results of new mortar with and without WGP: **a** TG-DTG curves; **b** quantitative analysis of TG results

According to Fig. 8.25b, the mass fractions of CH and nonevaporable water in the new mortar with WGP decreased 1.16% and 0.36%, respectively, compared with the new mortar without WGP. In cement paste containing SCM, nonevaporable water is contributed by cement hydration and replacement material [37]. It is reported that smaller amounts of additional water were required in pozzolanic reaction than cement hydration [42]. The decrease of the proportion of nonevaporable in the mortar containing WGP combined with the variation of C–S–H and CH contents proved the existence of pozzolanic reaction.

8.3.4.4 ^{29}Si NMR Characterisation Analysis

Agilent 500 MHz NMR was used to acquire solid-state ^{29}Si spectra from − 50 to − 150 ppm. The measurements were conducted in a 4 mm rotor at 71.4 kHz spinning rate and 30 s relaxation delays. The statistical deconvolution of ^{29}Si NMR spectra based on the Gaussian model was applied to obtain Q^0–Q^2 for unhydrated cement clinkers and C–S–H gel content analysis.

The ^{29}Si NMR spectra and corresponding deconvolution results of samples with and without WGP are shown in Fig. 8.26 and Table 8.8. The ^{29}Si NMR reflected the bonding relations between Si and O in cementitious materials [63]. A silicon chain can be represented by Q^n, and the number 'n' corresponds to the number of $(SiO_4)^{4-}$ units attached to the central $(SiO_4)^{4-}$ tetrahedral units. The complex degree of the silicate chain improved with the number of units, and the chemical shift became more negative [24]. Q^3–Q^6 corresponded to quartz and silicon-rich impurities. Here we only focus on the Si chains most directly related to Portland cement hydration, including Q^0, Q^1 and Q^2, whose chemical shifts were at around − 70 ppm, − 80 ppm and − 85 ppm, respectively. The Gaussian distribution area frictions of Q^3–Q^6 were deducted, and the remaining proportions were adjusted. Q^0 could be considered as the unreacted cement clinkers (C_2S and C_3S) [24]. Q^1 and Q^2 refer to the end-chain unit and middle-chain unit of C–S–H gel [24].

The peaks of soda-lime glass and unreacted cement clinkers overlapped [28]. In Fig. 8.26, the Q^0 peak of the sample containing WGP decreased, while the Q^1 and Q^2 peak improved compared with the sample powder without WGP, indicating that amorphous SiO_2 in WGP reacted with incompletely hydrated cement clinker to generate more C–S–H gel. In addition, WGP can also contribute to cement hydration to generate more C–S–H gel due to filler effect [26].

According to the relative intensities of Q^0, Q^1 and Q^2, the degree of hydration (DoH) of silicate phase, the polymerization degree (PD) of C–S–H and the mean chain length (MCL) of the silicate chain can be calculated [63]. The relative equations are as follows:

$$DoH = 100\% - Q^0 \tag{8.10}$$

$$PD = Q^1 + \frac{2 \times Q^2}{Q^1 + Q^2} \tag{8.11}$$

$$MCL = 2 \times \frac{Q^1 + Q^2}{Q^1} \tag{8.12}$$

The quantitative analysis results in Table 8.8 show that the hydration degree of silicate phase and polymerization degree of C–S–H in the sample with 20% WGP improved by 34.3% and 0.21, respectively, compared with the sample without WGP. In addition, the increase of the mean length of silicate chains in new mortar with WGP reflected the formation of denser and more stable C–S–H gel [59]. This explains why

Fig. 8.26 ^{29}Si NMR spectra of new mortar with and without WGP: **a** new mortar without WGP; **b** new mortar with WGP

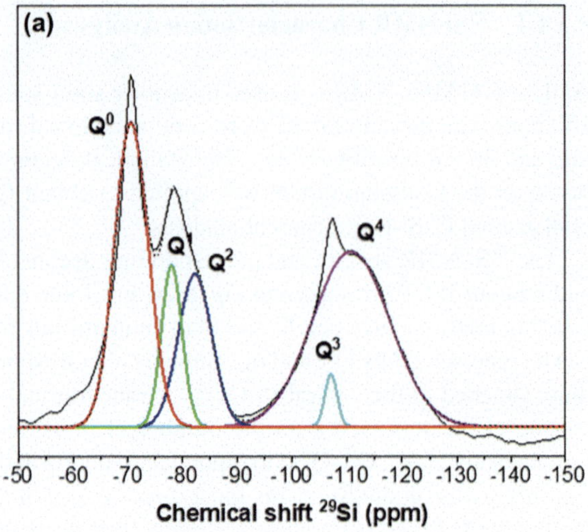

(a) New mortar without WGP

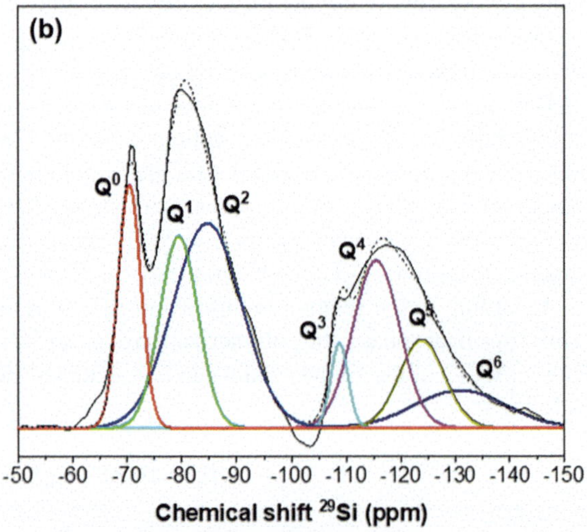

(b) New mortar with WGP

Table 8.8 Statistical deconvolution results of ^{29}Si NMR spectra

	Q^0 (%)	Q^1 (%)	Q^2 (%)	DoH (%)	PD	MCL
New mortar without WGP	55.2	17.8	27.0	44.8	1.38	5.03
New mortar with WGP	20.9	27.1	52.0	79.1	1.59	5.84

there were more hydration products and less pores/cracks in new ITZs containing WGP in image analysis. To summarize, the TG-DTG and ^{29}Si NMR test results supported the deductions in Sects. 8.3.3 and 8.3.4.2 about the functions of WGP in new ITZs.

8.3.5 Micromechanical Properties of ITZs

Nanoindentation test procedures follow Sect. 8.2.5.1. The nanoindentation test results with ITZ widths identified in Sect. 8.3.3 are shown in Fig. 8.27. The NCA adopted in this study was granite with the elastic modulus exceeding 160 GPa and the hardness over 25 GPa. It is worth noting that there was a clear drop in micromechanical properties on the edge of the modelled NCA, but not where the old mortar matrix adjacent to new ITZs. This could be explained that the elastic modulus and hardness of NCA chosen in this experiment were significantly higher than old mortar matrix. Old ITZ cannot provide enough restraint on the NCA edge. As a result, the value of the third column in Fig. 8.27a, b NAC region was evidently lower than the first two columns. It can also be found that the average elastic modulus and hardness of old ITZ were slightly lower than old mortar matrix, and the corresponding average elastic modulus and hardness ratios were 0.86 and 0.88, respectively. This is because of the 'wall effect' leading to a high water-cement ratio and low strength of old ITZ. Among them, the elastic modulus and hardness of old ITZ at 4 μm away from the surface of NCA were obviously lower than those far away from the surface of NCA. This corresponded to the local high water-cement ratio region near the NCA boundary observed in Fig. 8.24a.

As for MRAC without WGP, the average elastic modulus and hardness of new ITZ were 11.89 GPa and 0.29 GPa, respectively, which were slightly higher than those of the adjacent old mortar matrix (8.2 GPa and 0.2 GPa respectively). Although the 'wall effect' and release of pore water from the surface of RCA weakened the strength of the new ITZ, the lower initial water-cement ratio of the new ITZ still played a dominant role. Similar to the old ITZ, the micromechanical properties of the new ITZ were lower than those of the new mortar matrix (average elastic modulus ratio and hardness ratio were 0.69 and 0.71, respectively). In addition, the average elastic modulus and hardness of the new ITZ increased by 14.7% and 9.5%, respectively, compared with the old ITZ (10.37 GPa and 0.26 GPa), which confirmed the proposal to satisfy the strength requirements of RAC by reducing the water-cement ratio of the new mortar in real infrastructure construction.

As shown in Fig. 8.27e, f the average elastic modulus of the new ITZ with WGP and the adjacent old mortar matrix was 14.11 GPa and 12.93 GPa, respectively. The average hardness was 0.34 and 0.35 GPa, respectively. As expected, WGP participated in the pozzolanic reaction, generating more C–S–H gels with a high Si/Ca ratio, which enhanced the strength and compactness of the new ITZ. It is worth noting that the micromechanical properties of the old mortar adjacent to the new ITZ with WGP were also enhanced compared to the old mortar adjacent to the new ITZ

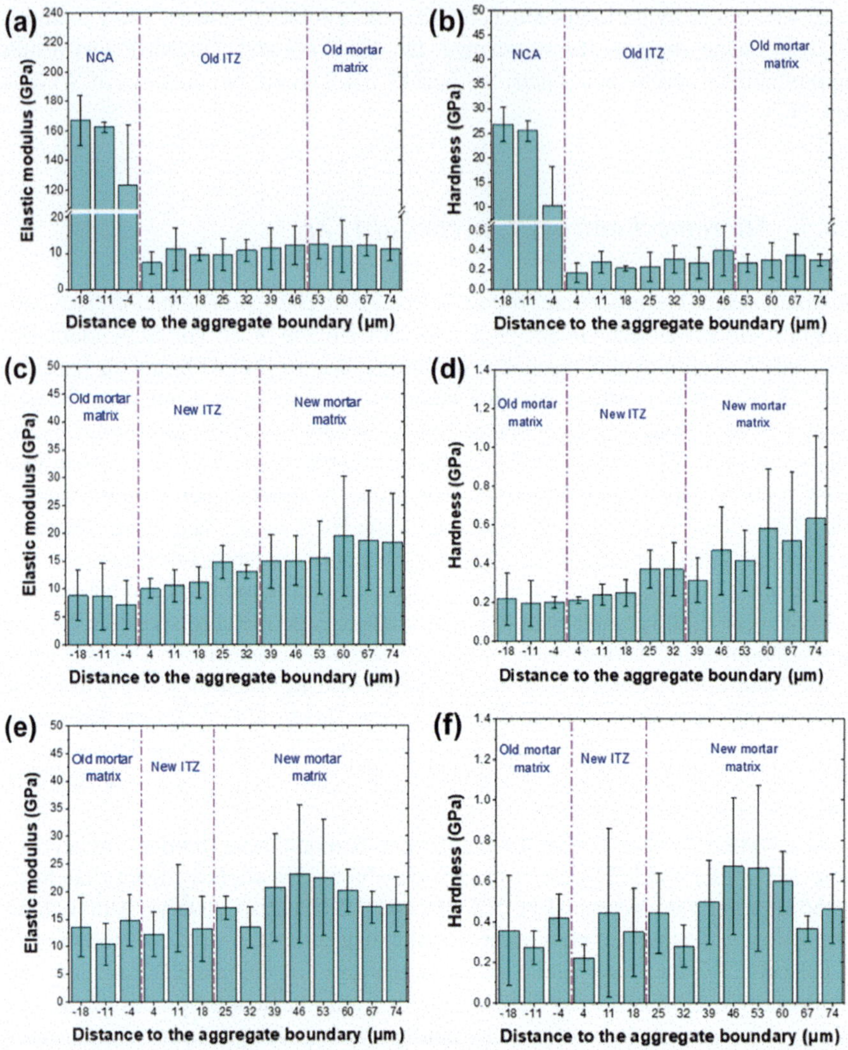

Fig. 8.27 Grid nanoindentation test results of modelled ITZs: **a** modulus of old ITZ; **b** hardness of old ITZ; **c** modulus of new ITZ without WGP; **d** hardness of new ITZ without WGP; **e** modulus of new ITZ with 20% WGP; **f** hardness of new ITZ with 20% WGP

without WGP (average elastic modulus and hardness increased by 57.9% and 70.8% respectively). This can be partly attributed to the dissemination effect of amorphous silica in the new mortar with WGP. SiO_2 in the new mortar penetrated into the adjacent old mortar region and proceeded the pozzolanic reaction, so that the local old mortar performance was improved and achieved a tighter bond between old and new mortar. Furthermore, the 'wall effect' was also evident in the MRAC with WGP. The average elastic modulus and average hardness ratio of the new ITZ to the new mortar matrix were 0.74 and 0.68, respectively. In addition, the data fluctuation of columns in the new ITZ containing WGP was more significant than old ITZ and new ITZ without WGP. A small amount of unreacted WGP particles may still hid under the observation surface. According to Bosque et al. [7], the nanoindentation elastic modulus of glass was up to 86.17 GPa, which was significantly higher than that of cement products. The heterogeneity of cement paste containing WGP could pose a certain degree of influence on the results of nanoindentation test.

The nanoindentation test results were collected and presented in Fig. 8.28 in the form of frequency distribution histograms and contour maps. The elastic modulus and hardness of ITZ roughly follows the normal distribution. Therefore, the Gaussian model was used to fit the frequency histogram of the nanoindentation test results. The elastic modulus and hardness of the old ITZ were mainly distributed around 5–15 GPa and 0.2 GPa. The elastic modulus and hardness of the new ITZ without WGP increased slightly compared with the old ITZ. The values in the new ITZ with WGP were scattered in a larger range, and the overall elastic modulus and hardness were higher than those of the old ITZ and the new ITZ without WGP. In the contour maps of new ITZs with and without WGP, there were points with higher elastic modulus and hardness than surrounding area. This revealed the existence of CH and glass clinkers in the sample.

8.4 Strengthening Mechanism of ITZs

Two types of modelled aggregate concrete samples were fabricated and three ITZs were simulated in this chapter. In order to conduct a more comprehensive and detailed comparative analysis, the properties of the three ITZs were investigated from microscopic morphology, chemical and micromechanical properties. The sketch of three ITZs are shown in Fig. 8.29.

The test results of the old ITZ were basically consistent with the performance of the traditional ITZs [54, 55]. The width of the old ITZ was around 50 μm. In addition, the old ITZ had a higher porosity than the old mortar matrix as shown in Fig. 8.29a. As a result, the cohesion of the old ITZ was the worst in the whole old mortar area. Meanwhile, due to extremely poor water absorption of granite, the free water in the mortar and attached on the surface of the natural aggregate was easy to accumulate on the boundary, forming a local cement paste region with a high water-cement ratio. These factors were shown in the nanoindentation tests as the lower average elastic modulus and hardness of the old ITZ than that of the mortar matrix. Moreover, the region closes to the natural aggregate surface along the old ITZ had the weakest mechanical properties. In some extreme cases [7, 16], a debonding phenomenon occurred at the ITZs. On the one hand, it may be due to the non-uniform shrinkage between the natural aggregate and the mortar, leading to gaps along the aggregate boundary [64]. On the other hand, the soft character of cement paste near the aggregate boundary was likely to be removed during the grinding and polishing process, forming a gap where the cement paste seemed to be completely

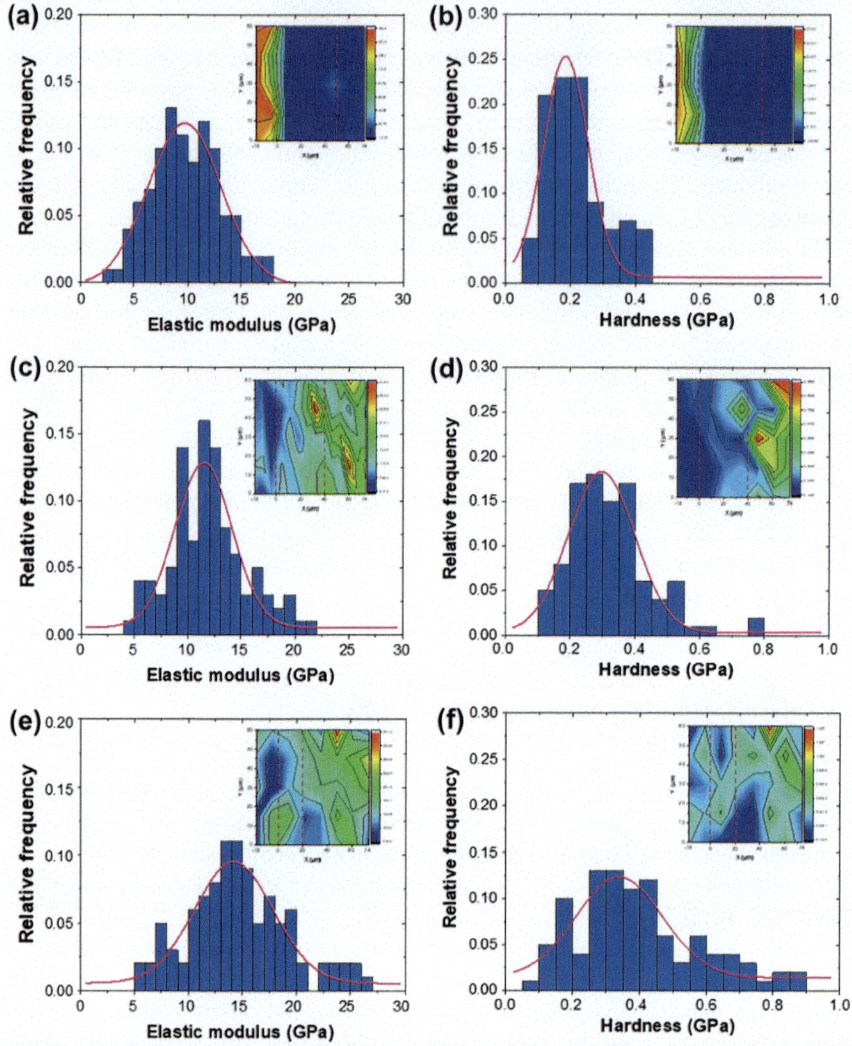

Fig. 8.28 Distribution of grid nanoindentation results in modelled ITZs: **a** old ITZ; **b** new ITZ without WGP; **c** new ITZ with WGP

peeled off from the aggregate under an electron microscope [42]. However, in some studies, there was also the case that the ITZ microstructure was not different from the cement paste and the micromechanical properties of ITZ were even stronger than the mortar matrix [5, 7]. The authors who came to these conclusions explained that the incompletely hydrated CH in the cement paste filled the pores within the ITZ and exhibited higher strength in nanoindentation tests [5, 7]. In fact, CH has almost no cohesion properties and it would not affect the final result of the ITZ being the weakest zone in ordinary concrete [10, 12]. In this study, the old mortar was cured

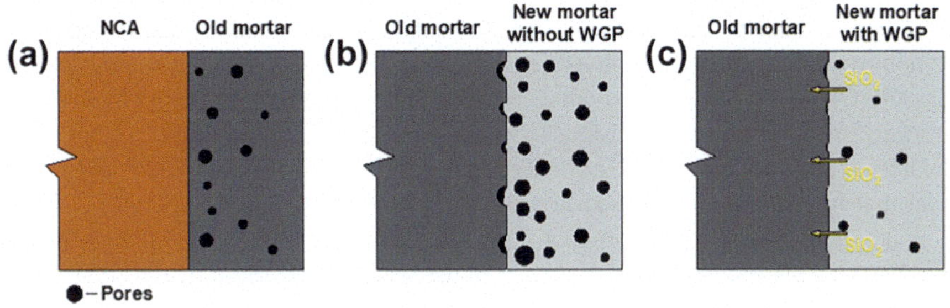

Fig. 8.29 ITZs microstructure general view and WGP enhancement mechanism: **a** old ITZ; **b** new ITZ without WGP; **c** new ITZ with WGP

for more than two months under the standard environment, and the cement paste achieved a higher hydration degree. Therefore, the difference in micromechanical properties between the old ITZ and the old mortar matrix was more obvious.

New mortar was prepared with a lower water-cement ratio than old mortar in this study. The effect of water-cement ratio on the mechanical properties and durability of RAC corresponds to the microscopic performance [35]. In Fig. 8.25, the number of effective points and Si/Ca ratio of C–S–H in the new ITZ without WGP was slightly higher than those in the old ITZ, indicating stronger C–S–H cohesion strength [57]. In the nanoindentation test, the new ITZ without WGP also exhibited higher elastic modulus and hardness than the old ITZ. The image analysis in Sect. 8.3.3 also shows that the width of the new ITZ was slightly lower than that of the old ITZ. These test results were all due to the low water-cement ratio reducing the distance between cement clinkers. However, in Fig. 8.23, the volume fraction of pores and cracks in the new ITZ was higher than that in the old ITZ. This seems to contradict the above conclusion of high cohesion strength of C–S–H in the new ITZ without WGP. For the new ITZ, in fact, due to the microstructure characteristics of the cement mortar, it was impossible to achieve the same degree of flatness on the old mortar as the modelled NCA surface by only grinding and polishing work [9]. There would inevitably be a large number of pores and gaps on the surface of the old mortar. In order to fill the pores on the surface of the old mortar, the new ITZs were likely to form more pores and cracks, which reduced the effective bonding area between the old and new mortar in Fig. 8.29b. This is consistent with the theory that when fresh concrete fills moulds, a large number of pores will be generated inside, and the air bubbles need to be removed by vibration. However, these pores may not be detected in nanoindentation tests. Because the regions with high porosity and large cracks were avoided in nanoindentation. In addition, when the indenter was stuck into a hole, the facility may not obtain a value or the load-indentation depth curve seriously deformed [65]. As a result, the nanomechanical properties of the new ITZ still remained at a high level, although possessed a high porosity.

For the new ITZ with WGP, all test results show a significant improvement in micro performance compared to the old ITZ and the new ITZ without WGP. Searching the previous research results and combining the conclusions obtained from the TG and

NMR test results in this study, it can be easily concluded that WGP accelerated cement hydration from both physical and chemical level to generate more high Si/Ca ratio C–S–H gel, which provided enough bonding strength between old and new mortar [56]. In addition, according to Fig. 8.23, it can be found that different from the new ITZ without WGP, the porosity of the new ITZ after adding WGP was lower than that of the old ITZ. In addition to the pozzolanic reaction induced by WGP to make the new ITZ more compact, free silica in the new ITZ with WGP released into the adjacent old mortar, participated in the secondary hydration, and refined the flatness of the old mortar surface, as shown in Fig. 8.29c. Therefore, the negative influence of the surface porosity of the old mortar on the microstructure of the new ITZ was alleviated. Meanwhile, it has to be admitted that the pozzolanic activity of WGP was poor, and it need a long time to finish pozzolanic reaction according to some studies [66]. This is somewhat reflected in the nanoindentation test results in Fig. 8.27e, f. The influence of incompletely reacted glass particles on the porosity and micromechanical properties of new ITZ cannot be ignored.

In general, for the traditional ITZ, that is, the old ITZ in this study, the bonding strength of cement mortar to natural aggregate can be evaluated only based on the compactness or micromechanical properties of ITZs. Because the strong cohesiveness of C–S–H gel corresponds to the high compactness. However, for the new ITZs in RAC, there could be conditions with high C–S–H cohesion strength but high porosity at the same time. It is unreliable to only rely on a single indicator to evaluate the bonding strength between old and new mortar. Besides, the weak bonding of C–S–H gel to unreacted clinkers should also be considered. In future research, it is necessary to comprehensively evaluate the bonding performance between old and new mortar through effective contact area and cohesion strength of C–S–H gel.

8.5 Conclusion

In this study, MAC and MRAC samples were fabricated to simulate regular ITZs. The performance of old and new ITZs was compared by using statistical techniques and multiscale experimental methods. Meanwhile, the improvement efficiency of WGP on the ITZs attached to modelled NCA and RCA was evaluated. The related conclusions can be drawn up as below:

(1) The regular natural aggregate-new mortar, natural aggregate-old mortar and old mortar-new mortar interfaces produced in this study can be clearly observed under an electron microscope. Modelled ITZ effectively reduced unnecessary workload as well as achieved satisfactory accuracy of test results. The sample preparation procedure in this study is sufficient to meet the testing requirements for BSE and nanoindentation tests.

(2) The 20% WGP can greatly reduce the width of ITZ, featured by the reduction of unreacted cement particles and the increase of hydration products. The volume fraction of cracks and pores in the new ITZ without WGP was 7.29%, which was

higher than 4.33% in the old ITZ, while the content of the reaction was similar, around 80.5%. In comparison, the proportion of pores and unreacted products within the new ITZ with WGP decreased to 4.30% and 10.27%, and the content of reaction products increased to 85.43%. The proportion of pores and cracks was greatly reduced in the ITZ and mortar matrix containing WGP, because the WGP participated in the pozzolanic reaction to generate more C–S–H gels, and the unreacted WGP played a role in physically filling pores.

(3) The volume fractions difference of reaction and unreacted products were extremely significant for the old and new ITZ without WGP compared with the corresponding mortar matrix (*P* values lower than 0.01), while for the new ITZ with WGP, only the porosity was significantly different from that in the new mortar matrix (*P* value was between 0.01 and 0.05). In the comparison between different ITZs, WGP can increase the difference significance in the reaction products.

(4) The cement hydration can be evaluated according to the uniform distribution of Ca and Si elements in EDS mapping. Moreover, the unreacted WGP can also be captured through the distribution of Si and Na elements. In the ITZs without WGP, the Si/Ca ratio of C–S–H was around 0.5. The Si/Ca ratio in the new ITZ without WGP was around 1.0, which was slightly higher than that of the old ITZ (0.6 on average). WGP can effectively promote hydration and pozzolanic reaction from physical and chemical level to generate a large amount of C–S–H with a Si/Ca ratio of around 1.5, which provided stronger cohesion strength for the ITZ.

(5) XRD and FTIR results show that the WGP consumed the CH and generated additional C–S–H gels. In addition, the deconvolution results of the FTIR spectrum between 700 and 1300 cm^{-1} indicate that the high content of soluble alkali metal oxides in the WGP improved the pH of the pore water, which accelerated the carbonisation speed of cement paste and slowed down the formation rate of ettringite.

(6) Nanoindentation tests well reflected the influence of the 'wall effect' on the micromechanical properties of ITZ. The average elastic modulus and hardness of ITZs were lower than those of the corresponding mortar matrix. The elastic modulus of ITZ and mortar matrix without WGP were between 5–20 GPa and 5–23 GPa, respectively, and the frequency density distribution roughly presents a standard normal distribution. However, two relatively independent components appeared between 10–20 GPa and 19–27 GPa in elastic modulus FDH of ITZ and mortar matrix containing WGP in MAC, implying the C–S–H gel and mixed phase of C–S–H and unhydrated or partially hydrated WGP, respectively. The data volume of the nanoscratch was higher than nanoindentation. Low-density C–S–H and high-density C–S–H were found in fracture toughness FDH of ITZ without WGP by deconvolution analysis. In contrast, the distribution of C–S–H in the mortar matrix without WGP was uniform. The ITZ and mortar matrix with WGP can also be separated into two components according to the fracture toughness FDH: C–S–H and mixed phase of C–S–H and WGP.

(7) In the comparative analysis of nanoindentation hardness and nanoscratch hardness, nanoindentation results of ITZ and mortar matrix without WGP were consistent with the nanoscratch ones. However, the nanoindentation hardness of ITZs and mortar matrix with WGP was slightly higher than nanoscratch one because the scratch area of the nanoscratch was much higher than that of nanoindentation. Therefore, nanoscratch is more likely to reflect the mixed nano/micromechanical properties of multiple phases. For construction materials with high homogeneity, nanoscratch can greatly improve testing efficiency. However, for extremely inhomogeneous materials, the accuracy of nanoindentation was higher based on enough indents.

(8) There were more exposed pores on the surface of the old mortar, which led to the fact that the bond between the old and new mortar cannot be as tight as that between old mortar and natural aggregate. Moreover, a large number of pores were formed in the new ITZ without WGP. After adding 20% WGP to the new mortar, amorphous silica improved the performance of the new ITZ as well as penetrated into the adjacent old mortar to accelerate the pozzolanic reaction. The surface structure of the old mortar was thus also improved. To evaluate the bonding strength between old and new mortar in RAC, not only the cohesion of C–S–H gel, but also the effective contact area should be considered.

References

1. W. Li, Z. Luo, Y. Gan, K. Wang, S.P. Shah, Nanoscratch on mechanical properties of interfacial transition zones (ITZs) in fly ash-based geopolymer composites. Compos. Sci. Technol. **214** (2021)
2. N. Arabi, H. Meftah, H. Amara, O. Kebaïli, L. Berredjem, Valorization of recycled materials in development of self-compacting concrete: mixing recycled concrete aggregates—Windshield waste glass aggregates. Constr. Build. Mater. **209**, 364–376 (2019)
3. Z. Tang, Y. Hu, V.W.Y. Tam, W. Li, Uniaxial compressive behaviors of fly ash/slag-based geopolymeric concrete with recycled aggregates. Cem. Concr. Compos. **104** (2019)
4. J.-X. Lu, P. Shen, H. Zheng, B. Zhan, H.A. Ali, P. He, C.S. Poon, Synergetic recycling of waste glass and recycled aggregates in cement mortars: physical, durability and microstructure performance. Cement Concr. Compos. **113**, 103632 (2020)
5. J. Xiao, W. Li, Z. Sun, D.A. Lange, S.P. Shah, Properties of interfacial transition zones in recycled aggregate concrete tested by nanoindentation. Cement Concr. Compos. **37**, 276–292 (2013)
6. W. Li, J. Xiao, Z. Sun, S. Kawashima, S.P. Shah, Interfacial transition zones in recycled aggregate concrete with different mixing approaches. Constr. Build. Mater. **35**, 1045–1055 (2012)
7. I.F. Sáez del Bosque, W. Zhu, T. Howind, A. Matías, M.I. Sánchez de Rojas, C. Medina, Properties of interfacial transition zones (ITZs) in concrete containing recycled mixed aggregate. Cement Concr. Compos. **81**, 25–34 (2017)
8. Z. Luo, W. Li, K. Wang, S.P. Shah, Research progress in advanced nanomechanical characterization of cement-based materials. Cement Concr. Compos. **94**, 277–295 (2018)
9. Z. Luo, W. Li, K. Wang, S.P. Shah, D. Sheng, Nano/micromechanical characterisation and image analysis on the properties and heterogeneity of ITZs in geopolymer concrete. Cem. Concr. Res. **152**, 106677 (2022)

10. A. Hosan, F.U.A. Shaikh, P. Sarker, F. Aslani, Nano- and micro-scale characterisation of interfacial transition zone (ITZ) of high volume slag and slag-fly ash blended concretes containing nano SiO_2 and nano $CaCO_3$. Constr. Build. Mater. **269**, 121311 (2021)
11. Z. Luo, W. Li, K. Wang, A. Castel, S.P. Shah, Comparison on the properties of ITZs in fly ash-based geopolymer and Portland cement concretes with equivalent flowability. Cem. Concr. Res. **143** (2021)
12. J. Xiao, W. Li, D.J. Corr, S.P. Shah, Effects of interfacial transition zones on the stress–strain behavior of modeled recycled aggregate concrete. Cem. Concr. Res. **52**, 82–99 (2013)
13. C. Fu, Y. Ling, K. Wang, An innovation study on chloride and oxygen diffusions in simulated interfacial transition zone of cementitious material. Cement Concr. Compos. **110**, 103585 (2020)
14. B.J. Zhan, D.X. Xuan, C.S. Poon, K.L. Scrivener, Characterization of interfacial transition zone in concrete prepared with carbonated modeled recycled concrete aggregates. Cem. Concr. Res. **136** (2020)
15. P. Vargas, O. Restrepo-Baena, J.I. Tobón, Microstructural analysis of interfacial transition zone (ITZ) and its impact on the compressive strength of lightweight concretes. Constr. Build. Mater. **137**, 381–389 (2017)
16. S. Diamond, J. Huang, The ITZ in concrete—a different view based on image analysis and SEM observations. Cement Concr. Compos. **23**(2–3), 179–188 (2001)
17. K. Lyu, W. She, C. Miao, H. Chang, Y. Gu, Quantitative characterization of pore morphology in hardened cement paste via SEM-BSE image analysis. Constr. Build. Mater. **202**, 589–602 (2019)
18. K. Lyu, E.J. Garboczi, W. She, C. Miao, The effect of rough versus smooth aggregate surfaces on the characteristics of the interfacial transition zone. Cem. Concr. Compos. **99**, 49–61 (2019)
19. H.S. Wong, M.K. Head, N.R. Buenfeld, Pore segmentation of cement-based materials from backscattered electron images. Cem. Concr. Res. **36**(6), 1083–1090 (2006)
20. H.S. Wong, N.R. Buenfeld, Euclidean distance mapping for computing microstructural gradients at interfaces in composite materials. Cem. Concr. Res. **36**(6), 1091–1097 (2006)
21. Z. Luo, W. Li, Y. Gan, K. Mendu, S.P. Shah, Applying grid nanoindentation and maximum likelihood estimation for N-A-S-H gel in geopolymer paste: investigation and discussion. Cem. Concr. Res. **135** (2020)
22. K.L. Scrivener, Backscattered electron imaging of cementitious microstructures: understanding and quantification. Cement Concr. Compos. **26**(8), 935–945 (2004)
23. G. Fang, M. Zhang, The evolution of interfacial transition zone in alkali-activated fly ash-slag concrete. Cem. Concr. Res. **129**, 105963 (2020)
24. J. Hjorth, J. Skibsted, H.J. Jakobsen, 29Si MAS NMR studies of portland cement components and effects of microsilica on the hydration reaction. Cem. Concr. Res. **18**(5), 789–798 (1988)
25. B. Lothenbach, K. Scrivener, R.D. Hooton, Supplementary cementitious materials. Cem. Concr. Res. **41**(12), 1244–1256 (2011)
26. A. Omran, A. Tagnit-Hamou, Performance of glass-powder concrete in field applications. Constr. Build. Mater. **109**, 84–95 (2016)
27. R.-U.-D. Nassar, P. Soroushian, Strength and durability of recycled aggregate concrete containing milled glass as partial replacement for cement. Constr. Build. Mater. **29**, 368–377 (2012)
28. M. Mirzahosseini, K.A. Riding, Influence of different particle sizes on reactivity of finely ground glass as supplementary cementitious material (SCM). Cement Concr. Compos. **56**, 95–105 (2015)
29. H. Du, K.H. Tan, Properties of high volume glass powder concrete. Cement Concr. Compos. **75**, 22–29 (2017)
30. W. Kunther, Z. Dai, J. Skibsted, Thermodynamic modeling of hydrated white Portland cement–metakaolin–limestone blends utilizing hydration kinetics from 29Si MAS NMR spectroscopy. Cem. Concr. Res. **86**, 29–41 (2016)
31. W. Kunther, S. Ferreiro, J. Skibsted, Influence of the Ca/Si ratio on the compressive strength of cementitious calcium-silicate-hydrate binders. J. Mater. Chem. A **5**(33), 17401–17412 (2017)

32. F. Qu, W. Li, Y. Guo, S. Zhang, J.L. Zhou, K. Wang, Chloride-binding capacity of cement-GGBFS-nanosilica composites under seawater chloride-rich environment. Constr. Build. Mater. **342**, 127890 (2022)
33. Z.Z. Ismail, E.A. Al-Hashmi, Recycling of waste glass as a partial replacement for fine aggregate in concrete. Waste Manage. **29**(2), 655–659 (2009)
34. I. García Lodeiro, D.E. Macphee, A. Palomo, A. Fernández-Jiménez, Effect of alkalis on fresh C–S–H gels. FTIR analysis. Cem. Concr. Res. **39**(3), 147–153 (2009)
35. C. Lu, Z. Zhang, J. Hu, Q. Yu, C. Shi, Effects of anionic species of activators on the rheological properties and early gel characteristics of alkali-activated slag paste. Cem. Concr. Res. **162**, 106968 (2022)
36. D. Mostofinejad, S.M. Hosseini, F. Nosouhian, T. Ozbakkaloglu, B. Nader Tehrani, Durability of concrete containing recycled concrete coarse and fine aggregates and milled waste glass in magnesium sulfate environment. J. Build. Eng. **29**, 101182 (2020)
37. T. Ye, J. Xiao, W. Zhao, Z. Duan, Y. Xu, Combined use of recycled concrete aggregate and glass cullet in mortar: strength, alkali expansion and chemical compositions. J. Build. Eng. **55**, 104721 (2022)
38. F. Qu, W. Li, K. Wang, S. Zhang, D. Sheng, Performance deterioration of fly ash/slag-based geopolymer composites subjected to coupled cyclic preloading and sulfuric acid attack. J. Clean. Prod. **321**, 128942 (2021)
39. P. Li, W. Li, T. Yu, F. Qu, V.W.Y. Tam, Investigation on early-age hydration, mechanical properties and microstructure of seawater sea sand cement mortar. Constr. Build. Mater. **249**, 118776 (2020)
40. Z. Luo, W. Li, Y. Gan, K. Mendu, S.P. Shah, Maximum likelihood estimation for nanoindentation on sodium aluminosilicate hydrate gel of geopolymer under different silica modulus and curing conditions. Compos. Part B: Eng. **198** (2020)
41. M. Khedmati, Y.R. Kim, J.A. Turner, Investigation of the interphase between recycled aggregates and cementitious binding materials using integrated microstructural-nanomechanical-chemical characterization. Compos. B Eng. **158**, 218–229 (2019)
42. M. Khedmati, Y.R. Kim, J.A. Turner, H. Alanazi, C. Nguyen, An integrated microstructural-nanomechanical-chemical approach to examine material-specific characteristics of cementitious interphase regions. Mater Charact **138**, 154–164 (2018)
43. X. Yan, H. Zhao, Y. Shen, H. Ding, Material removal mechanism and penetration depth modeling of CF/PEEK composites under nano-scratching. Compos. Sci. Technol. **227**, 109566 (2022)
44. E. Rayón, M.P. Arrieta, T. Pasíes, J. López, J.L. Jordá, Enhancing the mechanical features of clay surfaces by the absorption of nano-SiO_2 particles in aqueous media. Case of study on bronze age clay objects. Cem. Concr. Compos. **93**, 107–117 (2018)
45. Y. Yin, Y. Zhao, K. En Koey, Q. Tan, M.-X. Zhang, H. Huang, In-situ synthesized age-hardenable high-entropy composites with superior wear resistance. Compos. Part B: Eng. **235**, 109795 (2022)
46. Y. Kim, A. Hanif, M. Usman, W. Park, Influence of bonded mortar of recycled concrete aggregates on interfacial characteristics—porosity assessment based on pore segmentation from backscattered electron image analysis. Constr. Build. Mater. **212**, 149–163 (2019)
47. P.R. Rangaraju, J. Olek, S. Diamond, An investigation into the influence of inter-aggregate spacing and the extent of the ITZ on properties of Portland cement concretes. Cem. Concr. Res. **40**(11), 1601–1608 (2010)
48. H.S. Wong, N.R. Buenfeld, Determining the water-cement ratio, cement content, water content and degree of hydration of hardened cement paste: method development and validation on paste samples. Cem. Concr. Res. **39**(10), 957–965 (2009)
49. C. Camposeco-Negrete, Optimization of cutting parameters for minimizing energy consumption in turning of AISI 6061 T6 using Taguchi methodology and ANOVA. J. Clean. Prod. **53**, 195–203 (2013)
50. G.L. Golewski, An extensive investigations on fracture parameters of concretes based on quaternary binders (QBC) by means of the DIC technique. Constr. Build. Mater. **351** (2022)

51. J. Thomas, N.N. Thaickavil, P.M. Wilson, Strength and durability of concrete containing recycled concrete aggregates. J. Build. Eng. **19**, 349–365 (2018)
52. B. Lei, W. Li, Z. Tang, V.W.Y. Tam, Z. Sun, Durability of recycled aggregate concrete under coupling mechanical loading and freeze-thaw cycle in salt-solution. Constr. Build. Mater. **163**, 840–849 (2018)
53. G.L. Golewski, B. Szostak, Strength and microstructure of composites with cement matrixes modified by fly ash and active seeds of C-S-H phase. Struct. Eng. Mech. **82**(4), 543–556 (2022)
54. Y. Wei, W. Kong, Y. Wang, A. Sha, Multifunctional application of nanoscratch technique to characterize cementitious materials. Cem. Concr. Res. **140**, 106318 (2021)
55. J. Xu, B. Wang, J. Zuo, Modification effects of nanosilica on the interfacial transition zone in concrete: a multiscale approach. Cement Concr. Compos. **81**, 1–10 (2017)
56. A.-T. Akono, F.-J. Ulm, Microscopic toughness of viscous solids via scratching: from amorphous polymers to gas shale. J. Nanomech. Micromech. **7**(3), 04017009 (2017)
57. Y. Hu, Z. Tang, W. Li, Y. Li, V.W.Y. Tam, Physical-mechanical properties of fly ash/GGBFS geopolymer composites with recycled aggregates. Constr. Build. Mater. **226**, 139–151 (2019)
58. C.H. Chen, R. Huang, J.K. Wu, C.C. Yang, Waste e-glass particles used in cementitious mixtures. Cem. Concr. Res. **36**(3), 449–456 (2006)
59. M. Mejdi, W. Wilson, M. Saillio, T. Chaussadent, L. Divet, A. Tagnit-Hamou, Investigating the pozzolanic reaction of post-consumption glass powder and the role of portlandite in the formation of sodium-rich C-S-H. Cem. Concr. Res. **123**, 105790 (2019)
60. L. Li, J. Lu, P. Shen, K. Sun, L.E.L. Pua, J. Xiao, C.S. Poon, Roles of recycled fine aggregate and carbonated recycled fine aggregate in alkali-activated slag and glass powder mortar. Constr. Build. Mater. **364** (2023)
61. V. Corinaldesi, G. Moriconi, Influence of mineral additions on the performance of 100% recycled aggregate concrete. Constr. Build. Mater. **23**(8), 2869–2876 (2009)
62. L. Guo, J. Wu, H. Wang, Mechanical and perceptual characterization of ultra-high-performance cement-based composites with silane-treated graphene nano-platelets. Constr. Build. Mater. **240** (2020)
63. L. Chen, Y. Zhang, L. Wang, S. Ruan, J. Chen, H. Li, J. Yang, V. Mechtcherine, D.C.W. Tsang, Biochar-augmented carbon-negative concrete. Chem. Eng. J. **431** (2022)
64. G.L. Golewski, An assessment of microcracks in the interfacial transition zone of durable concrete composites with fly ash additives. Compos. Struct. **200**, 515–520 (2018)
65. C.G. Hoover, F.-J. Ulm, Experimental chemo-mechanics of early-age fracture properties of cement paste. Cem. Concr. Res. **75**, 42–52 (2015)
66. M. Mejdi, W. Wilson, M. Saillio, T. Chaussadent, L. Divet, A. Tagnit-Hamou, Hydration and microstructure of glass powder cement pastes—a multi-technique investigation. Cem. Concr. Res. **151**, 106610 (2022)

Chapter 9
Nano/Microscopic Investigation on the Interfacial Transition Zones in Modelled Geopolymer Concrete

9.1 Introduction

Fly ash-based geopolymer is different from the PC binder [1, 2]. Sodium aluminosilicate hydrate (N–A–S–H) gel has highly cross-linked, disordered pseudo-zeolitic structure, and calcium silicate hydrate (C–S–H)/C–A–S–H has tobermorite-like structure [3–5]. Besides, the use of sodium silicate solution in alkali-activated materials also brings the differences in ITZs when compared with the PC concrete where water is used. Therefore, it is necessary to conduct microstructural characterizations and quantitative investigations, such as nanomechanical testing for a better understanding of ITZ in geopolymer. The research will be more meaningful if it is based on a comparison with PC concrete. As the ITZ in cement-based materials generally has a very small width, such as the range of 15–50 μm reported [1, 2, 6, 7], advanced characterisation techniques are required for the characterisation. Micro/nanomechanical testing techniques enable internal comparison of cementitious materials at the microstructure level, such as the comparison of ITZ with paste. Micro- and nanoindentation have been applied to evaluate the properties of ITZs in geopolymer [8–10] and PC concrete [1, 7].

In the previous study [7], two close aggregates were found to have different ITZs surrounding them. The ITZ in the aside of one aggregate has a high amount of pore, while the ITZ beneath another aggregate is concentrated with calcium hydroxide. Previous study [1] on image analysis found that the innermost 0–10 μm ITZ taken from different sand grain locations had significantly different pore content, which ranged from 4.4 to 16.6%, with the coefficient of variation as high as almost 40%. It is important to investigate if such inhomogeneity is also significant in geopolymer. From a test perspective, if the ITZ is highly inhomogeneous, the micro/nano scale ITZ properties investigation or related comparative study between samples should consider the interferences caused by the random selection of local testing objects. Besides, highly inhomogeneous ITZs can damage concrete due to the non-uniform stress distribution and/or stress concentration under loading, thus exaggerating concrete

W. Li et al., *Nanocharacterization of Cementitious Composites*,
https://doi.org/10.1007/978-981-96-2917-6_9

failure and deterioration. Understanding the features of ITZs is crucial for unveiling the failure mechanisms and behaviour of concrete.

This chapter discusses the properties and heterogeneity of ITZs in fly ash-based geopolymer concrete and provide a micromechanical test strategy to characterise more complicated ITZs surrounding multiple aggregate particles with a feasible test duration. Modelled concrete specimens, produced by a polished coarse aggregate particle and fly ash-based geopolymer mortar, were prepared to provide neat aggregate-binder boundaries and ITZs with varied orientations. Microstructures of the ITZs were also examined using SEM under backscattered electron (BSE) imaging. These images were further analysed to give detailed information of pores, cracks, hydration products and unhydrated clinkers. Micromechanical properties of the ITZs were studied using nanoindentation and nanoscratch techniques. Nanoscratch can obtain a series of continuous data including adhesion strength [2, 9], abrasion resistance [11, 12], fracture toughness [8, 13, 14], hardness [15–17], etc. in a short time, which has unique advantages in the micromechanical characterization of ITZ. The results were combined to comprehensively reveal the properties difference of ITZs surrounding coarse aggregates. The possible mechanisms were discussed fundamentally.

9.2 Comparison on the Properties of ITZs in Fly Ash-Based Geopolymer and Portland Cement Concretes

9.2.1 Materials and Sample Preparation

Geopolymer was synthesized by alkali-silica activation of fly ash. Alkali silica solution used was the mixture of sodium silicate and sodium hydroxide solution. The composition of sodium silicate was 14.7% of Na_2O, 29.4% of SiO_2 and 55.9% of H_2O. The sodium hydroxide solution was prepared by dissolving sodium hydroxide pellets into water and cooled down before mixing with sodium silicate. General purpose Portland cement was mixed with water for PC. The chemical composition of fly ash and Portland cement is shown in Tables 9.1 and 9.2, respectively. River sand was used with a fineness modulus of 2.1. Small cylinders with a diameter of 25 mm were drilled out from the same limestone rock contained mainly calcite. The cylinders were cut into slices to act as aggregates. One side of the slices was manually ground by using 320, 600 and 1200 grits of abrasive paper and polished using 1 μm alumina for 20 min to achieve smooth surfaces.

Table 9.1 Chemical composition of fly ash

Oxide	Al_2O_3	CaO	SiO_2	Fe_2O_3	K_2O	MgO	Na_2O	MnO	P_2O_5	TiO_2	LOI
Weight (%)	25.21	1.73	64.55	2.85	1.47	0.41	0.48	0.07	0.19	0.91	1.54

Table 9.2 Chemical composition of Portland cement

Oxide	Al_2O_3	CaO	SiO_2	Fe_2O_3	K_2O	MgO	Na_2O	SiO_3	LOI
Weight (%)	4.78	64.18	19.67	3.10	1.11	0.91	0.11	2.37	3.77

For the geopolymer and PC, it is difficult to conduct a representative comparison since the raw materials and mix design parameters are different. When one mix design parameter or property of PC and geopolymer is controlled to be the same, others are still variable and may lead to different compressive analysis. Therefore, the comparative studies can only reveal the differences under some specific conditions to enrich the in-depth understanding of the two binders and concretes. In this study, the companion is based on the equivalent flowability instead of mix ratio parameters. The similar flowability was assumed to create interfaces with similar mix and cast condition and then allowed to reveal some interface properties differences (e.g. debonding) caused by the different nature of cementitious materials. Both paste and mortar have been tried for investigation. Paste is ideal for minimizing interference but causes significant microcracking in ITZs due to the shrinkage. As a result, mortars by sand to fly ash/Portland cement ratio of two were used to prepare modelled sample I for BSE, EDS and nanoindentation test. The w/c ratio of PC was 0.42. The according solution/fly ash, Na_2O/fly ash, and silica modulus were 0.53, 0.08 and 1.00 for geopolymer, respectively. A similar flow table test value of 168 mm was achieved by PC and geopolymer mortars according to ASTM C1437 (Standard test method for flow of hydraulic cement mortar). For modelled sample II, mortar exhibited the same morphology features with the paste. However, the very strong bond of geopolymer mortar to aggregate produced abundant residual paste and no aggregate surface for ITZs observation. Thus, paste samples with the same mix ratio as corresponding mortar samples were used to cast the modelled sample II for of PC and geopolymer in the ITZ SEM characterization.

For the first type of modelled sample, the circle aggregate (Φ25) were ground to a rectangle shape of around 15 mm × 10 mm. A Φ25 mould was oiled, and then a layer of cling film was sticked to the mould. The rectangle aggregate was then put inside the mould and geopolymer or PC mortar was poured to form the modelled sample I. For the modelled sample II, the clear vinyl tubing of 10 mm diameter was cut into short tubes with a height of around 5 mm. One side of the tubes was polished to flat, and pasted to the circle slice aggregates (SSD) using super glue. The mixed geopolymer and PC paste were cast into the tubes as described in Fig. 9.1c. After two minutes of vibration, the geopolymer samples were covered with cling film and cured under a temperature of 65 °C for 24 h in an oven. Afterward, there were subjected to the standard curing with PC samples until 28 days.

Modelled sample I was cast with an interface condition similar to that in the real concrete. Besides, the cling film adhered between the modelled concrete and mould could avoid restrain on matrix deformation caused by bonding, and minimize damage of interface from the demoulding process. After demoulding, samples were put into a Φ35 mould and coated by epoxy resin to preventing damage during the

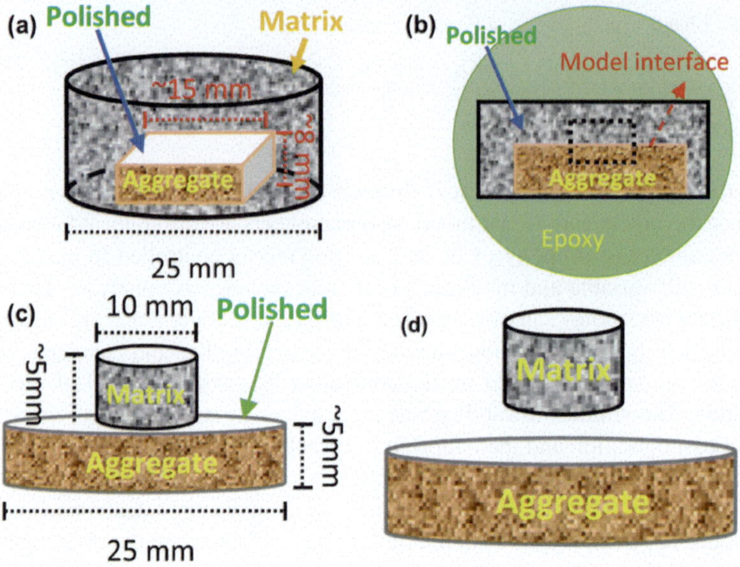

Fig. 9.1 Schematic diagram of the modelled interfaces: **a** modelled sample I; **b** surface for testing; **c** modelled sample II; **d** separated surface

surface preparation procedures. The sample as shown in Fig. 9.1a was cut from top to bottom to expose a surface as displayed in Fig. 9.1b. This surface with very neat interface in both the horizontal direction and the penetration direction was polished using 320, 600 and 1200 grits of abrasive paper and by alumina of 1 and 0.3 μm. The well prepared surface was used for the following microstructure and nanomechanical test. The modelled sample II was manually broken and separated into two parts of matrix and aggregate for interface observation as shown in Fig. 9.1d.

9.2.2 *Microstructure of Aggregate Surfaces*

Appearance of the aggregates after being separated from modelled samples II is shown in Fig. 9.2. In Fig. 9.2a, a circular area (PC area) was less bright than its surrounding areas, which was identified as the location of PC before separation. The details in Fig. 9.2b indicate that after separated, no remarkable PC paste was left on aggregate. In contrast, as displayed in Fig. 9.2c, a significant amount of residual geopolymer was observed on the aggregate surface. Additionally, separation of geopolymer sample was more difficult than for PC sample. Indeed, the PC paste sample was very easily removed while considerable force was still required for geopolymer paste sample. The above morphology features clearly indicate that geopolymer bond to the surface of the aggregate is better than that of PC.

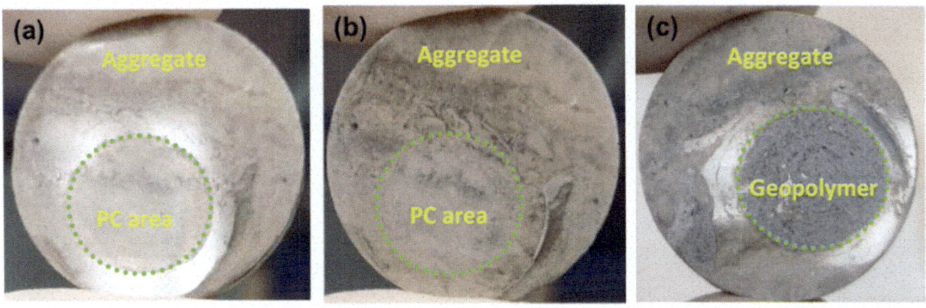

Fig. 9.2 Surface of aggregates after removing geopolymer or PC matrix: **a** PC; **b** PC; **c** Geopolymer

A scanning electron microscope (SEM, Zeiss Supra 55VP) was used for interfacial microstructure observation characterization at high magnification. The accelerating voltage was set as 10 kV. Within the circular area, the interfacial microstructure of geopolymer was observed at high magnifications based on attached paste. As shown in Fig. 9.3a, the geopolymer itself is highly heterogeneous, but a layer of paste rich in gel is found to adhere to the surface of aggregate, which was caused by both wall effect and slight bleeding. This layer of paste exhibited a very dense and uniform structure even observed at high magnifications 20,000×, which implies a well-formed interface. For the PC modelled sample, since there was no significant residual paste attached to aggregate, the surface of the bulk paste after separation could also be within the range of ITZs. Therefore, both the separated aggregate and paste were observed as exhibited in Figs. 9.4 and 9.5. In Fig. 9.4a, the residual PC paste presents a small size, thin thickness, and heterogeneous characteristics. At the magnification of 8000×, it is clear that abundant crystals can be found on paste very close to the surface of the aggregate, and some details can be observed at 20,000×. Moreover, some locations of the bulk paste shown in Fig. 9.5 are also found to be rich in calcium hydroxide and ettringite crystals. Crystals such as calcium hydroxide and ettringite generated from cement hydration, are able to develop into large size due to the high water to cement ratio (water films) in ITZ [5]. High porosity and weak bond between crystals and C–S–H gel induced a loose structure leading to the poor performance of ITZ in PC sample. For geopolymer, there were generally no newly formed crystals that could be detected by XRD [18–21]. Except for the possible zeolite at the ultra-fine size of several nanometers [22], a lot of the crystals existing in geopolymer would be introduced by the raw material fly ash. Due to the wall effect, the content of fly ash and its crystals can be assumed to be low in the vicinity of the aggregates. In addition, although the liquid to fly ash ratio can be expected to be high in the area near to aggregate, transformation of the possible nanoscale zeolite into harmful large scale crystals is not possible. Water content, temperature, and soluble silicate are important factors that dominate the formation of large size zeolites [23]. Water just accounts for a small proportion of the solution. The increase in water to fly ash ratio is lower than that of liquid to fly ash ratio in ITZ. Thus, the increase in water content is small and not be able to overcome the inhibition of crystallization from the soluble silicate in the sodium silicate solution. The above mechanism indicates

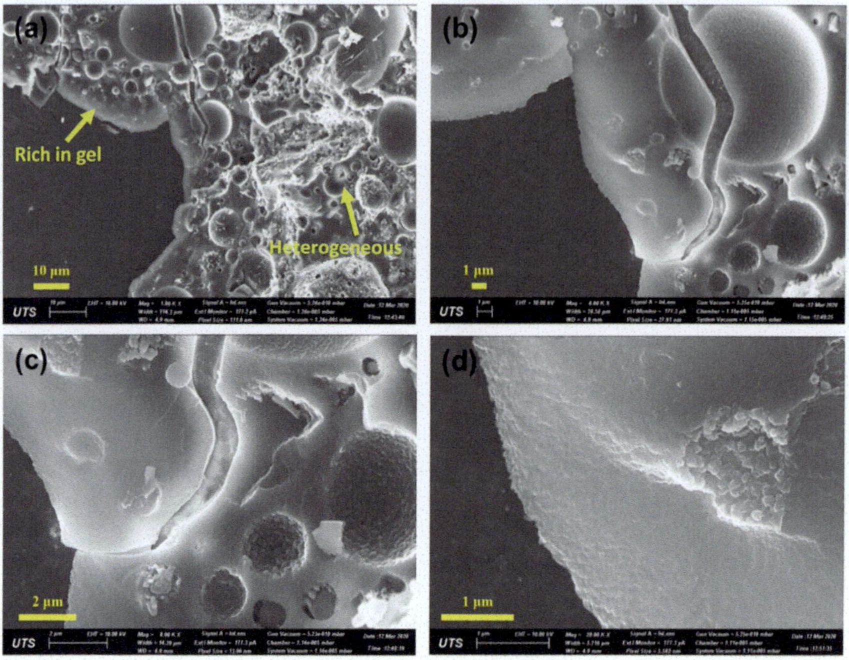

Fig. 9.3 Residual geopolymer on aggregate: **a** 1000×; **b** 4000×; **c** 8000×; **d** details 20,000×

that crystals and even porosity would not be the dominant factors affecting ITZ in geopolymer. Besides, the sodium silicate solution presents adhesive properties by nature, which could contribute to achieving better bonding in geopolymer than the PC. When reactive aggregate is used, the benefits of alkali solution on ITZ are further manifested by improved bonding due to chemical reactions [17, 24].

9.2.3 Constituents Distribution of ITZ

The distribution of constituents at different distances to the surface of the aggregates was investigated by statistical analysis of segmented BSE images. The distribution and width of the ITZs were determined by the statistical of 40 BSE images at a magnification of 500×. In each image, 17 of 5 μm-width bands were taken successively from the surface of the aggregate to paste. As shown in Fig. 9.6, the image segmentation was conducted based on the different grey values of the BSE image. Thereafter, the percentage of unreacted FA, reacted products and defects (e.g. pore, cracks) in each strip of geopolymer samples were obtained. Similar operations were also conducted on PC as shown in Fig. 9.7. The average distribution of the constituents in PC and geopolymer based on the statistic analysis of BSE images is provided in Fig. 9.8. Wall effect can be clearly observed. The contents of both FA and clinker

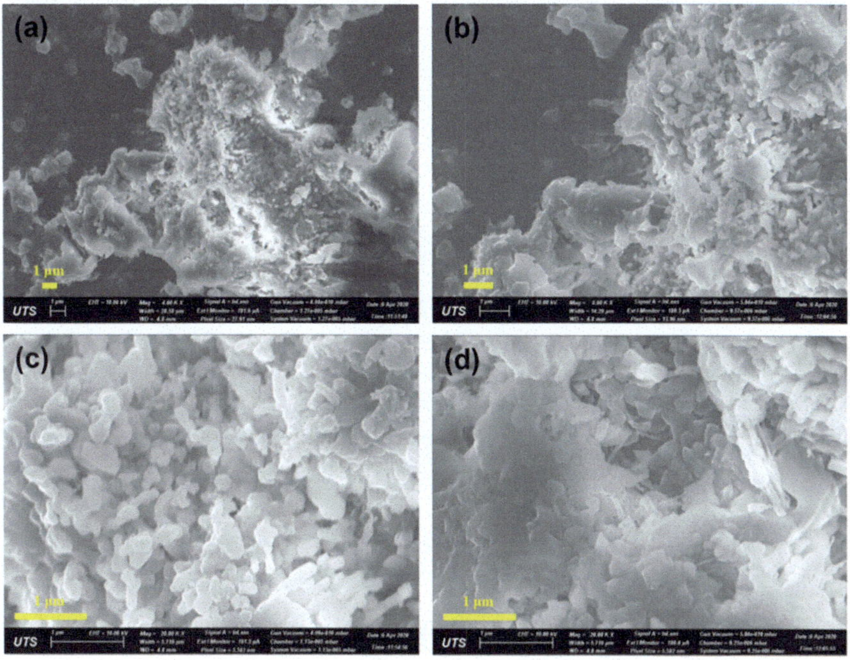

Fig. 9.4 Residual PC on aggregate: **a** 4000×; **b** 8000×; **c** details 20,000×; **d** details 20,000×

are substantially low at the area near to aggregate and then increases to an almost stable value. The zone before the stable stage is around 20 µm and 30 µm in the PC and geopolymer samples, respectively. Those regions were identified as the ITZ. The thickness of ITZs in the modelled PC sample of this study is in agreement with results obtained in real PC concrete [25, 26], which validates that this artificial interface is feasible for ITZ research. The average volume of reacted products in ITZ is similar for geopolymer and PC, 66.32% in the first 20 µm of PC and 65.78% in the first 30 µm of geopolymer. Although the width of ITZ in geopolymer is larger than in PC, the high gel content instead of crystalline phases in the reaction products, as well as the low proportion of pores and cracks, contribute to the better performance.

Due to the different in shrinkage and elastic modulus of aggregate and paste, microcracks are likely to form in ITZ, as observed in both geopolymer and PC samples. However, microcracks in PC are generally larger and more developed than in geopolymer, which is consistent with the feature at macroscope (Fig. 9.2). In addition to different gel shrinkage, higher microcracking in PC could also be partially explained by the crystals and large pores in ITZ of PC. Interfacial debonding was observed in real PC samples by Khedmati et al. [24, 27] and Diamond [1, 2]. The width of interfacial debonding was measured as 15–50 µm [28] by laser scanning microscopy, which was larger than the few microns cracks observed in this study.

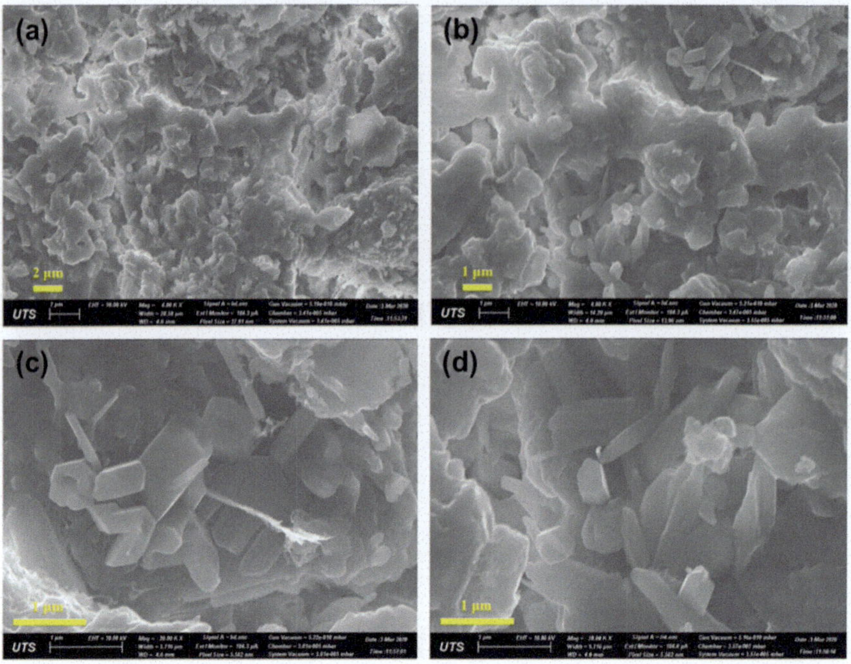

Fig. 9.5 Surface of bulk PC after removing of aggregate: **a** 4000×; **b** 8000×; **c** details 20,000×; **d** details 20,000×

9.2.4 Element Analysis of ITZs

EDS mapping and line scans were conducted for interfaces in geopolymer and PC samples to reveal the details of element distribution in ITZ. The EDS measurements were performed using a Zeiss EVO LS15 with an accelerating voltage of 15 kV. As shown in Figs. 9.9 and 9.10, the boundary between aggregate and ITZ can be distinctly identified according to the element distributions. Calcium is the dominant element in rock, which is far less in the alkali-activated low calcium fly ash geopolymer. The distribution of Al, Si, and Na can be found in paste, indicating N–A–S–H gel. Besides, different kinds of fly ash which are composed of different elements can be visibly identified. Figures 9.11 and 9.12 reveal the element distribution of ITZ in the PC sample. This area is found to be rich in elements of Ca and Si, which are the elements in the main hydration products such as C–S–H gel and CH. Elements of Fe and Al were also identified in some phases.

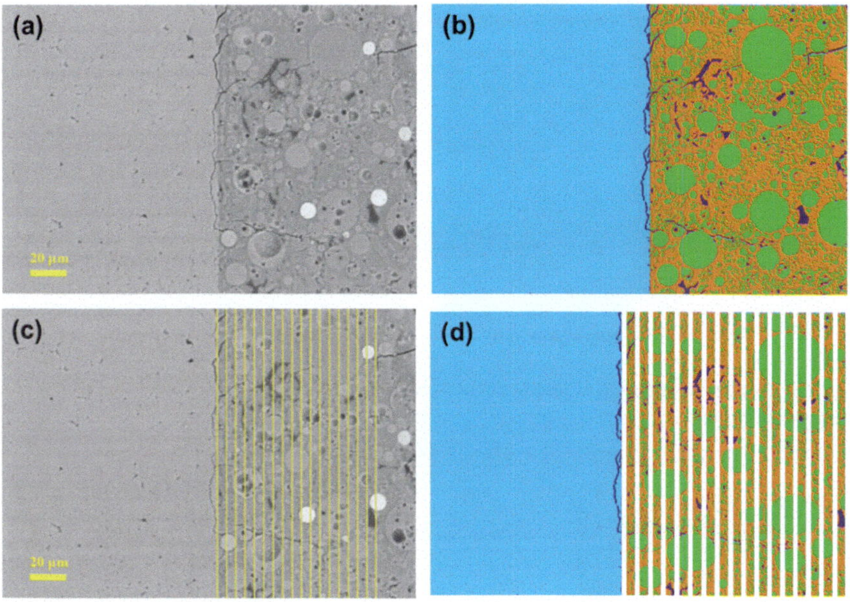

Fig. 9.6 Statistical analysis of ITZ in modelled geopolymer concrete: **a** BSE image; **b** segmented image; **c** taking strips from ITZ to paste; **d** statistic of strips

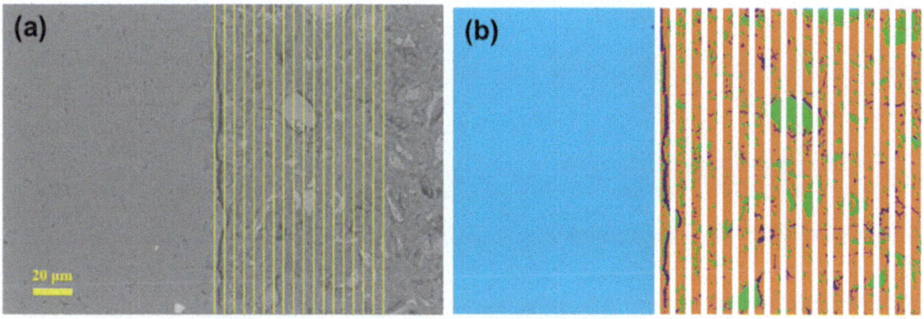

Fig. 9.7 Statistical analysis of ITZ in the modelled PC concrete: **a** taking strips from ITZ to paste; **b** statistic of strips

9.2.5 *Nanoindentation on ITZs*

Microindentation normally requires measuring the residual indent impression, while impression less than 20 μm is susceptible to imprecision according to ASTM E384-17 (Standard test method for microindentation hardness of materials). With the advances in testing techniques, depth-sensing nanoindentation with nanoscale indenter has been widely used in material characterization. The contact area during the test can be obtained by combining the depth detected and the geometry of the indenter, allowing

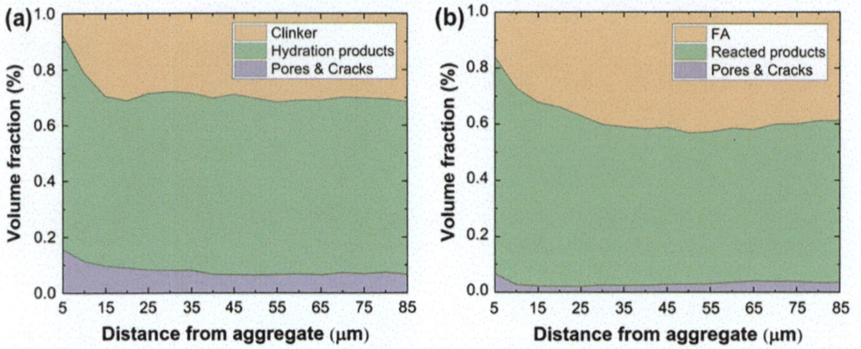

Fig. 9.8 Average distribution of microstructural constituents in ITZs: **a** ITZ in PC concrete; **b** ITZ in geopolymer concrete

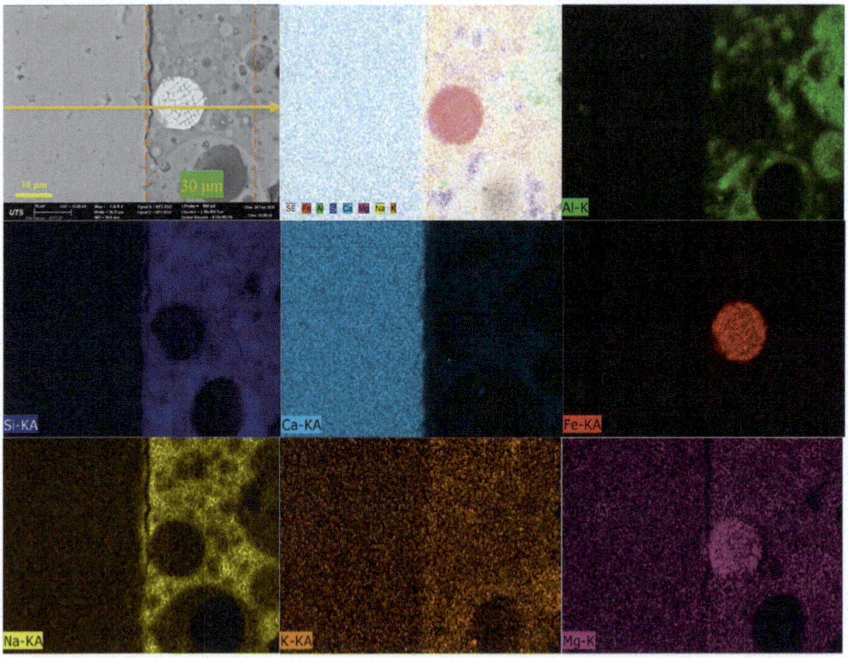

Fig. 9.9 EDS element mapping of geopolymer ITZ

much shallower indentation tests. In this study, the BSE results indicated that the width of ITZs in PC and geopolymer was around 20 μm and 30 μm, respectively. Thus, nanoindentation has advantages over the microindentation for investigation.

In previous studies such as [18], the grid nanoindentation investigations of ITZ were almost all at a small force, such as 1200 and 2000 μN. The average value of each column of the test results was used to reveal the variation of micromechanical

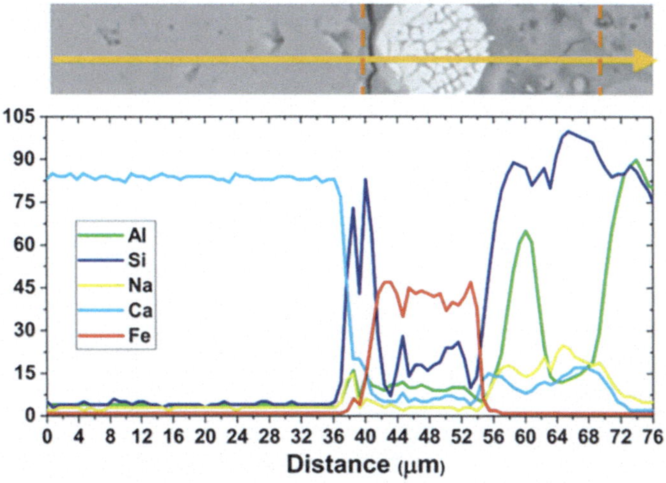

Fig. 9.10 EDS line scanning of the geopolymer ITZ

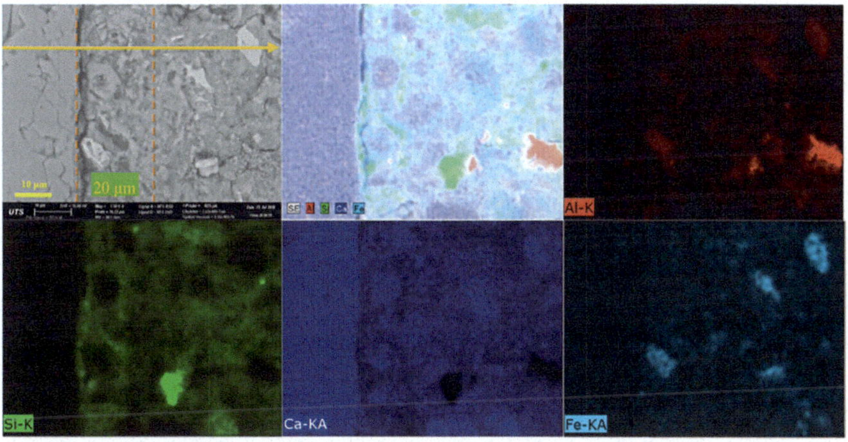

Fig. 9.11 EDS mapping of the PC ITZ

properties from ITZ to paste. Small force could make it possible to conduct dense indents with small spacing to reveal more abundant mechanical properties of ITZs. However, the shallow indentation usually provides the property of a specific phase in reaction products. For pores and cracks under low force loading, during the test they either do not provide a value, or cause significantly abnormal nanoindentation curves. Those indent points are generally omitted during analysis. Calcium hydroxide (CH) crystal is detrimental to the performance of ITZ owing to the loose bonding between CH and C–S–H gels, but have higher micromechanical properties than the gel phase [19, 20]. The loose bonding causes it easy to drop off from the polished surface.

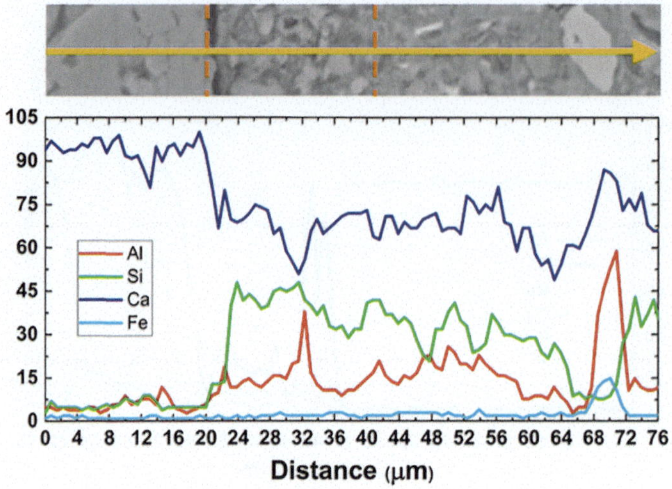

Fig. 9.12 EDS line scan of the PC ITZ

The shallow test would detect a lower fraction of crystals than that in real situation. Meanwhile the shallow indent points on CH phases only capture the high mechanical properties, but not the poor bonding properties. The large size and high amount of pores, cracks, and CH with weak bonding are indicators of the inferior properties of ITZ in PC concrete, but these adverse effects could not be well reflected during low load indent. Higher force with large penetration depth that results in the interaction of multiple phases would be more appropriate to reveal the relative properties of ITZ to paste as the phenomenon at the macro scale. However, to ensure the accuracy of result, the involved range should not exceed the width of ITZ. Moreover, the interferences from adjacent indent points as well as the restraint effect of aggregate should be avoided. The activated depth in nanoindentation test was considered as 3 to 4 times the indentation depth [13]. In ASTM E384-17, the minimum recommended spacing for Knoop and Vickers indentations are 2 or 2.5 times of the diagonal. In order to make the test range more reliable and controllable, the nanoindentation was conducted with constant penetration depth. The above two criteria were referenced to determine the depth of indentation penetration.

Two types of large depth tests with different numbers of nanoindentation columns on ITZ were conducted as shown in Fig. 9.13. In the first type, two columns of nanoindentation test points were set on the ITZ. Considering a smaller ITZ width of around 20 μm, the horizontal grid spacing and the distance between the surface of aggregate and nearest test points were set as 7 μm. The vertical grid spacing was 15 μm. A total of ten 4 × 9 grids were conducted on both geopolymer and PC to reveal the variation in mechanical properties with increased distance to the surface of aggregates. The corresponding nanoindentation depth was determined as 850 nm. The second type of test was conducted at a larger scale to compare the properties of ITZ with paste. As shown in Fig. 9.13b, just one column of indent data was conducted

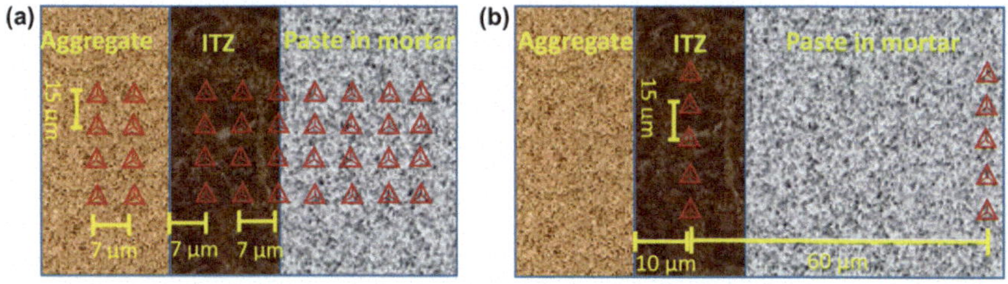

Fig. 9.13 Grid nanoindentation on modelled ITZ: **a** two columns of test indents on ITZ; **b** one column of test indents on ITZ

on ITZ, which allowed the test to achieve the maximum feasible interaction range within ITZs. The properties of paste were tested simultaneously by one column of test points at the distance of 70 μm from the surface of aggregate. After several trial tests, the nanoindentation depth of 1200 nm was adopted. Total 25 of these 5 × 2 grids were randomly selected along the modelled interface for test investigation. The locations close to sand grains were avoided.

The residual indent marks after the indentations are displayed in Fig. 9.14. After the test with a penetration depth of 850 nm, four rows of residual indent impressions can be observed on the aggregate in Fig. 9.14a. Figure 9.14b shows that the residual impressions on paste are smaller than that on aggregate. More clearly residual impressions are shown in Fig. 9.14c, d, which are from indents with the indent penetration depth of 1200 nm.

The nanoindentation test results for geopolymer and PC are presented in Figs. 9.15, 9.16, 9.17 and 9.18 and summarized in Table 9.3. Figures 9.15 and 9.16 provide the test results for geopolymer and PC under a penetration depth of 850 nm. The ITZ in geopolymer had a very similar micromechanical properties distribution to that of bulk paste. The average elastic modulus values of the nanoindentation test points within ITZ and bulk paste were around 25.70 GPa and 26.38 GPa, respectively. The elastic modulus ratio of ITZ to bulk paste was around 0.97. The corresponding average hardness values were around 1.87 and 1.93 GPa, with a hardness ratio of around 0.97. Even in PC concrete, there is still no distinct difference between the distributions of elastic modulus in ITZ and bulk paste. The average elastic modulus for ITZ and bulk paste is around 28.34 GPa and 29.37 GPa, respectively. The identifiable inferior property on ITZ compared to paste is only found in the hardness distribution histogram of PC. The average hardness is around 0.98 GPa for ITZ, while a larger value of around 1.20 is obtained for the paste. The average elastic modulus ratio and hardness ratio of ITZ to paste is around 0.96 and 0.82, respectively. Thus, except for hardness in PC, the micromechanical properties distribution of both materials did not show significant weakness in ITZ. Geopolymer exhibited a slightly lower elastic modulus but an evidently higher hardness compared to PC. In terms of aggregate, it is noted that the second column of test data on the aggregate in geopolymer presented obviously lower mechanical properties than the first column of data. This was not observed in the PC sample. It should be partly attributed to

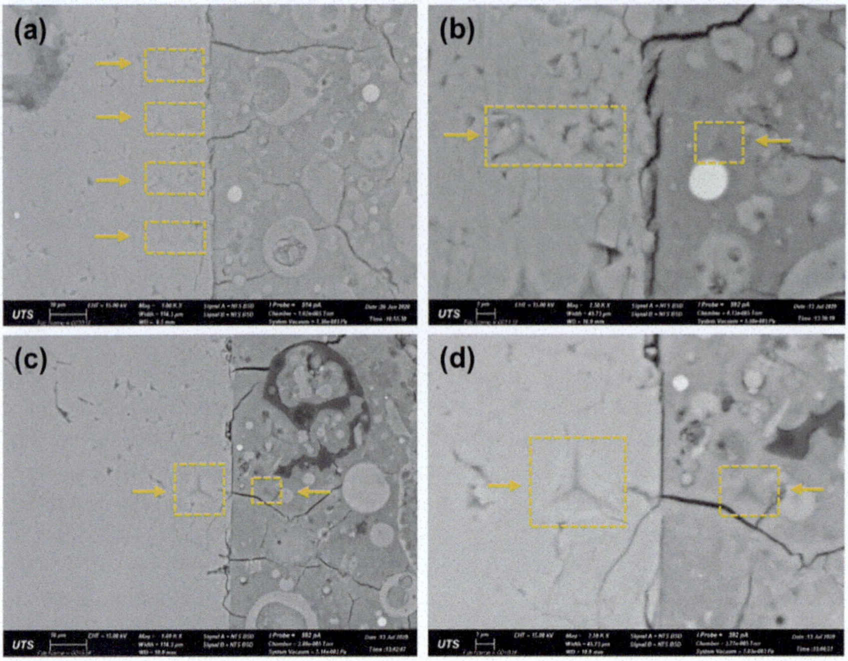

Fig. 9.14 Residual indent impression after nanoindentation: **a** penetration depth of 850 nm (1000×); **b** penetration depth of 850 nm (2500×); **c** penetration depth of 1200 nm (1000×); **d** penetration depth of 1200 nm (2500×)

the better bonding of geopolymer with aggregate. Figure 9.3 shows that geopolymer adheres tightly to aggregate. The shrinkage of the geopolymer is restrained by this strong bond, which sometimes leads to cracking of the brittle aggregate as typically shown in Fig. 9.7.

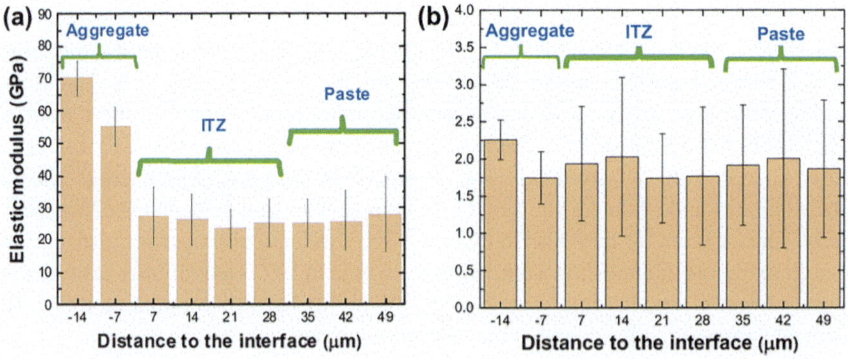

Fig. 9.15 Micromechanical properties of geopolymer from aggregate to paste (850 nm): **a** elastic modulus; **b** hardness

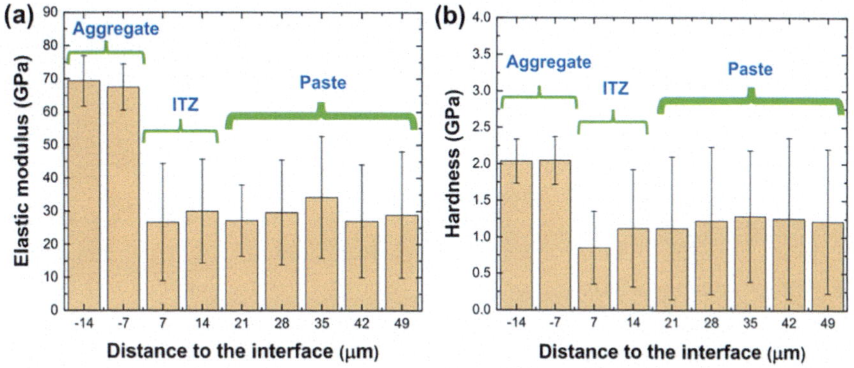

Fig. 9.16 Micromechanical properties of PC from aggregate to paste (850 nm): **a** elastic modulus; **b** hardness

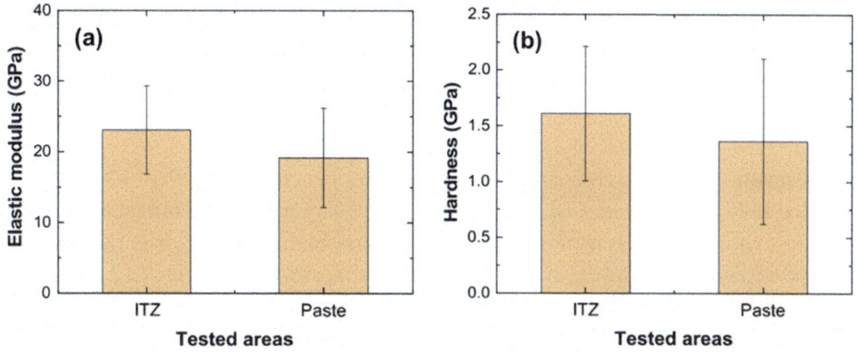

Fig. 9.17 Comparison on micromechanical properties of ITZ with paste in geopolymer (1200 nm): **a** elastic modulus; **b** hardness

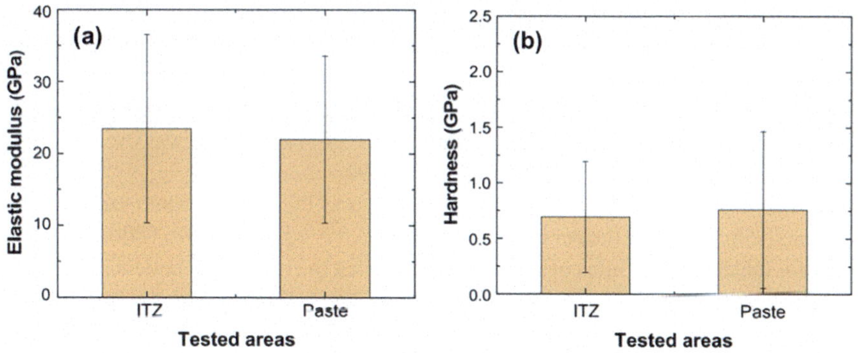

Fig. 9.18 Comparison on micromechanical properties of ITZ with paste in PC (1200 nm): **a** elastic modulus; **b** hardness

Table 9.3 Results for geopolymer and PC from penetration depth of 850 nm and 1200 nm

Objects	Test depth (nm)	M (GPa)	H (GPa)	σ_M	σ_H	M_{ITZ}/M_{Paste}	H_{ITZ}/H_{Paste}
Geo-ITZ	850	25.70	1.87	7.84	0.86	0.97	0.97
Geo-paste	850	26.38	1.93	9.58	0.99		
PC-ITZ	850	28.34	0.98	16.69	0.65	0.96	0.82
PC-paste	850	29.37	1.20	16.22	1.01		
Geo-ITZ	1200	23.08	1.61	6.22	0.60	1.21	1.18
Geo-paste	1200	19.15	1.36	7.02	0.74		
PC-ITZ	1200	23.46	0.69	13.11	0.50	1.07	0.91
PC-paste	1200	21.96	0.76	11.62	0.71		

Note M is the average elastic modulus and H is the average hardness. σ_M and σ_H are the standard deviation of elastic modulus and hardness, respectively. For the test with the penetration depth of 850 nm, the test data belong to ITZ or paste are determined as shown in Figs. 9.15 and 9.16. Then, M, H, σ_M and σ_H of ITZ or paste are obtained based on all of the test data within them. M_{ITZ}/M_{Paste} is the ratio of the average elastic modulus of ITZ to that of paste. H_{ITZ}/H_{Paste} is the corresponding ratio for hardness

The results for geopolymer and PC under a larger penetration depth of 1200 nm are provided in Figs. 9.17 and 9.18. For this test with a larger involved range, the results for geopolymer are different compared to the results obtained under lower penetration depth of 850 nm. Both the elastic modulus and hardness of ITZ in Fig. 9.17 are found to be larger than paste. The elastic modulus of ITZ and paste were 23.08 GPa and 19.15 GPa, respectively. Correspondingly, the hardness of ITZ and paste was 1.61 GPa and 1.36 GPa, respectively. The elastic modulus of ITZ is 1.21 times of the paste and the hardness is 1.18 times of the paste. In terms of the PC sample, the elastic modulus of ITZ and paste was around 23.46 GPa and 21.96 GPa, respectively, and the hardness was around 0.69 GPa and 0.76 GPa, respectively, for ITZ and paste. The hardness ratio of ITZ to paste was around 0.91 GPa, which still indicates the low hardness of ITZ compared to paste. However, the difference was reduced when compared with the ratio of around 0.82 obtained under the lower penetration depth of 850 nm. Additionally, the elastic modulus ratio of ITZ to paste was even slightly greater than 1.0, reaching the value of around 1.07.

As shown in Table 9.3, for both geopolymer and PC, both elastic modulus and hardness decrease when the penetration depth is increased from 850 to 1200 nm, especially for the elastic modulus of PC. This implies that the micromechanical properties are lower when a larger extent of multiple phase interaction is involved. Even under the depth of 1200 nm, the micromechanical properties obtained are still found to be larger than the corresponding gel phase in materials, N–A–S–H gel [29, 30] and LD C–S–H gel [25, 31, 32]. In addition to micromechanical properties, the standard deviation also was decreased, indicating more homogeneous results obtained during the larger penetration depth test. However, due to the limitation of ITZ width on the maximum test depth of nanoindentation, the materials are still heterogeneous for the indent test scale. The distinct difference between micromechanical properties of phases governs the intrinsic large standard deviation of this type of test, which cannot be reduced by increasing the number of indent data. For both the test result from 850 and 1200 nm, it is found that PC has slightly higher elastic modulus but significantly lower hardness than geopolymer. The hardness of PC is just 0.52 to 0.62 times (850 nm) and 0.43 to 0.56 times (1200 nm) of the hardness of geopolymer. Moreover, the standard deviation of PC elastic modulus is always significantly higher than that of geopolymer, indicating more heterogeneous modulus properties in PC, which is not consistent with the results reported in another study involving smaller penetration depth [33].

As illustrated previously, tests with a larger involved range should be more appropriate to reflect the micromechanical properties of ITZ. Based on the result from the maximum allowed depth of 1200 nm, the interface of alkali-activated low calcium fly ash geopolymer is found to have higher micromechanical properties than that of the paste. This phenomenon is not observed when considering smaller depth tests as reported in both this study and other studies [29]. The ITZs of geopolymer do not have many crystals and pores. The higher micromechanical properties could be attributed to the layer of dense paste rich in gel products shown in Fig. 9.3. The strong paste and low content of harmful large crystalline phase and pores contribute to the strong interfaces in geopolymer, which is consistent with the macro phenomenon shown in Fig. 9.2. The layer of geopolymer paste remaining on the surface of aggregate after the broken separation indicates that the ITZ should be stronger than the bulk paste. Thus, the large width of ITZ in geopolymer would be a favorable factor for its performance.

In terms of PC sample, ITZ elastic modulus is slightly higher, and hardness is slightly lower than the paste, which seems to conflict with the conventional concept that the ITZ is a significant weak part of PC concrete. In fact, the high micromechanical properties tested for ITZ are not groundless. As shown in Figs. 9.4 and 9.5, there is a large amount of CHs in ITZs, which greatly enhances the micromechanical properties of ITZ. The hardness is somewhat related to strength [34], which means that ITZ is still inferior to paste in some properties. The macro behaviour of ITZ is affected by many factors. The existence of large cracks at the surface of aggregate due to shrinkage could be a more influencing factor explaining the weak feature of ITZ as typically displayed in Fig. 9.7. Besides, the high content and large size of CHs with loose bonding make ITZ easy broken under loading. This kind of property cannot be effectively captured in the micromechanical test. Thus, ITZs of PC should be a weak region, which can be verified by the observation in Figs. 9.2a, b. Nevertheless, the 'weak features' of ITZs do not affect its microelastic modulus and microhardness significantly, which are close to the corresponding properties of the paste.

9.3 Nanoscratch on Mechanical Properties of Interfacial Transition Zones (ITZs) in Fly Ash-Based Geopolymer Composites

9.3.1 Raw Materials and Sample Preparation

Geopolymer was synthesized by alkali-activation of low calcium fly ash. The chemical composition of Class F fly ash is listed in Table 9.1. Alkali solution adopted was the mixture of sodium hydroxide solution and sodium silicate solution. Sodium hydroxide pellets were dissolved into water and cooled down. Thereafter, it was incorporated into commercial D grade sodium silicate (14.7% of Na_2O, 29.4% of SiO_2 and 55.9% of H_2O) to adjust the silica modulus (SiO_2/Na_2O) from the original value of 2 to the widely used values of 1 and 1.5. For geopolymer samples with different silica modulus, they have the same Na_2O/fly ash of 8%, water/solid ratio of 0.33. A sand/fly ash ratio of 2 was used to prepare mortar samples. The geopolymer samples with different silica modulus of 1 and 1.5 were denoted by the abbreviations of Geo-Ms1 and Geo-Ms1.5, respectively.

Coarse aggregate used was drilled from a limestone rock with calcite as the main component. The drilled φ25 mm long cylinder was cut into slices. The top and bottom sides of the slices were ground to parallel. These φ25 mm aggregate slices were further grounded by abrasive papers to around $15 \times 10 \times 8$ mm^3 cuboids. The top side of the cuboid aggregate slices was polished to smooth by using 1 μm alumina for 20 min and put in a φ25 mm mould as shown in Fig. 9.19a. For this type of mould, a layer of cling was stuck to it by using oil, which was designed to avoid the contact between geopolymer and mould, minimizing the damage of ITZ in the demoulding process. Afterwards, fresh geopolymer mortar was poured into the mould containing polished aggregate. The second type of modelled interface for ITZ observation was referenced from previous ITZ studies of Portland cement concrete [34, 35]. As displayed in Fig. 9.19c, one side of a φ10 mm vinyl tubing was polished to smooth and then glued to a polished surface of a φ25 mm aggregate slice to act as mould, which was poured with geopolymer paste. Both types of modelled samples were vibrated for 2 min and covered with cling. They were cured under 65 °C for 24 h in an oven and followed by standard curing until 28 days. The modelled samples obtained are provided in Fig. 9.19b, d. After the demoulding, the modelled sample I was coated with epoxy resin and cut from top to bottom to expose a surface with neat ITZ. The surface was grounded to flat by using 320, 600 and 1200 grits of abrasive paper and subjected to final polish with 1 μm alumina particles for 40 min and 0.3 μm alumina particles for 20 min. It was used for the nanoscratch test of ITZ. The modelled sample II was separated into two parts for the SEM observation of the interface between the surface of aggregate and residual geopolymer paste.

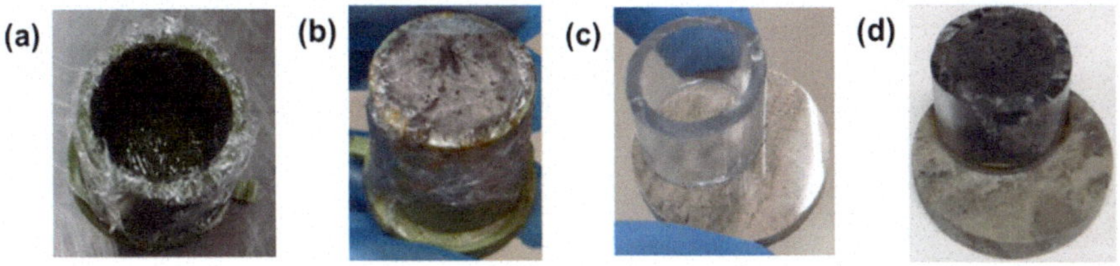

Fig. 9.19 Preparation of modelled samples for nanoscratch tests: **a** mould I; **b** modelled sample I; **c** mould II; **d** modelled sample II

9.3.2 Determining ITZ by BSE

Same to procedure introduced in Sect. 9.2.3, BSE images were segmented into three components of reacted products (green), fly ash (light red) and pores/cracks (blue) as revealed in Fig. 9.20. Pores on unreacted fly ash are sometimes unable to be precisely captured by grey values and incur inaccuracy when classified as pores manually. Therefore, they are still counted as fly ash in the part of BSE analysis. For a segmented image, fourteen consecutive 5 μm strips cover ITZ and partial paste were taken starting at the boundary of aggregate. A total of forty such BSE images were adopted for statistics of the average proportion of components in strips. The ITZ in concrete is considered to be caused by the "wall" effect [36–38]. It means that the amount of unreacted fly ash would decrease whereas the solution to ash ratio would increase in ITZ due to the presence of aggregate as a wall. This phenomenon is clearly presented in the statistical results of BSE images as shown in Fig. 9.21, in which the proportion of constituents were plotted in the middle of the position of each strip. For geopolymer with different silica modulus, the distribution of constituents generally shows a similar trend. However, the "wall" effect is less significant in geopolymer with a higher silica modulus of 1.5. Besides, geopolymer with a silica modulus of 1 presents more unreacted fly ash and slightly more pores/cracks, but fewer reacted products than the sample with a higher silica modulus of 1.5.

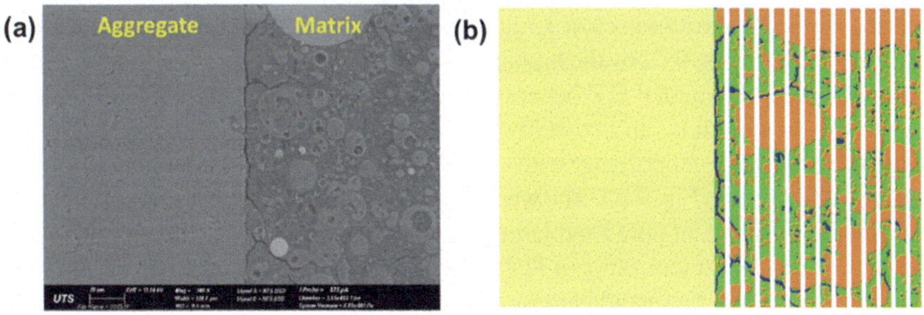

Fig. 9.20 Segmentation of BSE image based on grey value: **a** BSE image (500×); **b** segmented image

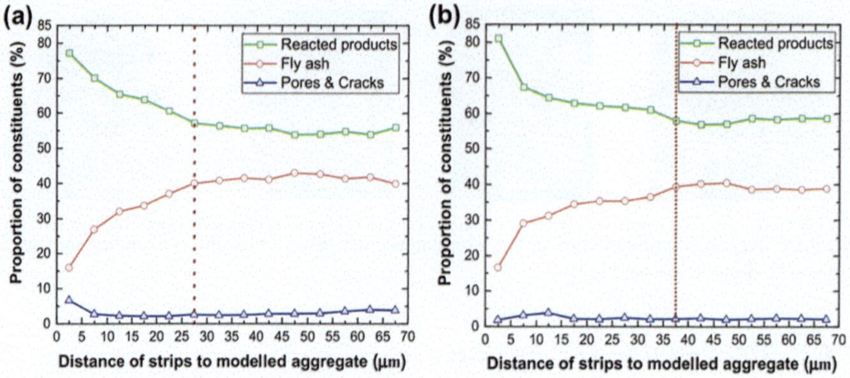

Fig. 9.21 Variation of constituents with increased distance to modelled aggregate (the dashed vertical lines indicate the boundary of ITZ and paste determined by BSE image analysis): **a** Geo-Ms1; **b** Geo-Ms1.5

For geopolymer with silica modulus of 1, the most significant part of ITZ was the first six strips, which corresponds to a width of around 30 μm. The increase of fly ash in geopolymer with silica modulus of 1.5 is less apparent. Nevertheless, it was identified that the increasing trend sustained until to 40 μm. Based on the results, the scratch test of ITZ was conducted at a distance of 15 μm to the boundary of aggregate.

9.3.3 Microstructure of ITZ

SEM observation of ITZ in modelled sample II was conducted by Zeiss Supra 55VP with an accelerating voltage of 10 kV. After separating bulk paste and aggregate of modelled sample II, abundant residual geopolymer paste was observed on the surface of aggregate as typically shown in Fig. 9.22a, b at low magnifications. The detail of the boundary was revealed in Fig. 9.22c, d, where a dense and homogeneous binder can be observed at the region close to aggregate. Although the homogeneous features were magnified in Fig. 9.22 by the higher flowability of the solution than that of solids, it should still exist in actual ITZ because of the small liquid to solid ratio in ITZ. The change in silica modulus in this study did not bring significant differences for the observed results. For geopolymer with silica modulus of 1.0, a similar phenomenon is typically displayed in Fig. 9.23. The weak ITZ properties in Portland cement concrete is attributed to abundant pores and large crystals [20], which are related to hydration and high water to cement ratio in ITZ. For geopolymer, the usage of sodium silicate instead of water has avoided these adverse factors. The high amount of residual geopolymer observed on aggregate actually indicates that geopolymer bonded tightly to aggregate. The interface would be sometimes stronger than the highly heterogeneous geopolymer bulk paste, which contains different types of smooth fly ash and

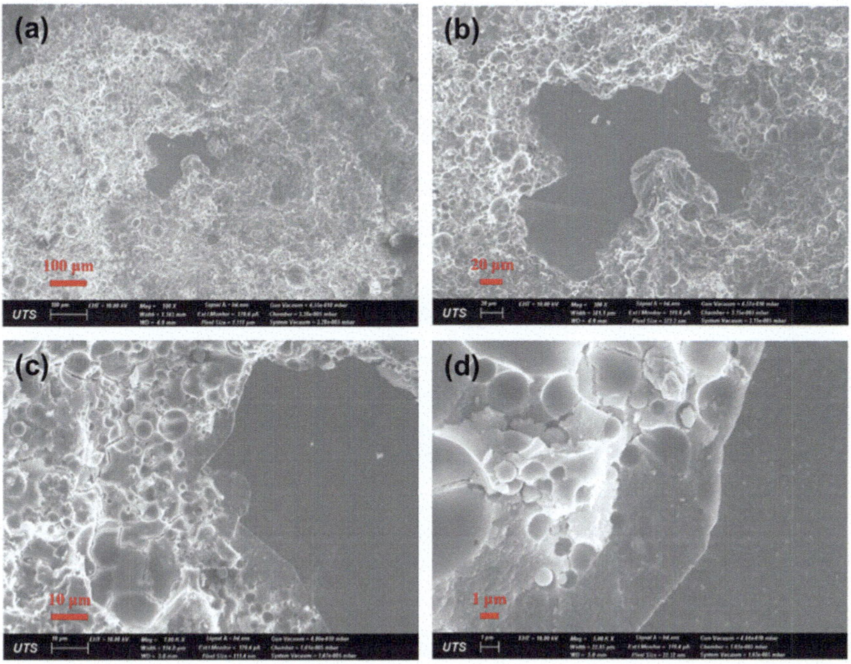

Fig. 9.22 Microstructures of ITZ in geopolymer with silica modulus of 1.5 on the aggregate surface: **a** 100×; **b** 300×; **c** 1000×; **d** 5000×

microscale crystals [16, 39]. For the modelled sample prepared, the bond between geopolymer and polished aggregate is weaker than that of geopolymer and field aggregate which has a rough surface. Thus, the same phenomenon of a good bond of geopolymer to aggregate could also occur in real geopolymer concrete.

9.3.4 Nanoscratch Test Results

9.3.4.1 Scratch Technique

Nanoscratch technique can provide a serials successive micromechanical data in a much shorter time than nanoindentation. Hardness and fracture toughness obtained from nanoscratch tests can help comprehensively analysis micromechanical properties of cement-based materials. Scratch test was performed on Agilent G200 Nano Indenter by using a hemispherically-tipped, conical stylus. The radius and apex angle of this tip are 5 μm and 120°, respectively. It has the same apex angle but a far small radius compared with the representative tip (radius of 200 ± 10 μm) in ASTM G171-03 (Standard Test Method for Scratch Hardness of Materials Using a Diamond Stylus) to facilitate the test on ITZ. For a real interface at the microscopic

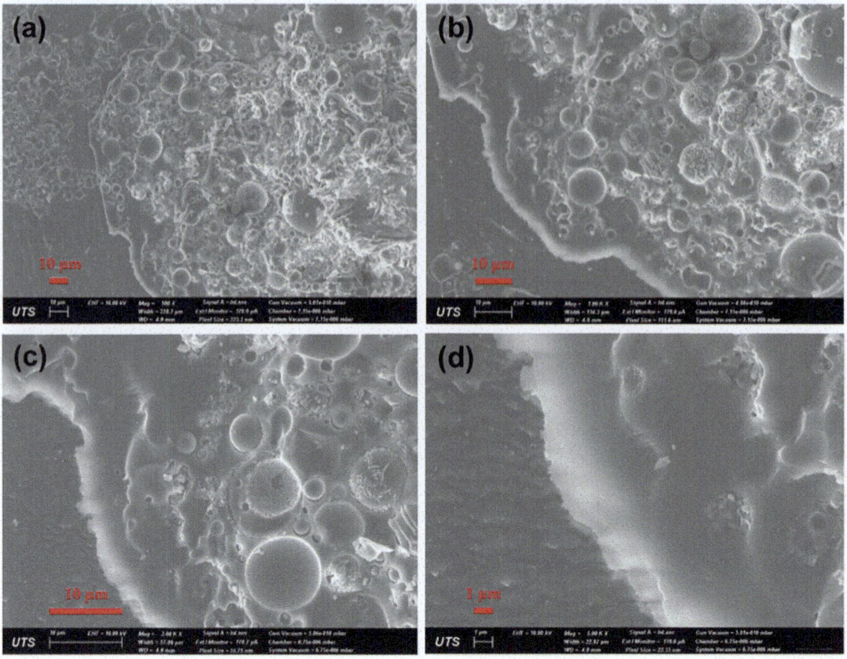

Fig. 9.23 Microstructures of ITZ in geopolymer with silica modulus of 1.0 on the aggregate surface: **a** 500×; **b** 1000×; **c** 2000×; **d** 5000×

scale, due to the irregular shape of aggregate, the starting point of ITZ is likely to be a curved surface rather than a straight plane. In this study, as shown in Fig. 9.24a, the neat aggregate boundary provided by modelled sample allows scratch to precisely test on "pure ITZ" to collect rich test data without incorporating properties results from aggregate and paste. According to the statistical results of BSE, the scratch tests on ITZ were set at 15 μm away from the boundary of aggregate. Tests were set on paste for comparison, which were 65 μm away from the boundary of aggregate. Regions with large cracks were avoided. For the first several ITZ tests, scratches with the same distance of 15 μm to the boundary were conducted on aggregate to help locate and observe scratch impression. The scratch paths were adjusted under a microscope to ensure that they were parallel to the aggregate boundary and started at the same level. The properties of ITZ and paste were both determined by twelve effective 100 μm scratches. In addition to the scratch test scheme I, parallel to the boundary of aggregate, as in Fig. 9.24a on individual objects, thirty three scratches based on the traditional scratch scheme II, perpendicular to the boundary, as given in Fig. 9.24b were conducted on geopolymer with a silica modulus of 1.5 to compare the test results.

To ensure the accuracy of the test, the normal load used should be moderately large. However, for the research here, an important factor limiting the normal load is the small width of the ITZ. The constraint effects of aggregate could possibly

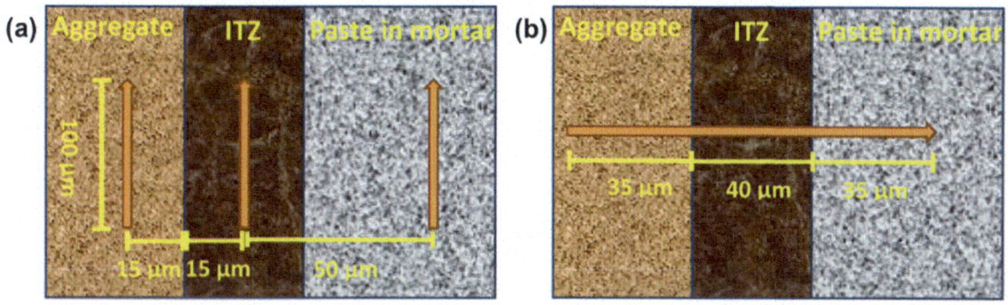

Fig. 9.24 Nanoscratch schemes for the ITZs: **a** scratch scheme I; **b** scratch scheme II

enhance the properties of ITZ significantly when the penetration depth is large. After a trial on a series of load levels, the final load for the scratch study of ITZ and paste was determined as 4 mN by referring to both ASTM G171-03 and the involved range of indentation test [25]. 2 mN and 8 mN scratch results on the more homogeneous modelled aggregate were also provided to help understand the effects of load levels. The tests on aggregate with different load levels were conducted at the location around 1500 μm away from the boundary of aggregate, which had the same scratch length and number as ITZ and paste tests. For all scratch tests, a constant scratch speed of 4 μm/s was applied. Prescratch and postscratch scans were always performed at a small profiling load of 20 μN. The data acquisition rate is one point per micron.

According to ASTM G171-03, the scratch hardness HS_p based on this kind of tip can be determined by Eq. (9.1). Scratch friction coefficient μ was obtained by Eq. (9.2), based on scratch lateral force F_{scr} and normal force P. The geometric transition depth d_{gt} provided in Eq. (9.3) means a critical depth. When the test depth d is larger than this value, the test would involve the conical portion of the indenter tip. In this study, d_{gt} was calculated as 669.87 nm. For the test with depth-sensing technique, similar to a previous study [28], the scratch width w was obtained based on the geometry of the indenter and the depth acquired as given in Eq. (9.4). The scratch width w_{gt} under geometric transition depth was calculated as 5 μm, equal to the tip radius.

$$HS_p = \frac{8P}{\pi \omega^2} \tag{9.1}$$

$$\mu = \frac{F_{scr}}{p} \tag{9.2}$$

$$d_{gt} = r\left(1 - \sin\left(\frac{\alpha}{2}\right)\right) \tag{9.3}$$

$$\omega = \left\{ \frac{2\sqrt{r^2 - (r-d)^2} d \le d_{gt}}{\omega_{gt} + 2\tan\left(\frac{\alpha}{2}\right) \times \left(d - d_{gt}\right) d > d_{gt}} \right. \tag{9.4}$$

where the value of α and r is 120° and 5 μm, respectively, representing the apex angle of the conical portion and the radius of the hemispherical portion of the tip adopted.

9.3.4.2 Effects of Load Level

The 2 mN, 4 mN and 8 mN scratch loads were applied on the modelled aggregate. The obtained results for hardness and friction coefficient were presented in the form of frequency density histogram as displayed in Figs. 9.25 and 9.26. Although histogram varies with its bin size and it is difficult to find an appropriate bin size [3, 40], the same bin size specified for all samples would be still feasible for the comparison purpose. In addition to histograms, the properties distribution was described as Gaussian mixture models, with the parameters determined by maximum likelihood estimation [4, 41]. The number of components for models was increased until it can match the histogram well. Then, the models were used to represent the frequency density histograms and plotted in the same figure to facilitate comparison.

As revealed in Figs. 9.25d and 9.26d, the results clearly indicate that both the hardness HS_p and friction coefficient μ increased with load level P. The average value

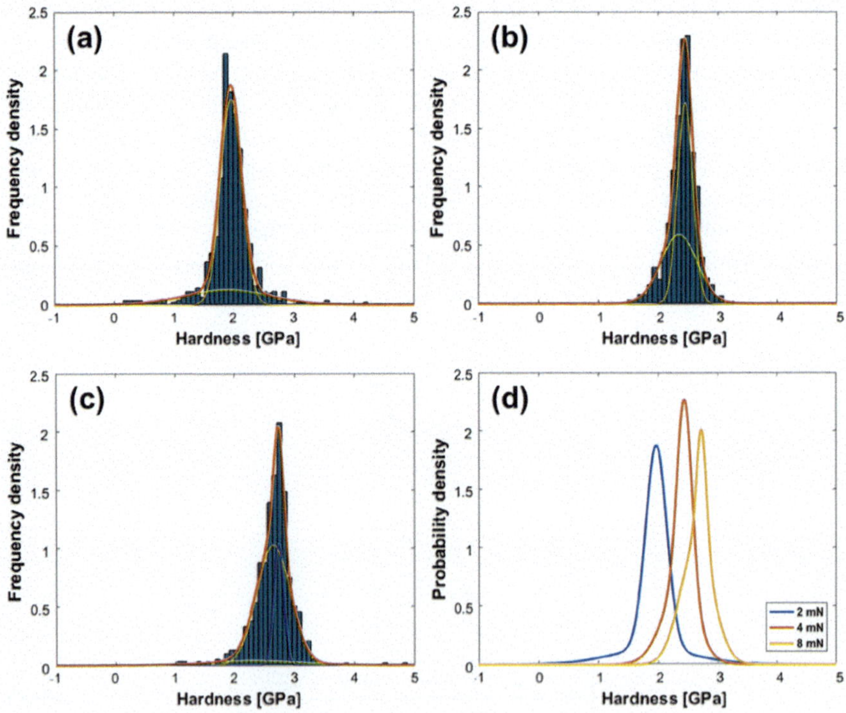

Fig. 9.25 Scratch hardness of aggregate at different load levels: **a** 2 mN (bin size = 0.08); **b** 4 mN (bin size = 0.08); **c** 8 mN (bin size = 0.08); **d** comparison on load levels

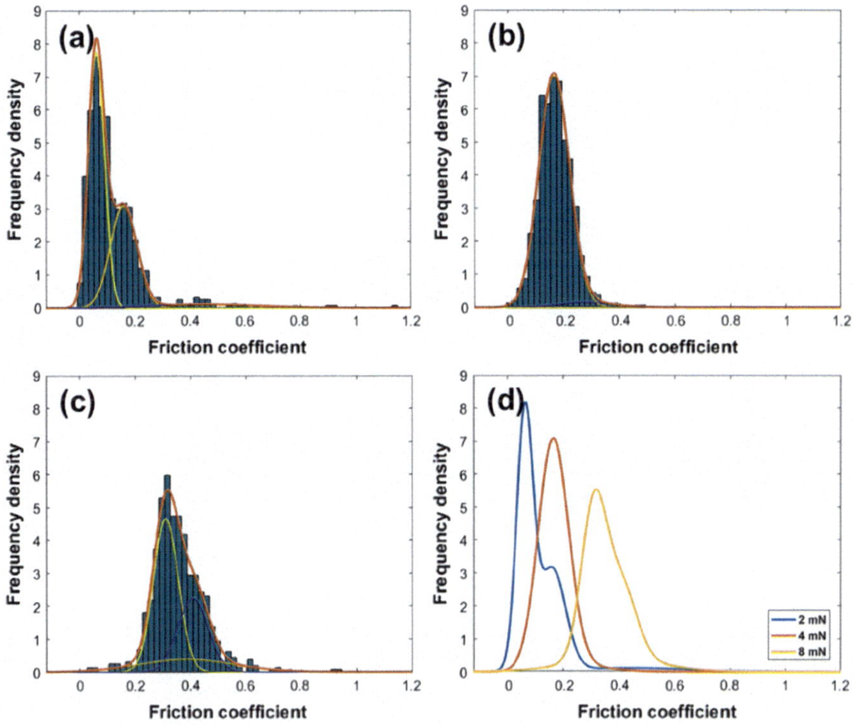

Fig. 9.26 Scratch friction coefficient of aggregate at different load levels: **a** 2 mN (bin size = 0.02); **b** 4 mN (bin size = 0.02); **c** 8 mN (bin size = 0.02); **d** comparison on load levels

and standard deviation of test result data are given in Table 9.4. Friction coefficients are very small values in the range of 0.12–0.35, while hardness is around one order of magnitude greater than the friction coefficient, which varies between 1.95 GPa and 2.64 GPa. Although both hardness and friction coefficient increased with the increase of normal load, their increasing trends are different. When the normal load was increased from 2 to 4 mN, the hardness and friction coefficient were increased by 23.08% and 41.67%, respectively. The corresponding increase from 4 to 8 mN is 10.00% and 105.88%, respectively. The load level is obviously more influential on the friction coefficient and causes less change in hardness. It could be attributed to the effects of surface defects under small load tests, which is especially significant for friction coefficient that governed by the topology of the contacted surface [42]. Since properties are dependent on test factors, the same scratch load and speed are set for the test of ITZ and paste.

Table 9.4 Average scratch hardness and friction coefficient of aggregate under different load levels

Load level (mN)	Scratch hardness (GPa)	Scratch friction coefficient
2	1.95 ± 0.37	0.12 ± 0.11
4	2.40 ± 0.23	0.17 ± 0.06
8	2.64 ± 0.33	0.35 ± 0.09

Table 9.5 Average hardness and friction coefficient of ITZ and paste with difference silica modulus

Samples	Scratch hardness (GPa)	Scratch friction coefficient
Geo-Ms1.5-ITZ	1.38 ± 0.75	0.31 ± 0.30
Geo-Ms1.5-paste	1.39 ± 0.83	0.31 ± 0.27
Geo-Ms1-ITZ	1.72 ± 1.42	0.32 ± 0.29
Geo-Ms1-paste	1.49 ± 1.46	0.42 ± 0.37

9.3.4.3 Scratch Hardness of ITZ and Paste

Based on the constant load of 4 mN under test scheme I, the average value and standard deviation of hardness and friction coefficient results of ITZ and paste are obtained as listed in Table 9.5. ITZ and paste are found to have lower hardness but higher friction coefficient and standard deviation than that of aggregate when the same scratch load and speed were applied. The high standard deviation is caused by the distinct hardness of different constituents. For geopolymer with different silica modulus, results reveal that ITZ of Geo-Ms1.5 sample has almost the same hardness and friction coefficient as its paste, whereas the ITZ of Geo-Ms1 shows distinctly higher hardness but lower friction coefficient than paste. The hardness and friction coefficient of Geo-Ms1 are similar to or larger than that of Geo-Ms1.5. For heterogeneous samples, the comparative study cannot be just based on the average value. For instance, some types of fly ash have great mechanical properties but account for a small proportion. The scratch tests can not reflect the real proportion of the fly ash very accurately and hence affect the comparison results based on the average value. This aspect can be known by the largest hardness obtained by scratch, which is 52.49 GPa for Geo-Ms1-ITZ, but just around 10 GPa or 35 GPa for other samples. Compared with unreacted fly ash, gel and defects are far more important constituents that would determine the macro performance of samples. Hence, deconvolution analysis was conducted for further discussion.

9.3.4.4 Properties Comparison of ITZ and Paste by Deconvolution Analysis

The scratch hardness of ITZ and paste in geopolymer samples with different silica modulus are displayed in Fig. 9.27. The deconvolution results for hardness are listed in Table 9.6. It can be observed that the Gaussian mixture models with 3 components match with corresponding frequency distribution histograms. Geopolymer is mainly composed of unreacted fly ash, N–A–S–H gel, crystals and defects (pores and cracks), where the proportion of the latter two is relatively small. Due to the large indenter tip adopted for scratch and the small characteristic size of some constituents, the deconvolution study is not expected to generate individual pure phases. However, the decomposition of Gaussian mixture models is still useful for the analysis of mechanisms. For the three components in each sample, the increased mechanical properties mean the increased test points from fly ash particles and crystals, and the decreased test points from gel and defects. The test points belong to the second and especially the third component should be mainly from fly ash particles. The test points in the first component with minimum hardness should be primarily from N–A–S–H gel. Compared with others, the first component is the least affected by fly ash/crystals and hence the key for comparative study. The results in Table 9.6 indicate that the first component of Geo-Ms1.5-ITZ and Geo-Ms1.5-paste has a similar value of around 0.94 GPa. However, the corresponding components in ITZ and paste of Geo-Ms1 present distinctly different hardness values, which are 1.10 GPa and 0.77 GPa, respectively. The discrepancy can also be clearly observed by the summarized distributions provided in Fig. 9.28.

In this study, 3 components were set for the convenience of comparison. To reveal more details, the number of components was increased for Geo-Ms1-paste. Further increasing the number of components was accompanied by a better match of probability density curve with frequency density histogram. When the number of components increased to 5, as shown in Fig. 9.29, the component with hardness of 0.77 was decomposed into two components with hardness of 0.58 GPa and 1.06 GPa, respectively. The deconvolution results for the 5 component model provided in Table 9.6 show that in addition to the small proportion, the second newly decomposed component with hardness of 1.06 GPa is similar to the minimum component in other test objects. The lowest hardness of 0.77 GPa obtained in 3 component models for Geo-Ms1paste should be caused by the first newly decomposed component with a small hardness of 0.58. This component should have a considerable amount of defects related test points (low strength gel). Defects such as some capillary pores in gel are not large enough to be detected by BSE analysis at the magnification of 500×, but would still affect scratch properties significantly.

Table 9.7 presents the percentage of test data from typical large scratch widths of larger than 5 μm ($H < 0.407$ GPa) and 4 μm ($H < 0.637$ GPa). Results in Table 9.7 indicate that the percentages are very close for Geo-Ms1.5-ITZ, Geo-Ms1.5-paste and Geo-Ms1-ITZ, which vary in the range of 1.66%-1.94% and 7.46%-8.65%, respectively. Lager proportions of 8.45% and 21.21% are found for Geo-Ms1-paste. For hardness data less than 0.407 GPa and 0.637 GPa in Geo-Ms1-paste, the proportions of data from the defects related phase ($H = 0.58$) can be estimated based on Eq. (9.5) and deconvolution results in Table 9.6, which were obtained as 89.56% and 84.27%, respectively. Thus, these results again indicate that the small hardness data in Geo-Ms1-paste are primarily from the defects related phase. It should be noticed that the defects related phase should still contain some normal gel points and the real differences in the proportion of defects between Geo-Ms1-paste and others may be

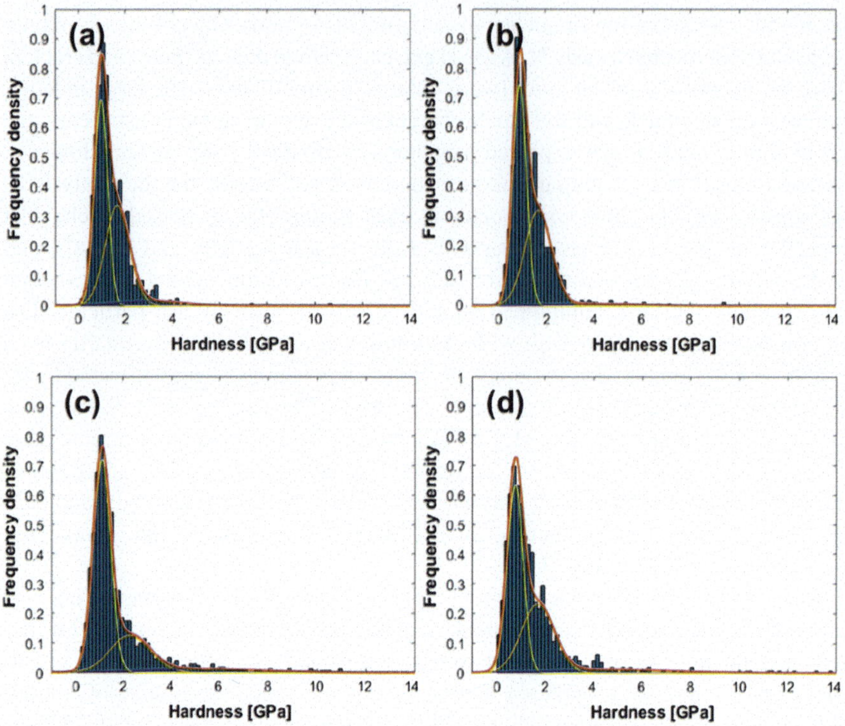

Fig. 9.27 Scratch hardness of ITZ and paste in geopolymer samples: **a** Geo-Ms1.5-ITZ (Bin size = 0.15 GPa); **b** Geo-Ms1.5-paste (Bin size = 0.15 GPa); **c** Geo-Ms1-ITZ (Bin size = 0.15 GPa); **d** Geo-Ms1-paste (Bin size = 0.15 GPa)

less significant as manifested in Table 9.7. After all, the contact area of nanoscratch test at the small hardness point is very large, and the distance between adjacent test points is only 1 μm. Still, it can be pointed that the low strength gel caused by small defects are more in Geo-Ms1-paste.

Based on the scratch hardness results and SEM observation above, it can be found that the ITZ of geopolymer samples studied is not weaker than that of the paste. Besides, the silica modulus of 1.5 used makes the difference between ITZ and paste less significant, enabling to make better use of both parts. The benefits brought by silica modulus on ITZ coincide with previous studies using siliceous aggregates [43, 44].

$$f = \pi_1 \Phi\left(\frac{m - \mu_1}{\sigma_1}\right) / \sum_{i=1}^{5} \pi_i \Phi\left(\frac{m - \mu_i}{\sigma_i}\right) \tag{9.5}$$

where f is the proportion of data from the defect related phase shown by the purple phase in Fig. 9.29. The m is the hardness value of 0.407 GPa or 0.637 GPa provided

Table 9.6 Deconvolution results for hardness probability density distribution

Samples	Components	Hardness (GPa)	Proportion (%)	Standard deviation
Geo-Ms1.5-ITZ	1	0.95	49.05	0.28
	2	1.65	45.01	0.53
	3	2.93	5.94	1.50
Geo-Ms1.5-paste	1	0.94	52.17	0.28
	2	1.72	44.64	0.56
	3	4.11	3.19	1.89
Geo-Ms1-ITZ	1	1.10	66.48	0.37
	2	2.27	26.09	0.83
	3	5.34	7.43	2.40
Geo-Ms1-paste	1	0.77	51.80	0.33
	2	1.71	39.87	0.68
	3	4.88	8.33	2.81
Geo-Ms1-paste	1	0.58	30.37	0.23
	2	1.06	32.01	0.32
	3	1.81	25.80	0.52
	4	3.30	9.20	1.09
	5	7.82	2.61	2.95

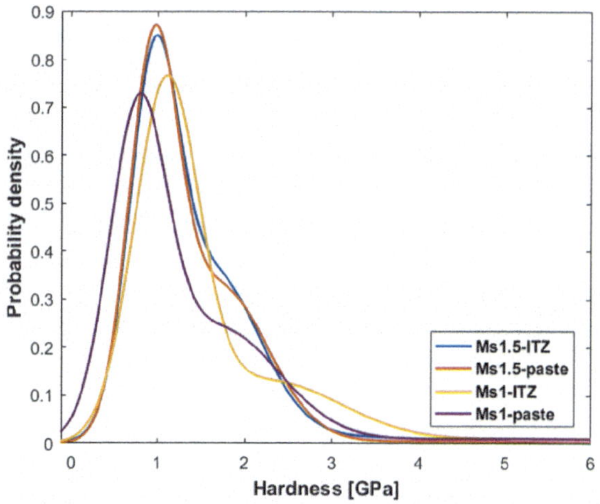

Fig. 9.28 The probability density distribution of scratch hardness

in Table 9.7 π_i, μ_i and σ_i are the proportion, average value and standard deviation of the i th component of the five components model.

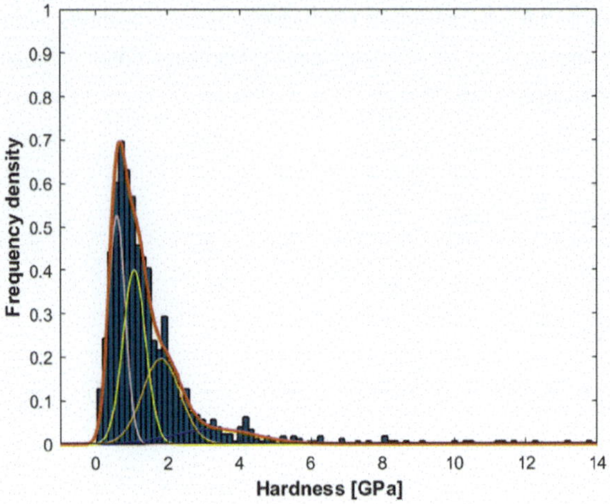

Fig. 9.29 Matching histograms by Gaussian mixture models with 5 components for Geo-Ms1-paste

Table 9.7 Percentage of test data from large scratch widths

Samples	$w > 5$ µm ($H < 0.407$ GPa), %	$w > 4$ µm ($H < 0.637$ GPa), %
Geo-Ms1.5-ITZ	1.66	8.65
Geo-Ms1.5-paste	1.82	7.88
Geo-Ms1-ITZ	1.94	7.46
Geo-Ms1-paste	8.45	21.21

9.3.4.5 Scratch Friction Coefficient of ITZ and Paste

The frequency density of scratch friction coefficient distribution is shown in Fig. 9.30. Compared with hardness, the distribution of friction coefficient is more concentrated, which mainly varies in the range of 0–1.5. The more concentrated distribution enables a small bin size to reveal more details. The 3-component models are unable to reflect the data well. Thus, the number of components is all increased to 5, which leads to an excellent match of models with histograms. These probability distribution curves are summarized for comparison in Fig. 9.31.

The distribution of friction coefficient does not show strong consistency with scratch hardness. Compared with others, Ms1-paste in Fig. 9.28 was found to have a substantial amount of low hardness data. The same phenomenon is not observed for its friction coefficient distribution. The results listed in Table 9.5 indicate that only Geo-Ms1-paste displays a significantly different average scratch friction coefficient, which shows larger values than others. As analyzed previously by the 5 components model, the distinct hardness probability distribution of Ms1-paste should be caused by the defects related component, which has an average hardness of 0.58 GPa. After clustering, hardness data belonging to this component were collected. For each hardness datum, by tracking the raw test data, the corresponding friction coefficient that tested at the same scratch location can be obtained. The hardness and

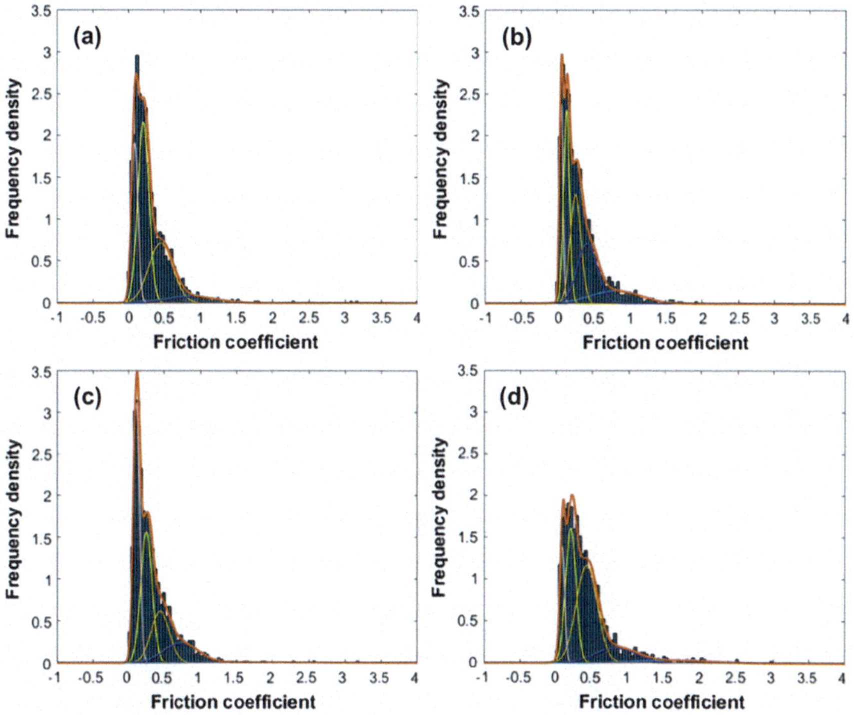

Fig. 9.30 Scratch friction coefficient of ITZ and paste in geopolymer samples: **a** Geo-Ms1.5-ITZ (Bin size = 0.04 GPa); **b** Geo-Ms1.5-paste (Bin size = 0.04 GPa); **c** Geo-Ms1-ITZ (Bin size = 0.04 GPa); **d** Geo-Ms1-paste (Bin size = 0.04 GPa)

Fig. 9.31 The probability density distribution of scratch friction coefficient

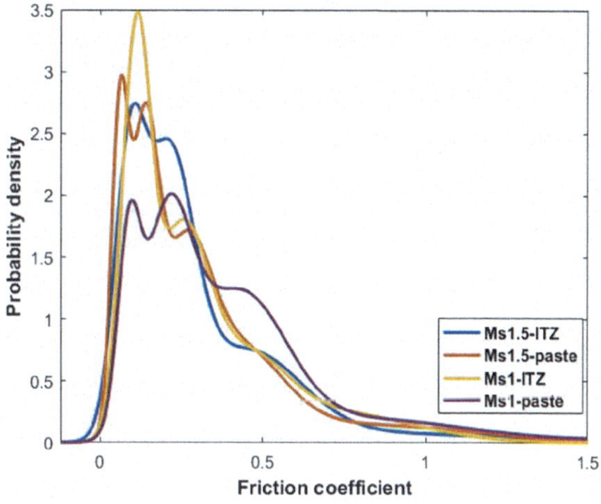

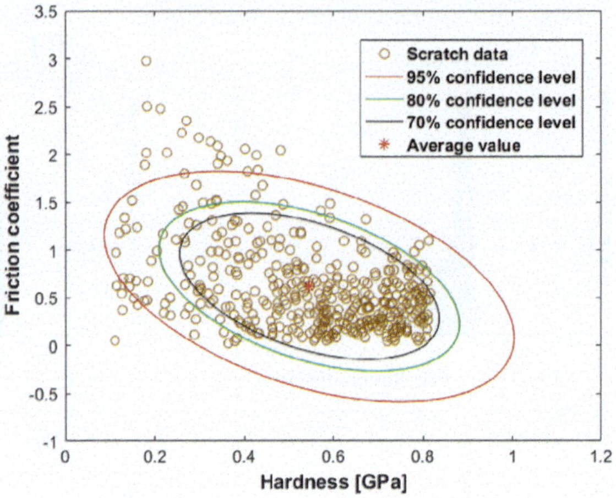

Fig. 9.32 Clustered hardness data and corresponding friction coefficient for the component with average hardness of 0.58 GPa

friction coefficient data are combined presented by scattering points in Fig. 9.32. For the clustered data, the hardness is 0.55 ± 0.19 GPa, while the friction coefficient is 0.62 ± 0.49. The hardness result indicates that although there is a slight deviation, the clustered data can still reflect the component with hardness of 0.58 GPa. The corresponding friction coefficient is found to have a very high average value and standard deviation. After removing the friction coefficient corresponding to the defects related component, the average value and standard deviation of remaining friction coefficient data are 0.31 and 0.22, respectively, which achieve similar values as other results in Table 9.5. Thus, the abnormally high friction coefficient of the Geo-Ms1-paste should be caused by the defects related component. A two-dimensional Gaussian distribution model [33, 45] was calculated based on these data. Since just one component is used, the model provided a good evaluation on average value but deviation on variance/covariance. Nevertheless, according to the confidence ellipses, it is not questionable to know that there is a negative correlation coefficient between hardness and friction coefficient. This phenomenon is in line with scratch results on different individual micro-constituents [46], where harder constituent has a lower friction coefficient. It is attributed to different mechanisms behind the tested properties [47]: hardness is governed by composition and morphology, while fiction is affected by the topology of the contacted surface. The defects related component is destined to have a large friction coefficient. Thus, small defects in gel that have a negative impact on macro performance has been identified by scratch technique via both scratch hardness displayed in Fig. 9.28 and scratch friction coefficient given in Table 9.5.

9.3.4.6 Effects of Scratch Direction

The results of the scratch test across aggregate, ITZ and paste of geopolymer with silica modulus of 1.5 are shown in Fig. 9.33. The result in Fig. 9.33a shows that there is a significant difference between the hardness of aggregate and binder (ITZ and paste). The average hardness of aggregate, ITZ and paste are 2.08 ± 0.73 GPa, 1.43 ± 0.92 GPa and 1.51 ± 1.15 GPa, respectively, which all show a larger standard deviation than the results collected from test scheme I. There is also a difference between the average hardness values obtained by these two different test schemes. For this test, it was set to have a similar number of test points on each object as the scratch test parallel to ITZ (scheme I). However, the wall effect leads to properties that change gradually with increased distance to aggregate. Compared with results all collected from a constant distance to the boundary of aggregate, the average results from this kind of test would be undoubtedly more variable. The around 33 test points at each distance make the results hard to overcome the heterogeneous features. The typical evidence is the different hardness of ITZ and paste on the same locations as that in the scratch scheme I, which are obtained as 1.28 ± 0.60 GPa and 1.53 ± 1.05 GPa, respectively, in scheme II for scratch tests 15 μm and 65 μm away from aggregate. When the scratch test reaches ITZ from aggregate, the hardness value decreases rapidly and then increases. Scratch hardness in the boundary of ITZ and paste also shows low value. There are some reasons such as interfacial cracks and very low content of unreacted particles that may be able to support the low hardness of ITZ at the location near to aggregate. Different from the first boundary, the low hardness at the boundary of ITZ and paste does not indicate a successful identification of boundary by scratch test, which would be mainly attributed to the unstable results from limited test points at each given distance. After all, the term 'transition zone' means gradually changing properties. Boundary defined between 'ITZ' and 'paste' is from the location where properties tend to be stable. In addition to unstable results, another inaccuracy in scratch test II is the constraint effect, which would enhance the properties of the part of ITZ that near to aggregate. Hence, although the average value of all hardness data on each object is not significantly different from that in test scheme I, the average results from scheme II should also be more unstable.

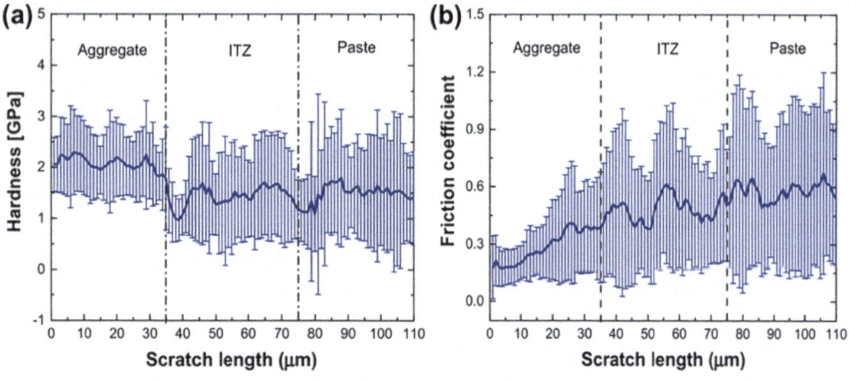

Fig. 9.33 Scratch test perpendicular to the boundary of modelled aggregates: **a** hardness; **b** friction coefficient

The friction coefficient along scratch is shown in Fig. 9.33b, which are 0.29 ± 0.21, 0.48 ± 0.34 and 0.58 ± 0.42 respectively for aggregate, ITZ and paste. The average value and standard deviation obtained are all larger than that in test scheme I. For aggregate, it may be caused by a high amount of defects at the location near to boundary resulted from polish. Compared with 0.31 in scratch scheme I, the friction coefficient of ITZ and paste tested at the distance of 15 μm and 65 μm to the surface of aggregate in scratch scheme II are larger values of 0.38 ± 0.25 and 0.56 ± 0.44, respectively. Although different directions may sometimes affect the friction coefficient even in the same location, the nearly one times of difference in the same paste still indicates the errors caused by the number of test data.

Thus, based on the same number of test points, scratch scheme I could provide a more reliable result for a comparison of ITZ with paste. Scratch scheme II with increased distance to aggregate has the potential to reveal the variation trend of properties but requires much more tests to eliminate the inaccuracy caused by insufficient test data on each given scratch distance. The number of test points should increase further if a smaller tip [41, 42] is adopted to obtain valid data without affected by constraint effect on the part of ITZ very close to aggregate.

9.4 Nano/Micromechanical Characterisation and Image Analysis on the Heterogeneity of ITZs in Geopolymer Concrete

9.4.1 Raw Materials and Sample Preparation

The modelled aggregates used were from calcite-dominated limestone rock. A φ25 cylinder rock sample was drilled out and cut into small slices. Then, φ25 slices were long time ground by 120 grit abrasive paper until become cuboids with a size of around 15 mm × 10 mm × 8 mm. The cuboids were further polished by 320, 600 and 1200 grits abrasive papers and 1 micron alumina to achieve smooth surfaces on the top, bottom and lateral sides. Class F fly ash was activated by alkali silicate solution to synthesise geopolymer. The chemical composition of fly ash is shown in Table 9.1. Sodium silicate purchased has the composition of 14.7% of Na_2O, 29.4% of SiO_2 and 55.9% of H_2O. Sodium hydroxide pellets were dissolved in water and mixed with commercial sodium silicate after cooled down to prepare alkali solution. The geopolymer mortar was synthesised with a solution/powder ratio of 0.57, Na_2O/fly ash of 0.08, silica modulus of 1.3 and sand/fly ash of 2. The mould used was a small φ25 cylinder mould. The internal surface of the mould was oiled and attached with a layer of plastic film to prevent the contact of geopolymer with mould and consequent demoulding damage on the interfaces.

After 5 min of mixing, fresh geopolymer mortar was poured into moulds. When the mortar reached half the height of a mould, the polished modelled aggregate was placed on the centre of the mould prior to the further casting. After casting, the

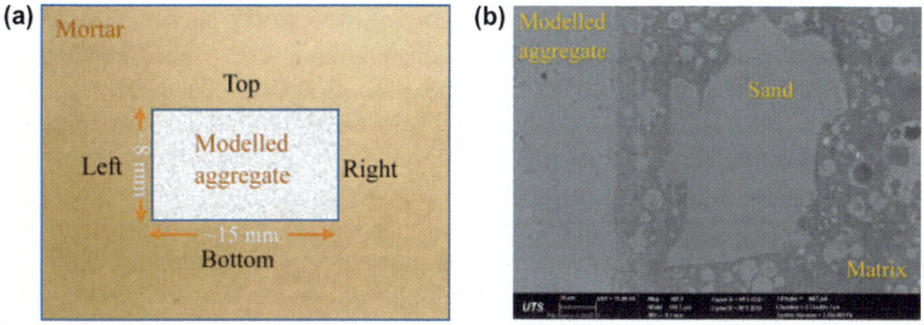

Fig. 9.34 Characteristics of the interface around natural aggregate and modelled aggregate: **a** modelled aggregate embedded in mortar; **b** comparison of different interfaces

prepared samples were vibrated on a vibration table for 1.5 min and then sealed by plastic film. The curing regime applied was heat curing under temperature of 65 °C for 24 h, followed by 22 °C curing to 28 days. Afterwards, the samples were demoulded and cut from top to bottom to expose the modelled aggregate and interfaces as shown in Fig. 9.34a. A sample with only slightly inclined aggregate was subjected to polish and used for the test. The surface was ground by 320, 600 and 1200 grit abrasive paper to flat and then polished by 1 μm and 0.3 μm alumina for 20 min each. The properties of ITZ at the top, bottom and lateral sides of the modelled aggregate were investigated in the following parts to reveal the properties difference caused by locations. The tests on the lateral side ITZ were about half on the left side and half on the right side. For nature aggregate with an irregular shape, it is sometimes difficult to strictly distinguish the location of ITZs, for instance, the top surface and the lateral surfaces of the sand particle as shown in Fig. 9.34b. Besides, it is difficult to accurately conduct micro/nanomechanical tests on the ITZs surrounding nature aggregates. Even if the starting point is on the ITZs, straight scratch tests with a length larger than tens of microns would sometimes still path through aggregate and paste. Modelled aggregate exhibited a very neat boundary, which can overcome these issues. The simplified mix condition also led to more even binders at each side and hence ensures the accuracy of the limited testing numbers.

9.4.2 *Variation in the Proportion of Constituents from ITZs to Paste*

BSE was operated under an accelerating voltage of 15 kV. BSE images were randomly taken along the interface between modelled aggregate and geopolymer at the magnification of 500×. Then, constituents of geopolymer were identified and classified in images by gray values threshold segmentation technique [48]. The pores on fly ash and the locations where fly ash dropped due to polish were filled and still segmented as the fly ash phase. The initial matrix adhered to modelled aggregate was divided into eighteen small bands of 5 μm. After averaging the results from fifty 500× BSE images, the proportion variation trend of constituents in consecutive bands with increased distance to aggregate was presented. To ensure the same length of the bands for different types of ITZ, the sample was rotated before taking BSE images to make the original top and bottom ITZs shown in Fig. 9.35a, b was also on the right side and analysed in the same way as shown in Fig. 9.35c, d.

The statistical results for the proportion of constituents are shown in Fig. 9.36. It can be found that the distribution of constituents is not the same for ITZs from different sides of the same aggregate, although the proportion difference is not very clear. The proportion of fly ash in the initial part of ITZ is higher in the top and bottom side ITZ than the lateral side ITZ, especially for the 5–10 μm and 10–15 μm bands. The same situation is also for bands far away from the aggregate, such as the bands with a distance more than 50 μm to the aggregates. For all the ITZs, the content of fly ash increases significantly in the first 15 μm. The band of 10–15 μm seems to have a slightly high amount of fly ash for bottom side ITZ and lateral side ITZ when compared with the expected increase trend. After that, the proportion of fly ash still increases remarkably for the lateral side ITZ, but rise with a low slope for the bottom side ITZ. For the top side ITZ, when the distance to the boundary of aggregate is more than 15 μm, the content of fly ash just increases slightly. According

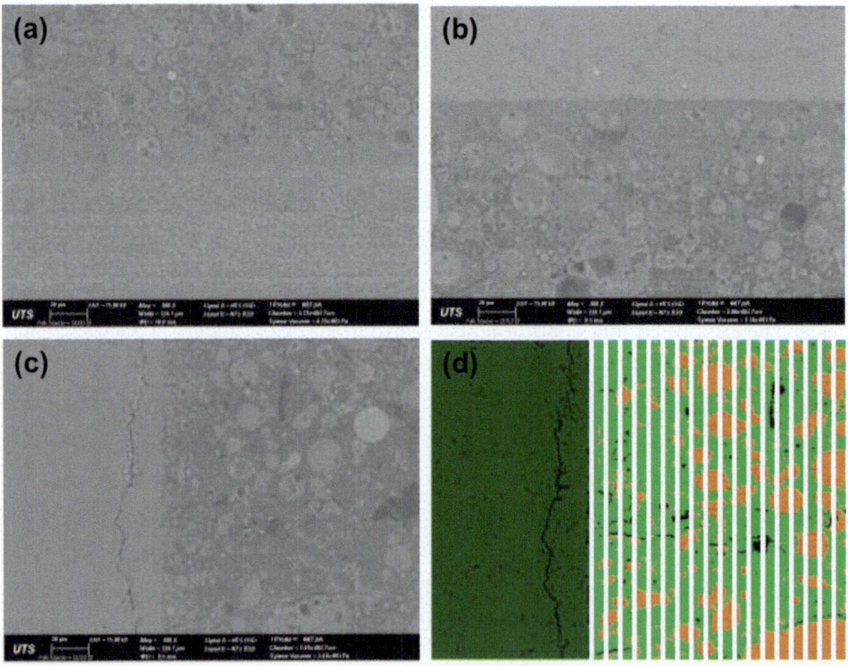

Fig. 9.35 ITZs at different locations surrounding aggregates: **a** ITZ at the top surface of aggregate; **b** ITZ at the bottom surface of aggregate; **c** ITZ at the right, lateral surface of aggregate; **d** proportion analysis based on bands

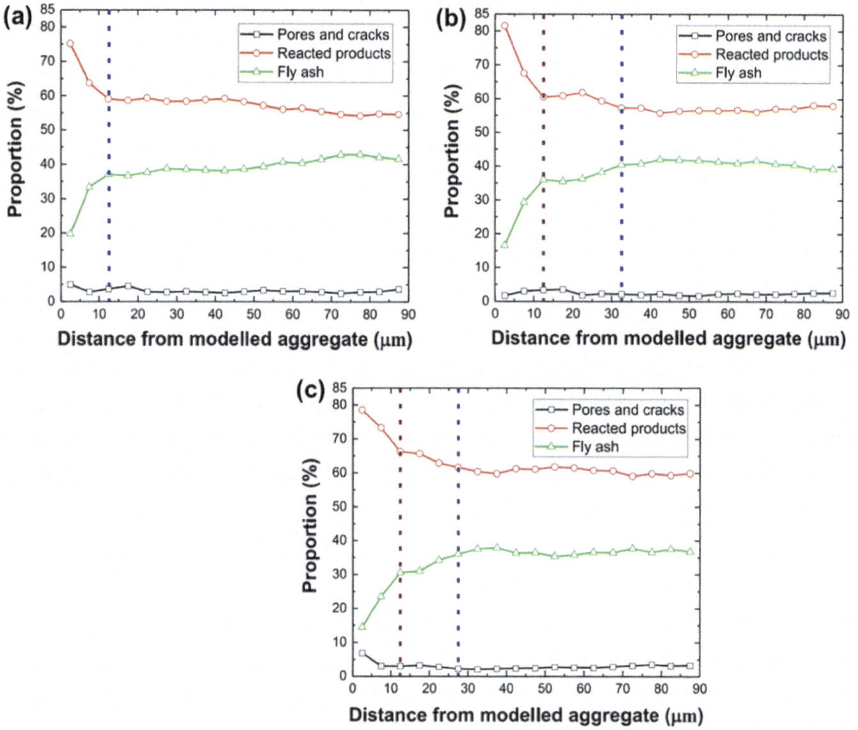

Fig. 9.36 Statistical results of constituent distributions in different ITZs: **a** top surface of aggregate; **b** bottom surface of aggregate; **c** lateral surface (left/right sides) of aggregate

to the significant increase part of the fly ash particles, the thickness of ITZs in the modelled geopolymer concrete was around 15 μm, 35 μm and 30 μm, respectively, for the top, bottom and lateral side ITZs. Generally, the effects of the aggregate wall on the proportion distribution of constituents are less significant for the bottom and especially the top side ITZ. In terms of the pores and cracks, there is only a very slight difference between the samples.

9.4.3 Variation in Number and Equivalent Diameter of Fly Ash from ITZ to Paste

In addition to the proportion, the number and diameter of fly ash in different ITZs were analysed as well. These two parameters cannot be studied via truncated bands as the traditional method above since it would break down the intact particles. Even based on a complete image, some independent objects in image analysis may be decomposed, while different objects may be merged as one object. To avoid them, the particles were firstly further filled. It was the particles that were no longer complete after reaction but still showed a clear outline. In this process, the interior part of fly ash that has reacted and classified into gel by the gray value was restored to fly ash according to the outline identified. The second aspect, clustering of independent particles can be clearly observed in Fig. 9.37a, where the dotted rectangle is the minimum rectangle that contains the object identified and the solid dot is the centroid of the object. For easy observation, only the rectangle of the ten objects with the largest areas was drawn, and the largest one was highlighted by a red rectangle. Since the first step consolidated the shape of the weak particles, morphological opening of the binary image was performed with a square-shaped structuring element. This step disconnected lots of particles as revealed in Fig. 9.37b and the eroding-dilating process ensured that the area of particles remained almost unchanged. Furthermore, tiny objects that fewer than 100 pixels were deleted to avoid their interference on results as manifested in Fig. 9.37c. Sometimes there were still particles that were combined and recognised as a single particle as revealed in the yellow rectangle of Fig. 9.37c, which were manually segmented by deleting several connected '255' gray value in the gray matrix as shown in Fig. 9.38. After that, the object in the yellow rectangle became two individual particles as displayed in Fig. 9.39a, c.

After dealing with the images, the distribution of number and equivalent diameter of fly ash with different distance to aggregate was investigated, where the equivalent diameter means the diameter of a circle that has the same area as the object. It should be noted that the equivalent diameter here is for the cross-section of unreacted and partially unreacted fly ash particles. It is assumed that these remaining particles can still reveal the relative difference between samples as before any reactions. Besides, the analysis based on cross-section underestimates the actual diameter of the three-dimensional particles [49], because most of the cross-sections would not pass through the centre of the sphere in this study. However, the equivalent diameter should be still helpful for internal comparison study if a large number of sections of fly ash are statistically analysed to decrease the impact from the random sections of spherical particles, and the difference between samples is not too small. The properties distribution were studied in the form of 10 μm × 9 bands and 30 μm × 3 bands as displayed in Fig. 9.39. The fly ash particles were assigned to a band according to their centroid. For instance, the properties of the third band of Fig. 9.39c represent the average properties of all fly ash particles whose centroid is in the range of 60 to 90 μm from the aggregates. Considering the possible significant impact from large particles, in addition to the average properties of all particles in each band, the average properties of the one largest particle and remaining particles after removing the largest one particle in each band were investigated for the 10 μm × 9 bands. For the 30 μm × 3 bands, the average properties of all particles, the three largest particles and remaining particles after removing the largest three particles in each band were studied. The final results presented are based on the average results of fifty images. Some schematic diagrams are shown in Fig. 9.39, where the largest particle in each band was marked by a red rectangle.

The variation in number and equivalent diameter of fly ash from ITZ to paste can be found in Figs. 9.40 and 9.41. The small band of 10 μm is used to understand the

Fig. 9.37 Image analysis of the number and diameter of fly ash: **a** binary image; **b** morphologically open the image; **c** delete small areas

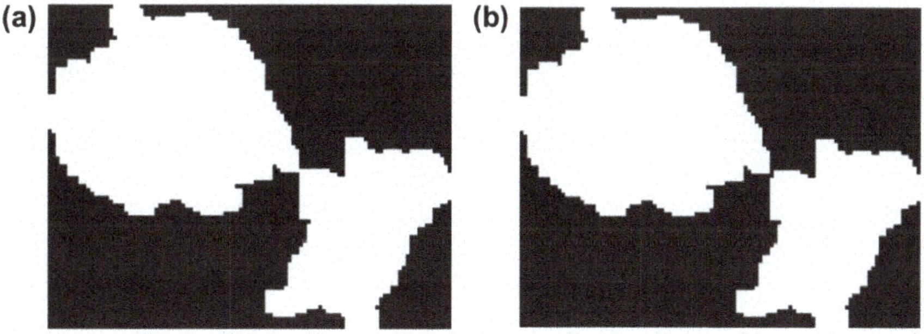

Fig. 9.38 Disconnection of special objects for image analysis: **a** original; **b** after disconnection

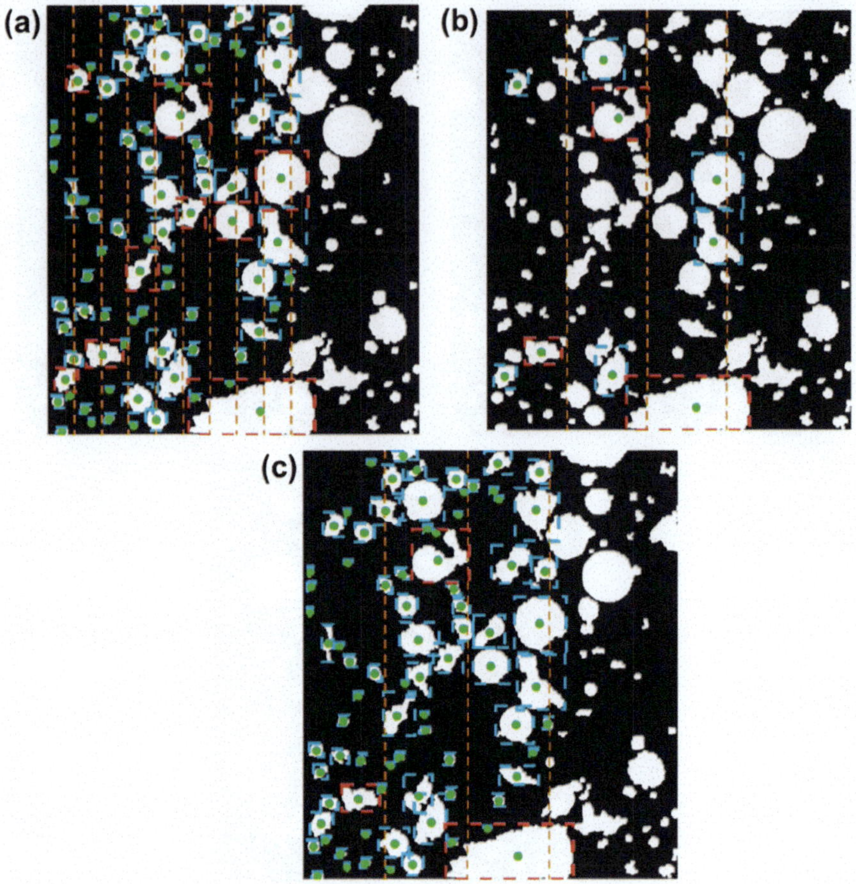

Fig. 9.39 Statistical of number and diameter of fly ash based on bands: **a** 10 μm × 9 bands; **b** 30 μm × 3 bands (largest three particles); **c** 30 μm × 3 bands

detailed distribution, whereas it would undoubtedly obtain less stable results than that of 30 μm bands. As shown in Fig. 9.40a, the results fluctuate significantly for the average equivalent diameter of all fly ash particles in each band. However, it still indicates that, after around 20 μm, the average diameter of fly ash is relatively larger in the lateral side region than that of the top and bottom side region. A more apparent difference between samples is observed for the number of particles. The 9 bands in the lateral side region always show fewer fly ash particles than the corresponding bands in the top and bottom side regions, especially for the first band with a distance of 0 to 10 μm away from aggregate. For all these regions at different sides, the significantly lower diameter and number values are generally just found for the first band, although there are sometimes still increasing trends for later bands. The average diameter is around 5.51–5.79 μm in the first band and increased to the range of 6.44–8.95 μm for the later bands. Similarly, the average number of fly ash particles in the first band is

3.84–6.60, where the lateral side ITZ has obviously fewer particles than the top side ITZ. The average number increased to 5.98–8.11 for later bands, with the difference between samples at each band less than 2.

In addition to all particles, Fig. 9.40c, d show the average diameter of the largest particle and average diameter of remaining particles after removing the largest particle in each band. Compared with the results of Fig. 9.40a, Fig. 9.40c presents a more remarkable increase in the diameter of fly ash with distance. Besides the first band, the second band also has average diameters generally smaller than later bands. Both the first and second band of the lateral side ITZs show smaller diameter when compared with the top and bottom side ITZ. Although both the average results in Fig. 9.40a, c indicate that the fly ash in initial 0–10 μm ITZ has a relatively small size, it does not mean that there are no large particles involved in this part. The first band of Fig. 9.40c is generally not lower than the later bands of Fig. 9.40a. Before averaging the results from fifty images for Fig. 9.40c, the largest diameter appeared is 18.97 μm, 15.27 μm, 17.06 μm in the first band and 30.23 μm, 34.59 μm, 25.67 μm in the second band for the top, bottom and lateral side ITZ, respectively. In addition,

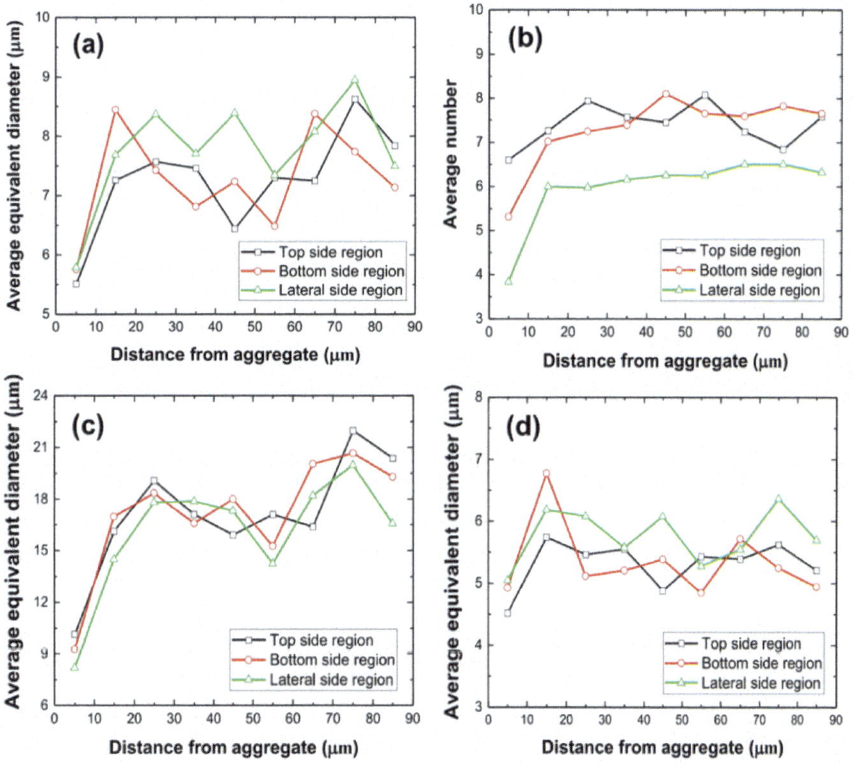

Fig. 9.40 The average equivalent diameter and number of fly ash particles in each band (10 μm × 9 bands): **a** all fly ash particles in each band; **b** all fly ash particles in each band; **c** the largest fly ash particle in each band; **d** after removing the largest fly ash particle in each band

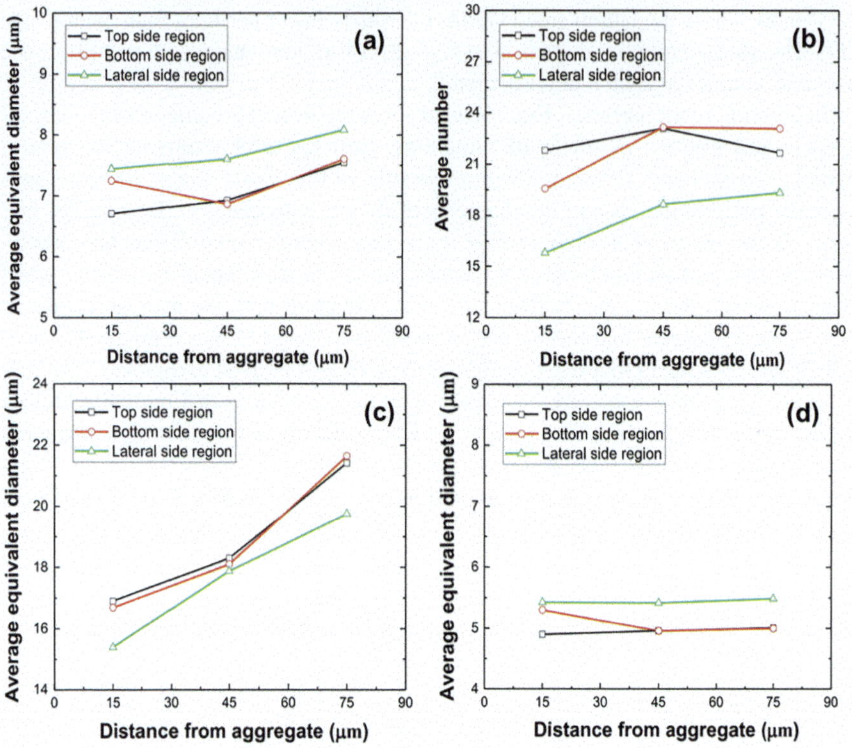

Fig. 9.41 The average equivalent diameter and number of fly ash particles in each band (30 μm × 3 bands): **a** all fly ash particles in each band; **b** all fly ash particles in each band; **c** the three largest fly ash particle in each band; **d** after removing the three largest fly ash particle in each band

as illustrated by a gray particle in Fig. 9.42, it is possible that part of large particles is in the initial ITZs but is not able to be displayed in the final results. After all, the band analysis of diameter/number is based on the centroid of fly ash particles. Proportion analysis would at least take into account the part that is in the band. That means the diameter/number analysis in the initial band would be more sensitive to wall effect than the proportion analysis when the particles size is significantly larger than the band size. If the largest fly ash particle is excluded as provided in Fig. 9.40d, the results display a less significant difference between the first and later bands than in Fig. 9.40a.

Figure 9.41 provides the statistical results from 30 μm bands. Figure 9.41a, b indicate that there are more fly ash particles in the top and bottom side regions, but their average diameter is smaller than the lateral side region. The average diameter of fly ash in the three bands is in the range of 6.71–7.44 μm, 6.87–7.61 μm, and 7.55–8.09 μm, respectively. The corresponding number of fly ash ranges from 15.81 to 21.81, 18.67 to 23.17, and 19.35 to 23.09, respectively, in these bands. For the results tested from different sides of aggregate, there is a general increasing trend for

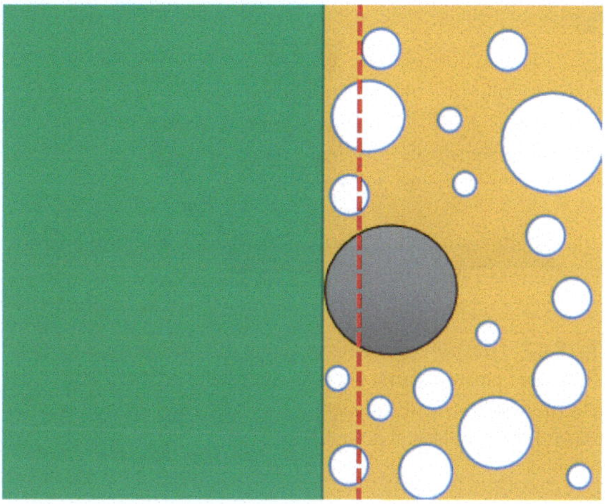

Fig. 9.42 Difference between proportion and number/diameter analysis by bands

both average diameter and number with the increased distance to aggregate, although deviations are also observed. The details presented in Fig. 9.41c, d imply that the smaller average diameter of the top and bottom regions is not caused by several large particles. Instead, the average diameter of the three largest fly ash particles of them is significantly larger than that of lateral ITZs. For the fly ash in all these sides, the diameter of the three largest fly ash particles increases remarkably with the distance to aggregate, resulting in the increased average diameter in Fig. 9.41a. Despite that the binary image for the number and diameter analysis was further filled compared with the image for area analysis, the results still reveal some similar phenomenon and can be analysed relevantly. For the larger proportion of fly ash and less significant wall effect in the top and bottom side shown in Fig. 9.36, the dominant reason should be the increased number of particles as displayed in Figs. 9.40b and 9.41b.

9.4.4 Micromechanical Property Tests

9.4.4.1 Test Procedure

Nanoscratch and nanoindentation tests were performed using the same instrument Agilent G200 Nano Indenter, but different tips. A Berkovich tip was used for the nanoindentation test, while a hemispherically-tipped, conical stylus was adopted for the scratch test to avoid errors caused by the orientation of the tip. The tip radius was 20 nm and 5 μm, respectively, for nanoindentation and nanoscratch. A comparison of these two test techniques can be found in Fig. 9.43 and the following test curves obtained from the same homogeneous object. Different from the traditional application of scratch for ITZs, the scratch investigation in this study is parallel to the boundary of aggregate.

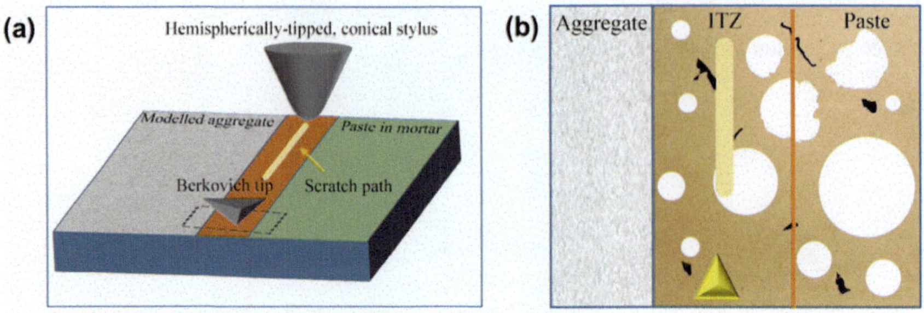

Fig. 9.43 Diagram of micromechanical properties test of ITZ in geopolymer concrete (white: fly ash; black: pores and cracks; yellow: reacted products) (for interpretation of the references to color in this figure legend, the reader is referred to the web version of this article): **a** scratch and indentation test on ITZ; **b** enlarged dotted area

Nanoscratch technique attracts more and more attention in the multifunctional characterisation of cement-based materials [42, 50, 51]. The typical scratch tests consist of prescratch, scratch and postscratch as shows in Fig. 9.44. In this study, scratch hardness is used. The scratch hardness value HS_p tested by a hemispherically-tipped, conical stylus can be calculated by Eq. (9.6) based on ASTM G171-03 (Standard test method for scratch hardness of materials using a diamond stylus). The scratch width w was determined by the depth detected and the geometry of the tip [52]. The constituent distribution shown in Sect. 9.4.3 indicates that in addition to the top ITZ, the significant part of ITZs is generally around 30 μm. Considering the large size of tip for scratch, the scratch test was set at 15 μm away from the aggregate, namely the location about the centre of ITZs. Moderately large normal force P is necessary for the accuracy of testing and obtaining more homogeneous properties, but too large force with large penetration depth d would make the results affected by constraint effect from the aggregate. By referring to the involved range of indentation test [32] and the requirements in ASTM G171-03, the normal force was finally determined as 4 mN, which ensured that even large depth from some soft phases would just activate the ITZ part but not significantly affected by the aggregate and paste part. The profiling loads in prescratch and postscratch scan were both taken as 20 μN. Total 18, 100 μm scratches were randomly conducted along each type of ITZ with the scratch speed of 4 μm/s and data acquisition rate of 1 point/μm. In addition to ITZ, the same scratch test was also conducted on paste far away from the polished aggregate and sand for comparison. Considering the features of low content of fly ash in ITZs and the fact that the mechanical property of fly ash is a less important aspect for the properties comparison, the paste regions without abundant large fly ash particles were chosen for the test. Due to the larger areas of paste than ITZ, the scratches were increased to 24 tests.

$$HS_p = \frac{8F_{N\max}}{\pi w^2} \tag{9.6}$$

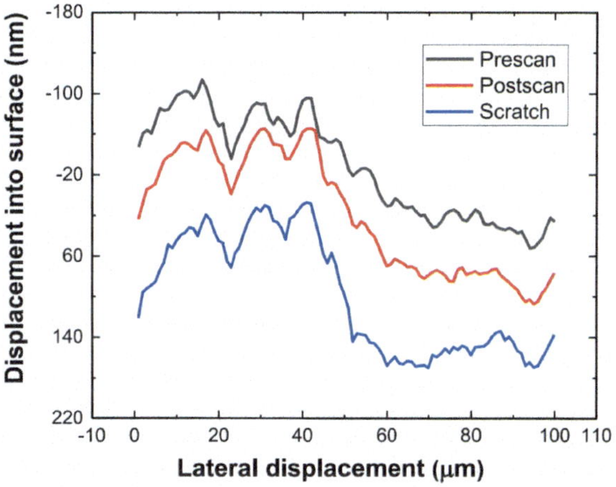

Fig. 9.44 Scratch test curves

$$w = \begin{cases} 2\sqrt{r^2 - (r-d)^2} & d \le d_{gt} \\ w_{gt} + 2\tan\left(\frac{\alpha}{2}\right) \times \left(d - d_{gt}\right) & d > d_{gt} \end{cases} \tag{9.7}$$

$$d_{gt} = r\left(1 - \sin\left(\frac{\alpha}{2}\right)\right) \tag{9.8}$$

where d_{gt} and w_{gt} are the geometric transition depth and the corresponding scratch width, which are calculated as 669.87 nm and 5 μm, respectively. The parameter r means the radius of the hemispherical portion of the tip, while α denotes the apex angle of the conical portion of the tip. For the tip used, their values are 5 μm and 120°, respectively.

The initial part of ITZ is the region most affected by the wall effect. Compared with nanoscratch, nanoindentation owing to the small tip allows test on the more significant ITZ part that is closer to the aggregate. The nanoindentation test points were set at the same distance of 10 μm to the aggregate boundary and conducted based on the depth-control model. The maximum feasible depth of 1200 nm that without affected by the constraint effect of aggregate was used to activate more homogeneous volume to compare the properties of ITZ [26]. The typical loading curve is shown in Fig. 9.45. After reaching 1200 nm, the load was sustained for 10 s, and then unloaded to 10% of the peak force to complete the test. Based on the curve, the modulus (E) and hardness (H) of each test point were calculated by Eqs. 9.9–9.11. A total of 125 effective indentation data were randomly tested from each type of ITZ. The spacing between adjacent test points was at least 15 μm. Due to the close distance to the aggregate, even if the maximum feasible depth is adopted for testing, the depth is still small and makes the results easy to be affected by individual phase. Fly ash due to the large mechanical properties and size is a significant factor that would interfere with the comparison result. Thus, the tests were on the gel-rich regions and large fly ash particles were intentionally avoided under the microscope. Nanoindentation was also conducted on bulk paste for comparison.

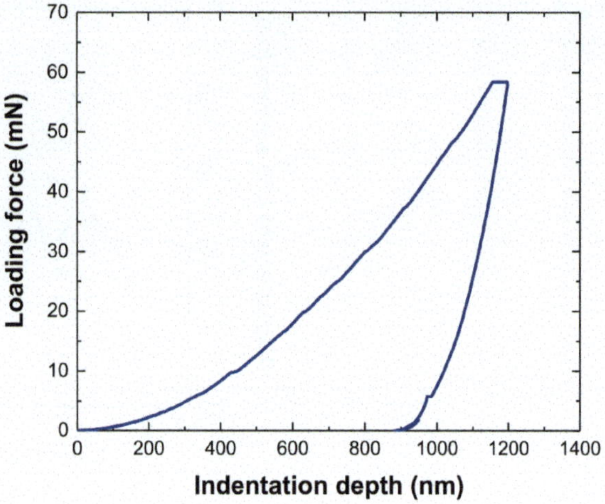

Fig. 9.45 Typical load and depth curve of nanoindentation test

$$H = \frac{P_{\max}}{A} \tag{9.9}$$

$$S = \left.\frac{dP}{dh}\right| h = h_{\max} = \frac{2}{\sqrt{\pi}} E_r \sqrt{A} \tag{9.10}$$

$$\frac{1}{E_r} = \frac{1 - v^2}{E} + \frac{1 - v_i^2}{E_i} \tag{9.11}$$

where P and h are the indentation load and indentation depth, respectively; S is the initial unloading stiffness; E_r is the reduced elastic modulus; E_i and v_i are the elastic modulus and Poisson's ratio of the indenter, respectively; E and v are the corresponding parameters for sample; A is the projected contact area. Constant Poisson's ratio of 0.2 was set for the testing.

9.4.4.2 Nanoscratch Results and Analysis

For a given type of ITZ, there are abundant scratch data acquired from different locations of it. After calculation, each test datum was transformed into scratch properties. The arithmetic mean of hardness data is provided in the third column of Table 9.8, which varies from 1.13 ± 0.94 to 1.59 ± 1.33 GPa. The top ITZ and lateral ITZ have the highest and the lowest hardness value, respectively. Owing to the small width of ITZ, the penetration depth is not allowed to be too large. Therefore, the scratch test result is still sensitive to the constituents. To further understand the hardness difference between the different ITZs, the properties data can be analysed via frequency density distribution histogram or models such as Gaussian mixture model (GMM). However, the components/peaks presented in GMM/histogram are sometimes spurious phases that do not precisely correspond to any real constituent of material. For the scratch investigation with a large tip, the test itself would be hard to obtain properties of individual phases. Most of the test data should be the result of multiple phase interaction. The data analysis is also a source of spurious phases [44]. It is necessary to use a reasonably small bin size for histogram or many components k for GMM, because they help to decrease the number of mixed-phases and display a more detailed property distribution, especially for the study of ITZs in the same sample that may just have minor differences. However, small bin size/large k generally results in sub-phases [32]. The spurious phases presented in the data analysis process are almost impossible to be avoided. For the scratch data analysis, the primary research purpose is only to compare different ITZs where spurious phases are allowed.

Since there may be sub-phases with negligible proportion in some samples, sometimes the effective phase number would be varied between samples and then affect the comparison results by GMM even if the same number of phase k is assigned for comparison. The bin size of histogram is not affected by sub-phases. Hence, the same bin size was set for the frequency density distribution histogram of samples to provide an equivalent condition for comparison. Very small bin sizes were tried, which should be able to reveal the details but not lead to a disordered shape of the histogram, namely, too many sub-phases. The final bin size used was 0.15. Despite the above advantage, histograms are hard to be compared directly. The previous studies [28, 53] show that although different methods, the GMM estimated by the maximum likelihood estimation (MLE) method can match frequency density distribution histogram by adjusting bin size or k [30]. Therefore, the number of components in GMM was increased to match the outline of histograms as shown in Fig. 9.46, where the one-dimensional normal distribution curves represent the components of GMM. Afterwards, the GMMs were plotted together in Fig. 9.47.

It should be mentioned that the GMM for paste with 5 components is not the optimal model with the maximum log-likelihood function value but the one with the second maximum value. For the given number of components of 5 and 6, the optimal model of paste is presented in Fig. 9.48. The deconvolution results are listed in Table 9.9. Compared with the model listed in Fig. 9.46d, the model in Fig. 9.48a matches the histogram slightly better at the region where the hardness is larger than 2 GPa but worse at the most prominent peak region. After increasing the number of components to 6 as displays in Fig. 9.48b, the model matches both regions well with the deconvolution results at the regions similar to the corresponding better matched 5 component model. However, for other models, a further increase in the number of components generally leads to spurious phases with a small proportion such as the bottom ITZ provided in Fig. 9.49. The small proportion spurious phase in the 6 components model of top and lateral ITZ just accounts for the negligible value of 0.70% and 0.73%, respectively. Different from the bottom ITZ, the components with the lowest hardness value in the top ($k_1 = 0.17$, $k_2 = 0.64$) and lateral ITZ ($k_1 = 0.34$, $k_2 = 0.53$) are no more corresponding to components in other samples and hence not useful for the comparison. Thus, the final comparison was primarily based on the 5 components models.

Table 9.8 Deconvolution results for hardness probability density distribution (5 components)

Test objects		k_{all}	k_1	k_2	k_3	k_4	k_5
Top-ITZ	$\mu \pm \sigma$	1.59 ± 1.33	0.71 ± 0.23	1.22 ± 0.23	1.74 ± 0.47	2.83 ± 0.93	6.91 ± 2.88
	f (%)	100	30.34	32.54	16.84	17.58	2.70
Bottom-ITZ	$\mu \pm \sigma$	1.44 ± 1.45	0.73 ± 0.24	1.21 ± 0.34	1.96 ± 0.59	3.21 ± 1.04	8.64 ± 2.49
	f (%)	100	48.58	24.08	17.53	7.42	2.38
Lateral-ITZ	$\mu \pm \sigma$	1.13 ± 0.94	0.46 ± 0.11	0.88 ± 0.25	1.59 ± 0.50	2.84 ± 1.00	8.93 ± 1.83
	f (%)	100	13.80	59.21	21.78	4.48	0.73
Paste	$\mu \pm \sigma$	1.36 ± 1.48	0.43 ± 0.11	0.71 ± 0.20	1.30 ± 0.43	2.89 ± 1.17	6.61 ± 2.98
	f (%)	100	12.47	39.89	30.65	13.36	3.64

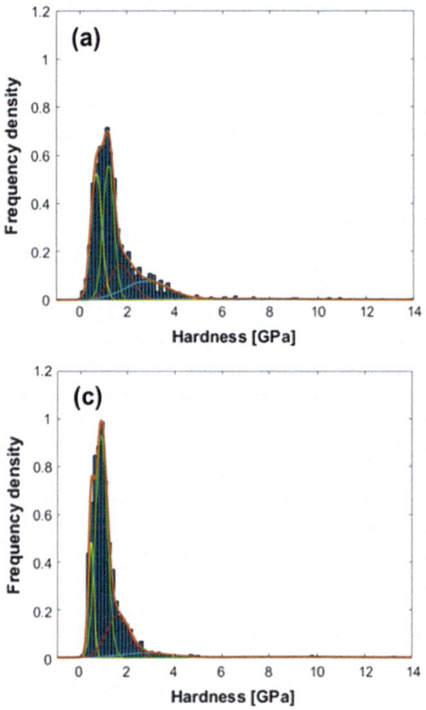

Fig. 9.46 Scratch hardness distributions of different ITZs and paste: **a** top-ITZ; **b** bottom-ITZ; **c** lateral-ITZ; **d** paste

Fig. 9.47 Summary of the probability density distribution of scratch hardness

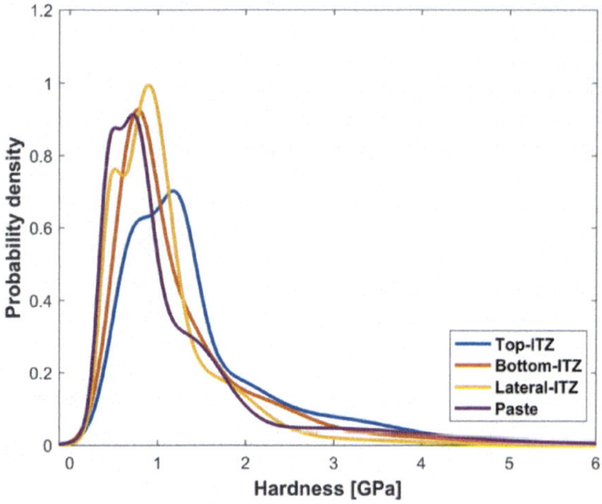

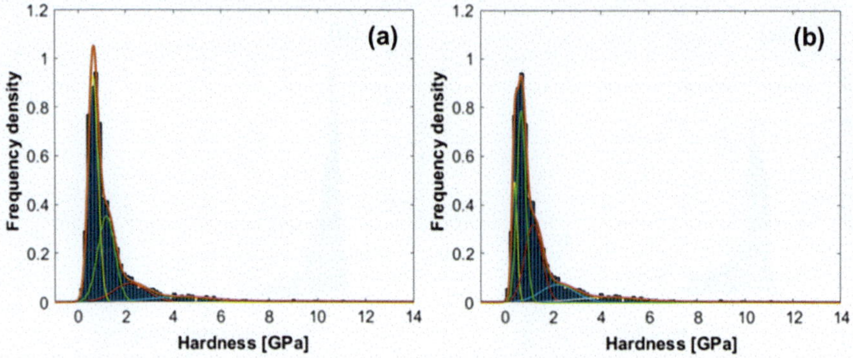

Fig. 9.48 The optimal model for paste with 5 and 6 components: **a** k is 5; **b** k is 6

Table 9.9 Optimal results for GMM models with increased number of components

Test objects		k_1	k_2	k_3	k_4	k_5	k_6
Paste_5	$\mu \pm \sigma$	0.61 ± 0.20	1.16 ± 0.36	2.18 ± 0.70	4.28 ± 1.33	8.82 ± 2.81	–
	f (%)	46.67	32.01	13.14	6.60	1.59	–
Paste_6	$\mu \pm \sigma$	0.42 ± 0.11	0.70 ± 0.19	1.21 ± 0.37	2.24 ± 0.71	4.34 ± 1.31	8.84 ± 2.80
	f (%)	13.74	36.70	29.23	12.42	6.33	1.58
Paste_7	$\mu \pm \sigma$	0.31 ± 0.02	0.76 ± 0.23	1.29 ± 0.22	1.81 ± 0.55	3.01 ± 1.01	8.37 ± 2.63
	f (%)	1.79	53.12	13.64	19.73	9.17	2.56

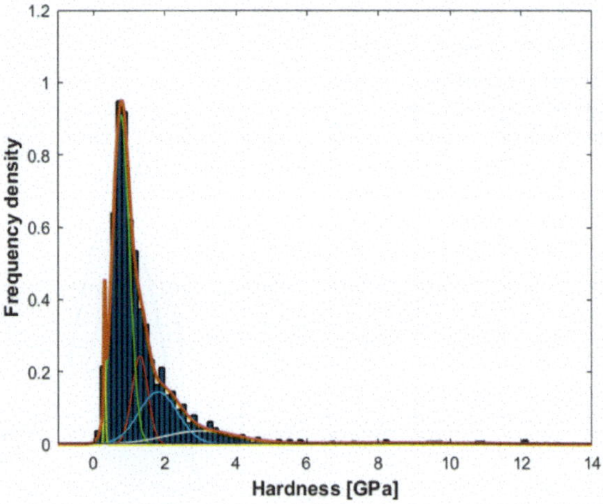

Fig. 9.49 Small proportion spurious phase of the bottom ITZs when using 6 components

For histograms provided in Fig. 9.46, there is only one remarkable peak. Although affected by other components, the peak should be mainly from the test points on the reacted products, which have the maximum volume fraction as shown in Fig. 9.36. The properties differences of samples can be observed from Fig. 9.47. It is evident that the initial part of the lateral side ITZ and paste curves is coincident, while bottom ITZ and especially top ITZ shows low probability density at this low hardness value part. The end part of the curves indicates that the top ITZ has the highest proportion of large hardness components, while the lateral side ITZ has the lowest proportion. These hardness distribution results show a consistent trend as the constituent distribution provided by BSE, where the lateral ITZ has more reacted products and a lower amount of fly ash particles than the bottom and especially the top ITZ.

A detailed comparison can be realised by analysing the components information. Although the peak of histograms and GMM are significantly different, the deconvolution results listed in Table 9.8 show that the difference sometimes can be understood by the different proportion of similar components. The first two components have similar hardness value for the top and bottom ITZ, whereas the top ITZ has a lower proportion of k1 and a higher proportion of k_2, which leads to a broader peak region. For the lateral ITZ and paste, in addition to the peak at around 0.7 GPa, there is a remarkable component with the hardness of around 0.4 GPa. When the phase is not separated from others in the deconvolution analysis, it leads to slightly small value of k_1 and k_2, as shown in Table 9.9 for the optimal model of paste with 5 components. Even if there is also a phase with the hardness of 0.31GPa for the bottom ITZ with 6 components provided in Table 9.9, the proportion of the phase is far smaller than the corresponding paste sample. Similarly, the proportion of k_1 with hardness of 0.17 GPa in the 6 components model of top ITZ is a negligible value of 0.70%. Thus, the low hardness phase is a significant difference between different samples. This low hardness phase exists merely in paste and the lateral ITZ should be formed by test points on low strength products. For instance, porous gel would have low hardness, while the pores are not large enough to be detected by BSE analysis at the magnification of 500× and still lead to a similar fraction of defects as shown in Fig. 9.36. There is a component with an average hardness of around 0.7 GPa for all samples as provided in Table 9.8, which is within the big peak of histogram as shown in Fig. 9.46. It may correspond to the mixed phase that is primarily reacted product gel. The later components would be more and more affected by crystals and fly ash.

9.4.4.3 Nanoindentation Results and Analysis

In addition to hardness, nanoindentation test is able to simultaneously obtain elastic modulus properties. The nanoindentation test results for ITZs and paste are shown in Fig. 9.50 and Table 9.10. The results show a relative strength trend similar to the average scratch hardness results, but there are differences in some details. The scratch hardness in the third column (kall) of Table 9.8 decreases progressively in the order of top ITZ, bottom ITZ, paste and lateral ITZ. In terms of the nanoindentation results, instead of the lateral ITZ, it is the paste that reaches the minimum values.

In general, the micromechanical properties of the top and bottom ITZ are close, and the lateral ITZ is close to the paste. Although the visible large fly ash particles/ defects were intentionally avoided and the maximum feasible depth was adopted to activate more homogeneous properties, the comparison results based on Table 9.10 would be still somewhat affected by the individual component of fly ash. Compared with others, the top ITZ has a slightly larger standard deviation and the results in Fig. 9.50 have several data that show very high mechanical properties. However, these data just account for a very small proportion. For the histograms with the same bin size, their peaks still clearly reveal the trend that the lateral ITZ and paste has lower mechanical properties than the top and bottom ITZ. For the nanoindentation data, the results represent the properties of mixed phase generally containing gel and not large size inclusions. This point is manifested by the small standard deviation of nanoindentation hardness than that of scratch hardness k_{all}. Therefore, the nanoindentation results are more closely linked to the gel-related phase in the deconvolution analysis, although the values are not close due to the different test technique, indenter and degree of involvement of other phases. The top and bottom ITZ have similar k1 but just different proportions in the scratch test. It should be the reason for the close elastic modulus and hardness value obtained by nanoindentation. Different from the top and bottom ITZ, scratch results indicate that the lateral ITZs and paste have a considerable amount of a small hardness phase. The nanoindentation at large penetration depth would detect them and hence result in low average micromechanical properties. The nanoindentation results coincide with the nanoscratch results in this aspect and indicate that the properties of ITZ are different at different locations surrounding the aggregate. Besides, both investigations found that all ITZs are not inferior to the bulk paste.

9.5 Conclusion

The properties of ITZs in modelled aggregate geopolymer concrete are investigated by micro/nanomechanical techniques and image analysis in this chapter. The main conclusions are drawn up as follows:

(1) After removing geopolymer from aggregate boundary surfaces, residual paste was observed on the surfaces of the polished aggregate. However, no obvious residual PC paste was found to adhere to the aggregate. Besides, the force needed to remove geopolymer was much higher than that for PC. It indicates that geopolymer has stronger bonding than that of PC to aggregate surface.
(2) Morphology features showed that a layer of dense gel-rich paste formed on the initial interfacial zone of geopolymer, while a large amount of CH crystals were found on the ITZ of PC. ITZ caused by wall effect was more distinct in geopolymer composites with silica modulus of 1. Based on the ranges in fly ash content, the size of ITZ was determined as around 30 μm and 40 μm for Geo-Ms1 and Geo-Ms1.5, respectively. The ITZ in the modelled

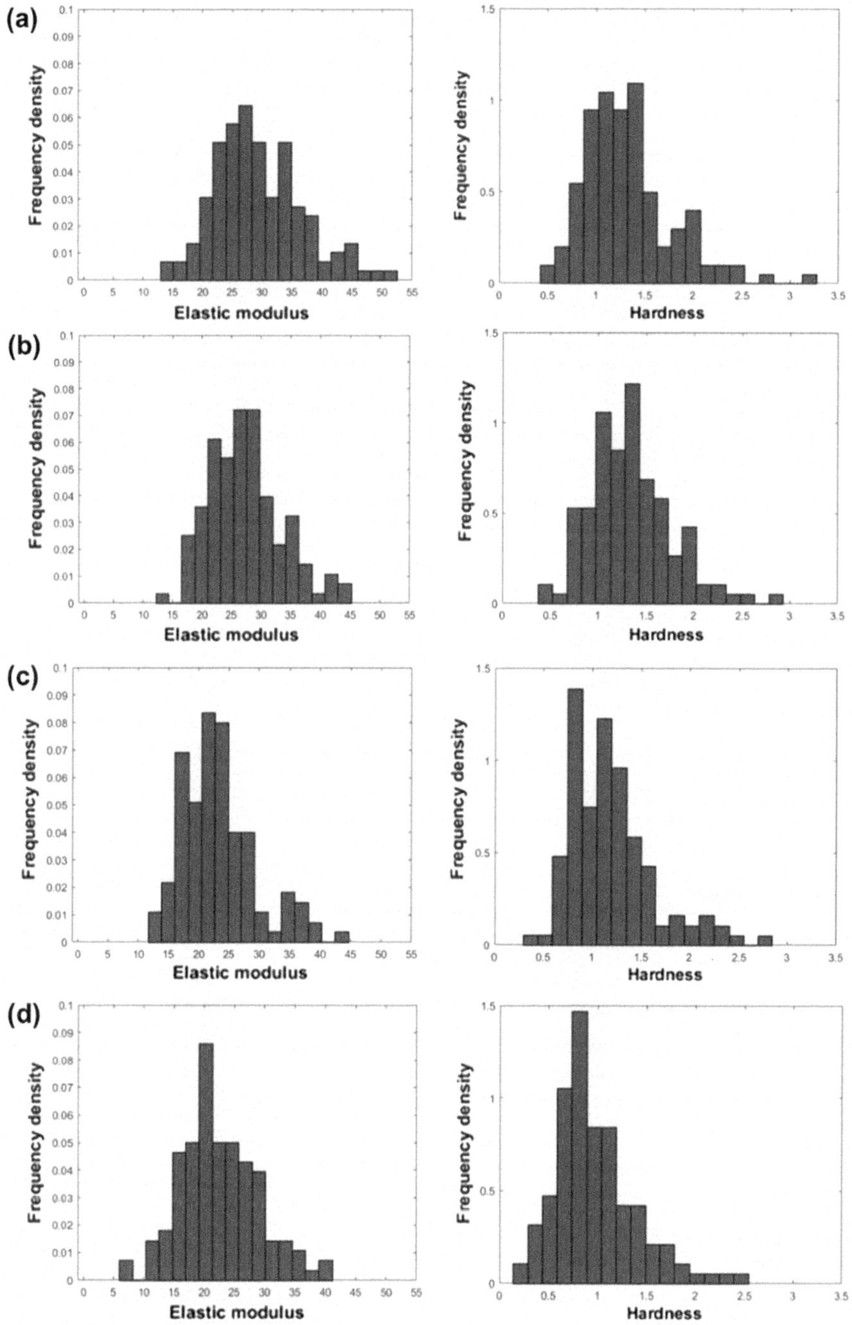

Fig. 9.50 Elastic modulus and hardness of ITZs/paste investigated by nanoindentation: **a** top ITZ; **b** bottom ITZ; **c** lateral ITZ; **d** paste

Table 9.10 Nanoindentation test results for ITZs and paste

Test objects	*M* (GPa)	*H* (GPa)
Top-ITZ	29.39 ± 7.35	1.32 ± 0.46
Bottom-ITZ	27.31 ± 6.13	1.33 ± 0.43
Lateral-ITZ	23.24 ± 6.02	1.19 ± 0.43
Paste	22.55 ± 6.40	0.98 ± 0.42

geopolymer concrete exhibited dense microstructures and was more uniform than geopolymer paste.

(3) Grid nanoindentation under the indent depth of 850 nm shows that the elastic modulus ratio and hardness ratio of ITZ to bulk paste are both around 0.97 in geopolymer. Grid nanoindentation indent under the depth of 1200 nm revealed more homogeneous properties. The elastic modulus and hardness of ITZ in geopolymer are around 1.21 and 1.18 times that of the bulk paste. The higher micromechanical properties of ITZs in geopolymer can be attributed to the dense and uniform gel-rich paste on the surface of aggregate, which contributes to the strong interfacial bonding properties.

(4) Scratch hardness and scratch friction coefficient depended on the normal load applied. When slightly increased the normal load from 2 to 4 mN and 8 mN, the hardness was increased by 23.08% and 10.00%, respectively, while the friction coefficient was increased by 41.67% and 105.88%, respectively.

(5) ITZ and paste of Geo-Ms1.5 showed very close scratch hardness and scratch friction coefficient. The scratch friction coefficient of ITZ and paste were 0.31 ± 0.30 and 0.31 ± 0.27, respectively. For Geo-Ms1, the scratch friction coefficient was 0.32 ± 0.29 for ITZ and 0.42 ± 0.37 for paste.

(6) Compared to the scratch test parallel to aggregate boundary, perpendicular scratch that continuously crosses aggregate, ITZ and paste presented the potential to reveal the detailed variation of properties with increased distance to aggregate. However, based on a similar number of test points, the properties results obtained from perpendicular scratch were less stable and reliable. Parallel scratch uniquely applicable to the modelled interface can quickly provide useful information to reflect the mechanical properties of ITZ.

(7) The top and bottom boundary regions (0–90 μm away from the aggregate) present a higher proportion and number of fly ash than the lateral boundary region, which makes the disrupted distribution of fly ash particles caused by aggregates become less obvious for the bottom and top ITZs. Besides the top region, the average number of fly ash particles generally presents a clear increasing trend with the increased distance from the aggregate.

(8) For all the ITZs, the average diameter of fly ash particles is significantly low in the initial 0 to 10 μm range of ITZ, which has an average diameter of around 5.51–5.79 μm and increased to the range of 6.44–8.95 μm for the later 10 μm width bands. The results from 30 μm bands show that the lateral side ITZ exhibits a higher average diameter of fly ash particles but a lower average diameter of the three largest fly ash particles than the top and bottom ITZs in

each band. The increase of diameter with distance from aggregate is primarily caused by the several largest fly ash particles.

(9) The deconvolution of scratch results indicate that compared with the top and bottom ITZs, the lateral side ITZs and bulk paste have a special low hardness component. The nanoindentation results from gel-rich region showed a similar phenomenon that the top and bottom ITZs have higher elastic modulus and hardness than the lateral side ITZs and the bulk paste. The elastic modulus and hardness by indentation ranged from 22.55–29.39 GPa to 0.98–1.33 GPa, respectively, where the top ITZs achieved the highest value, while bulk paste displayed the lowest value. The micromechanical properties of all types of ITZs are higher than that of the bulk paste.

(10) The bleeding of alkali-silicate solution in geopolymer concrete is less possible than that in the PC concrete due to the higher viscosity and smaller density difference with solids. The vibration process could lead to local compaction and intensified microstructure for the top and bottom side binders. This process is also accompanied by the accumulation of fly ash, while the increased small size particles with high activity in a small space may further enhance the micromechanical properties at the top and bottom ITZs. Therefore, the micromechanical properties of bottom ITZs are higher than lateral ITZs in the modelled geopolymer concrete.

References

1. S. Diamond, Considerations in image analysis as applied to investigations of the ITZ in concrete. Cement Concr. Compos. **23**(2–3), 171–178 (2001)
2. S. Diamond, J. Huang, The ITZ in concrete—a different view based on image analysis and SEM observations. Cement Concr. Compos. **23**(2–3), 179–188 (2001)
3. W. Li, Z. Luo, Y. Gan, K. Wang, S.P. Shah, Nanoscratch on mechanical properties of interfacial transition zones (ITZs) in fly ash-based geopolymer composites. Compos. Sci. Technol. **214** (2021)
4. Z. Luo, W. Li, K. Wang, A. Castel, S.P. Shah, Comparison on the properties of ITZs in fly ash-based geopolymer and Portland cement concretes with equivalent flowability. Cem. Concr. Res. **143** (2021)
5. A.R. Brough, A. Atkinson, Automated identification of the aggregate-paste interfacial transition zone in mortars of silica sand with Portland or alkali-activated slag cement paste. Cem. Concr. Res. **30**(6), 849–854 (2000)
6. J.J. Chen, L. Sorelli, M. Vandamme, F.J. Ulm, G. Chanvillard, A coupled nanoindentation/SEM-EDS study on low water/cement ratio Portland cement paste: evidence for C-S-H/$Ca(OH)_2$ nanocomposites. J. Am. Ceram. Soc. **93**(5), 1484–1493 (2010)
7. R.J. Detwiler, P.J.M. Monteiro, H.R. Wenk, Z. Zhong, Texture of calcium hydroxide near the cement paste-aggregate interface. Cem. Concr. Res. **18**(5), 823–829 (1988)
8. S. Du, Y. Jiang, J. Zhong, Y. Ge, X. Shi, Surface abrasion resistance of high-volume fly ash concrete modified by graphene oxide: macro- and micro-perspectives. Constr. Build. Mater. **237** (2020)
9. G. Fang, M. Zhang, The evolution of interfacial transition zone in alkali-activated fly ash-slag concrete. Cem. Concr. Res. **129** (2020)

10. A. Hajimohammadi, T. Ngo, J. Vongsvivut, Interfacial chemistry of a fly ash geopolymer and aggregates. J. Clean. Prod. **231**, 980–989 (2019)
11. Y. Ding, J.G. Dai, C.J. Shi, Fracture properties of alkali-activated slag and ordinary Portland cement concrete and mortar. Constr. Build. Mater. **165**, 310–320 (2018)
12. A. Hajimohammadi, T. Ngo, J.L. Provis, T. Kim, J. Vongsvivut, High strength/density ratio in a syntactic foam made from one-part mix geopolymer and cenospheres. Compos. Part B: Eng. **173** (2019)
13. C.G. Hoover, F.J. Ulm, Experimental chemo-mechanics of early-age fracture properties of cement paste. Cem. Concr. Res. **75**, 42–52 (2015)
14. J. Hu, Z. Qian, D. Wang, M. Oeser, Influence of aggregate particles on mastic and air-voids in asphalt concrete. Constr. Build. Mater. **93**, 1–9 (2015)
15. A. Elsharief, M.D. Cohen, J. Olek, Influence of aggregate size, water cement ratio and age on the microstructure of the interfacial transition zone. Cem. Concr. Res. **33**(11), 1837–1849 (2003)
16. A. Karimzadeh, M.R. Ayatollahi, Investigation of mechanical and tribological properties of bone cement by nano-indentation and nano-scratch experiments. Polym. Testing **31**(6), 828–833 (2012)
17. W.K.W. Lee, J.S.J. van Deventer, Chemical interactions between siliceous aggregates and low-Ca alkali-activated cements. Cem. Concr. Res. **37**(6), 844–855 (2007)
18. J.J. Kim, E.M. Foley, M.M. Reda Taha, Nano-mechanical characterization of synthetic calcium-silicate-hydrate (C-S-H) with varying CaO/SiO_2 mixture ratios. Cem. Concr. Compos., 65–70 (2013)
19. P.G. Allison, C.A. Weiss, R.D. Moser, A.J. Diaz, O.G. Rivera, S.S. Holton, Nanoindentation and SEM/EDX characterization of the geopolymer-to-steel interfacial transition zone for a reactive porcelain enamel coating. Compos. B Eng. **78**, 131–137 (2015)
20. M. Husem, The effects of bond strengths between lightweight and ordinary aggregate-mortar, aggregate-cement paste on the mechanical properties of concrete. Mater. Sci. Eng. A **363**(1–2), 152–158 (2003)
21. A.T. Akono, J. Chen, S. Kaewunruen, Friction and fracture characteristics of engineered crumb-rubber concrete at microscopic lengthscale. Constr. Build. Mater. **175**, 735–745 (2018)
22. V.S. Harutyunyan, E.S. Abovyan, P.J.M. Monteiro, V.P. Mkrtchyan, M.K. Balyan, A.P. Aivazyan, X-ray diffraction investigations of microstructure of calcium hydroxide crystallites in the interfacial transition zone of concrete. J. Am. Ceram. Soc. **86**(12), 2162–2166 (2003)
23. R.F. Gibson, A review of recent research on nanoindentation of polymer composites and their constituents. Compos. Sci. Technol. **105**, 51–65 (2014)
24. M. Khedmati, Y.R. Kim, J.A. Turner, Investigation of the interphase between recycled aggregates and cementitious binding materials using integrated microstructural-nanomechanical-chemical characterization. Compos. B Eng. **158**, 218–229 (2019)
25. J. Liu, Q. Zeng, S. Xu, The state-of-art in characterizing the micro/nano-structure and mechanical properties of cement-based materials via scratch test. Constr. Build. Mater. **254** (2020)
26. G. Mallikarjunachari, P. Ghosh, Analysis of strength and response of polymer nano thin film interfaces applying nanoindentation and nanoscratch techniques. Polymer **90**, 53–66 (2016)
27. M. Khedmati, Y.R. Kim, J.A. Turner, H. Alanazi, C. Nguyen, An integrated microstructural-nanomechanical-chemical approach to examine material-specific characteristics of cementitious interphase regions. Mater Charact **138**, 154–164 (2018)
28. Q. Li, Y. Li, L. Zhou, Nanoscale evaluation of multi-layer interfacial mechanical properties of sisal fiber reinforced composites by nanoindentation technique. Compos. Sci. Technol. **152**, 211–221 (2017)
29. Z. Luo, W. Li, Y. Gan, K. Mendu, S.P. Shah, Applying grid nanoindentation and maximum likelihood estimation for N-A-S-H gel in geopolymer paste: investigation and discussion. Cem. Concr. Res. **135** (2020)

30. J.L. Provis, A. Palomo, C. Shi, Advances in understanding alkali-activated materials. Cem. Concr. Res. **78**, 110–125 (2015)
31. M.J. Mac, M.H.N. Yio, G. Desbois, I. Casanova, H.S. Wong, N.R. Buenfeld, 3D imaging techniques for characterising microcracks in cement-based materials. Cem. Concr. Res. **140** (2021)
32. A. Ureña, J. Rams, M.D. Escalera, M. Sánchez, Characterization of interfacial mechanical properties in carbon fiber/aluminium matrix composites by the nanoindentation technique. Compos. Sci. Technol. **65**(13), 2025–2038 (2005)
33. J. Xiao, W. Li, Z. Sun, D.A. Lange, S.P. Shah, Properties of interfacial transition zones in recycled aggregate concrete tested by nanoindentation. Cement Concr. Compos. **37**(1), 276–292 (2013)
34. K. Wu, H. Shi, L. Xu, G. Ye, G. De Schutter, Microstructural characterization of ITZ in blended cement concretes and its relation to transport properties. Cem. Concr. Res. **79**, 243–256 (2016)
35. Y. Wei, X. Gao, S. Liang, A combined SPM/NI/EDS method to quantify properties of inner and outer C-S-H in OPC and slag-blended cement pastes. Cement Concr. Compos. **85**, 56–66 (2018)
36. K.L. Scrivener, Backscattered electron imaging of cementitious microstructures: understanding and quantification. Cement Concr. Compos. **26**(8), 935–945 (2004)
37. H. Peng, C. Cui, C.S. Cai, Y. Liu, Z. Liu, Microstructure and microhardness property of the interface between a metakaolin/GGBFS-based geopolymer paste and granite aggregate. Constr. Build. Mater. **221**, 263–273 (2019)
38. T. Ji, Q. Gao, W. Zheng, X. Lin, H.-C. Wu, Interfacial transition zone of alkali-activated slag concrete. ACI Mater. J. **114**(3), 347 (2017)
39. M. Kuroda, T. Watanabe, N. Terashi, Increase of bond strength at interfacial transition zone by the use of fly ash. Cem. Concr. Res. **30**(2), 253–258 (2000)
40. J.A. King, I. Miskioglu, D.D. Wright-Charlesworth, C.D. Van Karsen, Nanoscratch testing to assess the fiber adhesion of short-carbon-fiber composites. J. Appl. Polym. Sci. **103**(1), 328–335 (2007)
41. B. Nematollahi, J. Qiu, E.H. Yang, J. Sanjayan, Microscale investigation of fiber-matrix interface properties of strain-hardening geopolymer composite. Ceram. Int. **43**(17), 15616–15625 (2017)
42. J. Xu, D.J. Corr, S.P. Shah, Nanoscratch study of the modification effects of nanoSiO_2 on C-S-H gel/cement grain interfaces. J. Mater. Civ. Eng. **29**(9) (2017)
43. Y. Wei, W. Kong, Y. Wang, A. Sha, Multifunctional application of nanoscratch technique to characterize cementitious materials. Cem. Concr. Res. **140** (2021)
44. M. Zhou, Y. Jiang, Y. Sui, Microstructure and properties of interfacial transition zone in ZTA particle–reinforced iron composites. Appl. Phys. A: Mater. Sci. Process. **125**(2) (2019)
45. Z. Zhang, H. Wang, Y. Zhu, A. Reid, J.L. Provis, F. Bullen, Using fly ash to partially substitute metakaolin in geopolymer synthesis. Appl. Clay Sci. **88–89**, 194–201 (2014)
46. X.Y. Zhuang, L. Chen, S. Komarneni, C.H. Zhou, D.S. Tong, H.M. Yang, W.H. Yu, H. Wang, Fly ash-based geopolymer: clean production, properties and applications. J. Clean. Prod. **125**, 253–267 (2016)
47. W. Zhu, P.J.M. Bartos, Application of depth-sensing microindentation testing to study of interfacial transition zone in reinforced concrete. Cem. Concr. Res. **30**(8), 1299–1304 (2000)
48. R. San Nicolas, S.A. Bernal, R. Mejía De Gutiérrez, J.S.J. Van Deventer, J.L. Provis, Distinctive microstructural features of aged sodium silicate-activated slag concretes. Cem. Concr. Res. **65**, 41–51 (2014)
49. Z. Zhang, J.L. Provis, A. Reid, H. Wang, Fly ash-based geopolymers: the relationship between composition, pore structure and efflorescence. Cem. Concr. Res. **64**, 30–41 (2014)
50. J. Xu, D.J. Corr, S.P. Shah, Nanomechanical investigation of the effects of nanoSiO_2 on C-S-H gel/cement grain interfaces. Cement Concr. Compos. **61**, 7–17 (2015)
51. S. Zhao, E. Van Dam, D. Lange, W. Sun, Abrasion resistance and nanoscratch behavior of an ultra-high performance concrete. J. Mater. Civ. Eng. **29**(2) (2017)

52. Y. Yan, S. Chang, T. Wang, Y. Geng, Scratch on polymer materials using AFM tip-based approach: a review. Polymers **11**(10) (2019)
53. J.L. Provis, G.C. Lukey, J.S.J. Van Deventer, Do geopolymers actually contain nanocrystalline zeolites? a reexamination of existing results. Chem. Mater. **17**(12), 3075–3085 (2005)

Chapter 10
Challenges and Perspectives

10.1 Introduction

Concrete has been playing a significant role in construction for more than 300 years since 1900 [1]. As a resource-intensive material, concrete production annually consumes a large amount of natural sand, stone, and Portland cement [2, 3]. To achieve carbon neutrality as well as save natural resources, numerous researchers have attempted to recycle construction and industry wastes, such as crushed waste concrete, fly ash, and waste glass products, as raw materials or SCMs in concrete, [4, 5]. Crashed waste concrete can be used as coarse aggregate in new concrete after crushing and sieving, namely recycled coarse aggregate (RCA) [6]. However, it is difficult to obtain RCA with the same quality as the original nature aggregate because the old mortar inevitably adheres to the RCA surface [7, 8]. Although the properties of RAC are not as good as NAC, its economic value and environmental benefits are substantial [9, 10]. The prospect of recycling waste concrete for use on road, pavements, conduit and some non-structural components is still great [11–13]. The addition of fly ash and slag can not only bring environmental benefits, but also improve the long-term strength and durability of concrete. However, applying a high amount of fly ash would bring adverse effects on the strengths, especially the early age strength, of concrete [14]. Fly ash can be activated by alkali solutions to unleash its potential. Fly ash-based geopolymer [15] are promising construction materials that can maximize fly ash usage and develop clinker-free concrete. Proper design of fly ash-based cementitious materials can make its mechanical properties and durability comparable to Portland cement materials [16]. Despite the great advantages and potential of geopolymer materials, the understanding and application of geopolymer is far less than that of Portland cement concrete. In addition to general engineering properties, great efforts are needed in basic research to gain more scientific insights into geopolymer materials.

W. Li et al., *Nanocharacterization of Cementitious Composites*,
https://doi.org/10.1007/978-981-96-2917-6_10

In order to explain the failure mechanism and durability of RAC and geopolymer concrete, it is necessary to analyse the performance of concrete in depth at the microscopic level, especially the microstructure, micromechanical properties and chemical composition of ITZs and paste matrix [17–19]. In recent decades, the rapid development and popularization of cutting-edge technology have brought new horizons to the research of cement-based materials. One of the very noticeable advances is understanding and modifying of cement-based materials from small scales of micro and nanoscale. Micro and nanoscale characterization techniques such as nanomechanical test techniques and microstructure observation techniques can reveal the inner mechanisms of materials' macro behaviours. Concrete is a typical multiphase heterogeneous material. With the rise of sustainable and low-carbon ingredients (waste materials, admixtures and nanoparticles), the heterogeneity of concrete is further exacerbated. This makes it difficult to understand the failure and degradation process of modern concrete materials in complex environments. Therefore, it is very necessary to identify the microscopic properties of ITZ and paste matrix in these new-generation concrete materials through advanced micro and nanoscale characterization techniques. In-depth mechanism analysis is an important means of finding out engineering problems in various cases and providing strategies for improving materials. Scientific understanding and theory are the essential basis for the wide application of materials, especially the construction materials closely related to life and property safety.

This chapter discusses limitations and challenges of popular nano-characterization techniques for cementitious materials. In order to obtain accurate and stable test data, the sample preparation process and experimental parameters need to be strictly controlled. Mature nano-characterization techniques can be used to evaluate the performance of green and low-carbon materials such as RAC and geopolymer concrete.

10.2 Cementitious Sample Surface Preparation Methods

Surface preparation methods for cementitious samples are essential for ensuring accurate and reliable experimental results in various studies, including micromechanical testing and microstructural analysis. However, these methods face several challenges and have future directions for improvement as shown in Fig. 10.1.

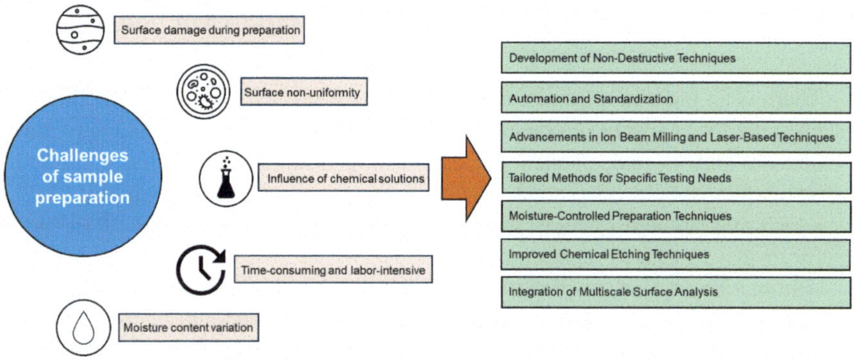

Fig. 10.1 Challenges and perspectives of sample preparation

10.2.1 Challenges of Cementitious Sample Surface Preparation Methods

1. The hardness of cement-based gel and geopolymer gel is approximately 1–2 GPa. Mechanical methods like grinding and polishing can introduce microcracks or distortions in the cement matrix, which may affect the mechanical properties of the material. When the polish speed is too fast and there is no effective cooling measure, the heat generated during grinding may cause thermal damage or changes in the hydration phases, impacting the material's behavior in tests.
2. Achieving a consistent, flat, and smooth surface can be difficult due to the heterogeneity of cementitious materials. Variations in the size of aggregate particles and differences in porosity can lead to uneven surfaces, which may affect testing results. In addition, multiple phases can cause difficulties in obtaining accurate measurements in micromechanical testing, such as inconsistent elastic modulus and hardness at clinker-gel interfaces.
3. Some preparation techniques may alter the microstructure of cementitious materials, potentially leading to an inaccurate representation of the original sample's microstructure. Techniques like acid etching, commonly used to reveal surface features, may dissolve certain phases of the cement matrix, leading to artifacts in microstructural observations.
4. Surface preparation methods, especially those requiring multiple steps like polishing with progressively finer abrasives, can be very time-consuming and labor-intensive. Advanced preparation methods, such as ion beam milling, while effective, are expensive and require highly specialized equipment and skilled technicians.
5. Moisture content significantly influences the surface properties of cementitious materials, and improper handling during preparation can lead to drying or hydration, affecting test results. The need to preserve the moisture content during surface preparation can add complexity, especially in samples intended for durability or porosity studies.

10.2.2 Future Directions in Surface Preparation Methods

1. There is an increasing need for non-destructive surface preparation methods to preserve the integrity of cementitious samples, particularly for microstructural analysis and micromechanical property testing. Some advanced characterization techniques should be promoted such as X-ray computed tomography (XCT) which may allow for better visualization of internal structures without the need for invasive surface preparation.
2. Automation in surface preparation methods could reduce the time and labor involved, leading to more consistent results. Robots and automated polishing systems can ensure uniformity and repeatability. In addition, Standardizing preparation protocols for specific tests would improve comparability between different studies and laboratories.
3. Ion beam milling, which is already used in some fields like electron microscopy, could become more widely applied in cementitious material studies due to its precision in removing material without damaging the microstructure. Laser-based methods, which can ablate material with minimal thermal damage, could be refined to provide efficient and non-destructive surface preparation for cementitious materials.
4. Developing surface preparation methods tailored to specific tests (e.g., mechanical, chemical, or microstructural analysis) could enhance the reliability of test results. For instance, samples intended for mechanical testing may require different surface treatments than those used for microscopy or durability testing.
5. Future methods may incorporate moisture-controlled environments during preparation, ensuring the sample's hydration state remains consistent throughout the process. This is particularly important for tests focused on permeability, shrinkage, or long-term durability.
6. Developing more selective chemical etching techniques could help in revealing microstructural details without excessively altering or damaging the cementitious matrix. Innovations in chemicals or etching protocols may allow for better differentiation of hydration phases without compromising structural integrity.
7. Combining multiple preparation techniques with advanced imaging technologies may provide multiscale surface analysis, offering a comprehensive understanding of the material's properties. Enhanced imaging could provide insights into both the surface and internal features simultaneously, reducing the need for invasive preparation methods.

These future directions point towards a more integrated, precise, and efficient approach to cementitious sample surface preparation, which will ultimately contribute to more accurate testing and better performance predictions for cement-based and geopolymer materials.

10.3 Microscopic Characterization Methods for Sustainable Cementitious Materials

10.3.1 Limitation of Characterization on Low-Carbon Cementitious Materials

Low-carbon cementitious materials are increasingly important in the construction industry as efforts to reduce greenhouse gas emissions intensity. These materials typically incorporate SCMs like fly ash, slag, silica fume, or natural pozzolans to replace a portion of traditional Portland cement, thereby lowering the overall carbon footprint. Microscopic characterization plays a crucial role in understanding the microstructure, composition, and performance of these low-carbon alternatives. This understanding facilitates the optimization of material properties, ensuring durability, strength, and sustainability.

Characterizing the microstructure and micromechanical properties of cementitious materials is crucial for understanding their performance, durability, and failure mechanisms. However, the inherent complexity of these materials presents numerous challenges in accurately capturing their microstructural features and mechanical behavior at small scales. The field continues to evolve, with future directions focusing on improving the precision and relevance of these characterization methods.

Cement-based materials are inherently heterogeneous, consisting of different phases like unhydrated cement particles, C–S–H, calcium hydroxide, and various voids and cracks. This heterogeneity becomes severer after the addition of low-carbon additives or waste materials. The various phases in low-carbon cementitious materials often overlap at the microstructural level and makes it difficult to isolate individual phases. Techniques like SEM and XRD can identify these phases, but precise phase separation and analysis remain challenging. In addition, conventional characterization tools like optical microscopy and SEM have resolution limits that may not capture the finest details of microstructural features, such as the intricate morphology of C–S–H or nanoscale cracks. Techniques that can reach higher resolutions, such as TEM, are limited by their complex sample preparation processes and may introduce artifacts that affect accuracy.

Characterizing micromechanical properties like hardness, stiffness, and fracture toughness at the microscale is difficult due to the influence of sample preparation, phase interactions, and size effects. Nanoindentation is widely used for this purpose, but results can be highly sensitive to surface roughness, measurement location, and interaction between different microstructural phases. Preparing samples for nano and micromechanical property tests, especially for techniques like nanoindentation, nanoscratch and peak-force QNM etc. often introduces artifacts. Mechanical grinding, polishing, or cutting can induce microcracks or alter the local microstructure, leading to inaccurate results. In addition, cement-based and geopolymer materials are sensitive to environmental conditions such as humidity, temperature and carbonation, which can affect their microstructure and mechanical properties over time. Preserving the sample's moisture content during preparation is another challenge, particularly for studies focusing on hydration products or porosity. Moreover, characterization techniques performed under different environmental conditions may yield varying results, complicating the interpretation of long-term durability and behavior. Real-time or in-situ characterization methods are still evolving, and there are challenges in maintaining representative environmental conditions while conducting time-dependent tests.

10.3.2 Perspectives of Microscopic Characterization Methods

1. In-situ characterization methods, such as environmental SEM (ESEM) and in-situ nanoindentation, allow researchers to observe microstructural evolution and mechanical properties under controlled environmental conditions (e.g., moisture, temperature, loading). Real-time characterization of hydration, microcrack formation, and phase changes can provide deeper insights into the long-term behavior of cement-based materials.
2. 3D imaging techniques such as X-ray computed tomography (XCT) and focused ion beam (FIB) tomography are becoming more refined, allowing for detailed visualization of internal microstructures without invasive sample preparation. These methods enable the study of porosity, crack networks, and aggregate distribution in three dimensions, offering a better understanding of how these features affect macroscopic behavior.
3. Future research will likely focus on integrating characterization methods across different length scales, from nanoscale (using techniques like atomic force microscopy, AFM) to macroscale (using mechanical testing and XCT). Such multiscale approaches can bridge the gap between local microstructural properties and global mechanical performance, providing a more comprehensive understanding of material behavior.
4. Nanoindentation is widely used for assessing micromechanical properties like hardness and modulus, but advancements are needed to improve accuracy, especially in heterogeneous cementitious materials. Future developments may include more localized phase-specific indentation, automated mapping to cover large areas, and coupling nanoindentation with other techniques (e.g., AFM) for better context on phase behavior.
5. With the growing use of high-resolution imaging and large datasets, machine learning and data analytics techniques can help process and interpret complex microstructural information. Machine learning could also be applied to predict microstructure-property relationships, allowing for the design of tailored cementitious materials with desired mechanical properties.
6. There is increasing interest in non-destructive techniques like XCT and digital image correlation (DIC), which can monitor mechanical performance without altering the sample. These methods are particularly valuable for studying

progressive damage, such as crack propagation or shrinkage, in cementitious materials under loading or environmental stresses.

7. New techniques for surface functionalization, such as chemical or physical treatments that specifically highlight certain phases or hydration products, could improve phase identification in microscopy. These methods would allow for more accurate quantification of individual phases and their contributions to mechanical properties.
8. Microstructural simulations, based on real characterization data, will play a significant role in future research. Simulations can model the behavior of different phases, predicting how microstructural changes (e.g., hydration, cracking) affect macroscopic properties. Combining experimental results with advanced modeling techniques like finite element analysis (FEA) or discrete element modeling (DEM) will help validate experimental findings and explore untested scenarios.
9. With the increasing use of nanoparticles (e.g., nanosilica, carbon nanotubes) in cement-based materials to improve strength and durability, there will be a greater focus on characterizing the interaction of these additives with the cement matrix. Advanced techniques that can capture nanoscale interactions and their effects on overall mechanical properties will be crucial for optimizing the performance of modified cementitious materials.
10. Data analysis plays a pivotal role in extracting, interpreting, and modeling complex data from microscopic characterization methods of cementitious materials. It allows researchers to better understand microstructural properties, predict material behavior, and ultimately design optimized concrete for various applications, including smart, high-performance, and low-carbon materials. Advanced computer technologies, such as artificial intelligence, machine learning and deep learning, can be used to process data more concisely and efficiently to obtain the proportions and contributions of each component in cementitious materials.

Microstructure and micromechanical property characterization of cementitious materials are advancing rapidly, yet several challenges remain, particularly related to heterogeneity, sample preparation, and environmental effects. Future directions point towards improved in-situ, non-destructive, and multiscale characterization techniques, enhanced by machine learning and simulation tools. These developments will provide more accurate, high-resolution insights into the complex behavior of cementitious materials, leading to better predictive models and material designs.

10.4 Nano/Micromechanical Testing Strategies for Complicated ITZs

The ITZs in real concrete have a three dimensional and complex structure. Micro/nanomechanical testing needs to be conducted on polished section sample. Even in this case, the ITZ is still highly heterogeneous. The heterogeneity of ITZ in a cutting cross-section sample of concrete could be simply understood from three increased levels where the higher level contains the lower level. Firstly, for a local region, ITZ consists of different micro-constituents. The properties of each type of micro-constituent could also be variable. Then, for an aggregate particle, the properties of surrounded ITZ are different between different local regions. In terms of the polished regular aggregate in this research, it can be understood from the two aspects of the difference of ITZ in different sides of the aggregate such as top, bottom, etc., and the inner difference of ITZ in a specified side. Finally, the different aggregate particles in the same concrete sample would have different ITZs. In an investigation, it is challenging to conduct so many tests that can overcome the level three heterogeneities and obtain the overall ITZ properties in a concrete sample. Owing to the small test scale, micro to nanoscale tests would even be significantly affected by the level one heterogeneity. The limited tests on the inhomogeneous object would make the results only local properties, which sometimes may be highly variable and lead to unreliable conclusions. Depending on the research purpose, a feasible investigation can be realised by strategies such as decreasing the complexity of the tested concrete system and improving existing test efficiency.

In the previous study [20], the heterogeneity of ITZ in the modelled samples was greatly reduced, which just contains the first level and part of the second level of heterogeneity as shown in Fig. 10.2a. The low complexity system allows a quick comparison of ITZs formed by different cementitious materials. It is especially meaningful when the test techniques are impossible to overcome the high heterogeneity that would significantly interfere with the comparison results. However, the simplified sample just provides basic ITZ information. In this study, the heterogeneity was increased to level two, which reveals that the top, bottom and lateral side ITZs have different properties. Above this level, the ITZ surrounding different aggregate particles after mixing would be highly complex since it is affected by far more factors, such as the difference in local mix state, aggregate characteristics (e.g. shape, size, orientation, etc.), and distance between aggregates.

Even for BSE images study, an area scan technique, one of the major obstacles reported for qualitative investigation is the very time-consuming procedure to obtain sufficient statistical results [20]. Thus, the complicated ITZ is hard to be appropriately evaluated by general discrete point micro/nanomechanical testing techniques. Indeed, the nanoindentation test in conventional instrument takes 6 to 7 min for a single indent. Despite faster tests, the test by modulus mapping and PeakForce QNM are limited to very small test areas. When aims to deeply understand the properties of complex ITZ in concrete or wants to incorporate more ITZ information for the comparison, it is necessary to find more efficient testing strategies to cope with the

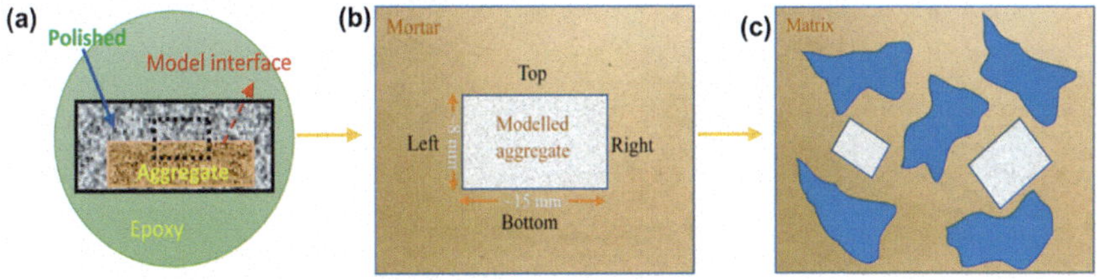

Fig. 10.2 Different heterogeneity of ITZ provided by modelled samples: **a** level I and partial II; **b** level II; **c** level II

largely increased workload. The modelled aggregate has a regular linear boundary, making it possible to apply the line scan technique, nanoscratch to test the ITZ. Nanoscratch is able to obtain massive test points from a long range in a very short time. For the investigation of a given long range of ITZ, the total time (e.g. testing, data analysis/image segment and statistic, etc.) needed for scratch technique is even lower than BSE. Thus, scratch technique combined with regular polished aggregate is a potential method to provide information to evaluate more complicated ITZ with acceptable time-consuming. Meanwhile, nanoindentation can be used to investigate the relatively important regions of the same sample or a more simplified modelled sample with the same binder.

The relatively complicated ITZ systems can be constructed with some real aggregate particles and a certain proportion of polished aggregate particles. They can be mixed the similar way with normal concrete. The polished aggregate within the system can be used as test objects for quick mechanical information acquisition of ITZ. It should be noted that the orientation of aggregate would change after mixing. Compared with the modelled aggregate perpendicular to the exposed test surface, the inclined modelled aggregate would be closer to the test surface on one side, which may result in constraint effect for the micromechanical penetration test. In addition, the section method for inclined aggregate would overestimate the actual width of ITZ because the apparent ITZ obtained is not from the planes perpendicular to the aggregate [21]. Correspondingly, the micromechanical test location on apparent ITZs surrounding the inclined modelled aggregate would correspond to the real ITZ that has a closer distance (d' in blue color) to aggregate, as revealed in Fig. 10.3. In addition, the involved ITZs in the penetration direction have a continuously changed distance to aggregate. Small force and depth as that in scratch can decrease the negative impact, but not all. For natural aggregates, the complex shape would cause or exacerbate similar test biases, making them almost unavoidable. For the regular modelled aggregate, the biases can be removed by testing aggregates with only small inclinations. A typical method is to use nondestructive three-dimensional imaging techniques such as X-ray computed tomography (X-ray CT) [22, 23]. Based on the known distribution, another cutting may be applied to find a test surface with more qualified aggregate particles. It may also be feasible to identify the orientation based on multiple cuts and the variation of section size of regular aggregate with the known three-dimensional size.

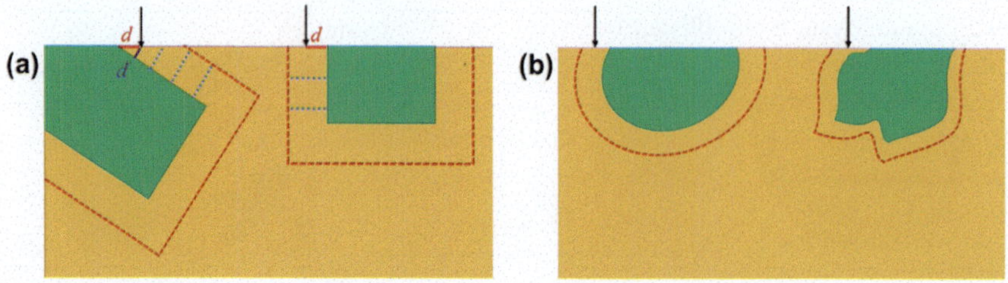

Fig. 10.3 Test biases for significantly inclined modelled aggregate and natural aggregate (side view) (for interpretation of the references to color in this figure, the reader is referred to the web version of this article): **a** modelled aggregate; **b** natural aggregate

The test based on polished aggregate may be different from that of natural aggregate. Given that the current micro/nanomechanical testing techniques are generally not even able to efficiently evaluate the ITZ around a single real coarse aggregate particle, use of polished aggregate could at least overcome the high heterogeneity and provide more information for complex ITZ. Especially that, most of the research is to compare ITZ formed by different binders. The surface of the tested aggregate is equally simplified in both samples, which may not significantly affect the comparison conclusion and would have higher accuracy than the comparison based on insufficient local properties. In addition to the nanoscratch, the polished surface also facilitates other research such as the BSE image segmentation analysis, EDS line scan and grid nanoindentation tests of ITZs. The complexity of the tested ITZ can be controlled by adjusting the amount of aggregate and variable parameters given to modelled aggregates such as size, shape, etc. In a more complicated case, the natural aggregate with few small polished surfaces may be used for testing.

10.5 Summary

Mechanical grinding and polishing for microscopy can introduce microcracks or alter the local microstructure, potentially leading to inaccurate observations. For cementitious materials, they are highly heterogeneous, containing multiple phases like C–S–H, calcium hydroxide, and aggregate particles. This complexity makes it difficult to obtain representative microscale measurements and isolate specific phases for analysis. Moreover, Different phases often overlap at the microscale, making it challenging to accurately separate and characterize individual phases using conventional imaging techniques. For example, traditional microscopic techniques such as SEM may lack the resolution needed to capture fine-scale details of the microstructure, especially for features like nanoscale cracks or hydration products.

Advancements in nanoindentation and in-situ nanoindentation, with phase-specific localization and automated mapping, will improve the accuracy of micromechanical property measurements in heterogeneous materials. Some machine learning and data analysis methods can be applied with microscopic characterization techniques to process large datasets. This is contributed to link microscale properties to overall material performance, which is crucial for optimization of material design. With increasing demands for new concrete technologies, nano-characterization will serve as a cornerstone of innovation in concrete research, enabling the design of more sustainable, resilient, and high-performance materials for concrete construction.

References

1. J. Xiao, W. Li, Y. Fan, X. Huang, An overview of study on recycled aggregate concrete in China (1996–2011). Constr. Build. Mater. **31**, 364–383 (2012)
2. W. Li, C. Long, V.W.Y. Tam, C.-S. Poon, W.H. Duan, Effects of nano-particles on failure process and microstructural properties of recycled aggregate concrete. Constr. Build. Mater. **142**, 42–50 (2017)
3. P. Zhang, K. Wang, Q. Li, J. Wang, Y. Ling, Fabrication and engineering properties of concretes based on geopolymers/alkali-activated binders—a review. J. Clean. Prod. **258**, 120896 (2020)
4. B. Wang, L. Yan, Q. Fu, B. Kasal, A comprehensive review on recycled aggregate and recycled aggregate concrete. Resour. Conserv. Recycl. **171** (2021)
5. J. Xiao, W. Li, C. Poon, Recent studies on mechanical properties of recycled aggregate concrete in China-a review. Science China Technol. Sci. **55**(6), 1463–1480 (2012)
6. Z. Tang, W. Li, V.W.Y. Tam, Z. Luo, Investigation on dynamic mechanical properties of fly ash/slag-based geopolymeric recycled aggregate concrete. Compos. Part B: Eng. **185** (2020)
7. V.W. Tam, C.M. Tam, Diversifying two-stage mixing approach (TSMA) for recycled aggregate concrete: TSMAs and TSMAsc. Constr. Build. Mater. **22**(10), 2068–2077 (2008)
8. Z. Tang, W. Li, V.W.Y. Tam, C. Xue, Advanced progress in recycling municipal and construction solid wastes for manufacturing sustainable construction materials. Resour. Conserv. Recycl, X 6 (2020)
9. J. Liu, K. Ma, J. Shen, J. Zhu, G. Long, Y. Xie, B. Liu, Influence of recycled concrete aggregate enhancement methods on the change of microstructure of ITZs in recycled aggregate concrete. Constr. Build. Mater. **371**, 130772 (2023)
10. J. Liu, Q. Zeng, S. Xu, Is scratch test proper to characterize microstructure and mechanical properties of cement-based materials? The effects of loading level and routine. Cem. Concr. Res. **133** (2020)
11. G. Fang, M. Zhang, Multiscale micromechanical analysis of alkali-activated fly ash-slag paste. Cem. Concr. Res. **135**, 106141 (2020)
12. Y. Hu, Z. Tang, W. Li, Y. Li, V.W.Y. Tam, Physical-mechanical properties of fly ash/GGBFS geopolymer composites with recycled aggregates. Constr. Build. Mater. **226**, 139–151 (2019)
13. B. Lei, W. Li, Z. Tang, V.W.Y. Tam, Z. Sun, Durability of recycled aggregate concrete under coupling mechanical loading and freeze-thaw cycle in salt-solution. Constr. Build. Mater. **163**, 840–849 (2018)
14. D. Adak, M. Sarkar, S. Mandal, Effect of nano-silica on strength and durability of fly ash based geopolymer mortar. Constr. Build. Mater. **70**, 453–459 (2014)
15. D.K. Chanda, S.R. Chowdhury, M. Bhattacharya, A.K. Mandal, N. Dey, A.K. Mukhopadhyay, Intelligently designed fly-ash based hybrid composites with very high hardness and Young's modulus. Constr. Build. Mater. **158**, 516–534 (2018)
16. X.Y. Zhuang, L. Chen, S. Komarneni, C.H. Zhou, D.S. Tong, H.M. Yang, W.H. Yu, H. Wang, Fly ash-based geopolymer: clean production, properties and applications. J. Clean. Prod. **125**, 253–267 (2016)
17. K. Lyu, E.J. Garboczi, W. She, C. Miao, The effect of rough vs. smooth aggregate surfaces on the characteristics of the interfacial transition zone. Cem. Concr. Compos. **99**, 49–61 (2019)

18. K. Lyu, W. She, C. Miao, H. Chang, Y. Gu, Quantitative characterization of pore morphology in hardened cement paste via SEM-BSE image analysis. Constr. Build. Mater. **202**, 589–602 (2019)
19. Z. Zhang, J.L. Provis, X. Ma, A. Reid, H. Wang, Efflorescence and subflorescence induced microstructural and mechanical evolution in fly ash-based geopolymers. Cement Concr. Compos. **92**, 165–177 (2018)
20. S. Diamond, J. Huang, The ITZ in concrete-a different view based on image analysis and SEM observations. Cement Concr. Compos. **23**, 179–188 (2001)
21. Z. Luo, W. Li, K. Wang, A. Castel, S.P. Shah, Comparison on the properties of ITZs in fly ash-based geopolymer and Portland cement concretes with equivalent flowability. Cem. Concr. Res. **143** (2021)
22. M.J. Mac, M.H.N. Yio, G. Desbois, I. Casanova, H.S. Wong, N.R. Buenfeld, 3D imaging techniques for characterising microcracks in cement-based materials. Cem. Concr. Res. **140** (2021)
23. S. Das, P. Yang, S.S. Singh, J.C.E. Mertens, X. Xiao, N. Chawla, N. Neithalath, Effective properties of a fly ash geopolymer: synergistic application of X-ray synchrotron tomography, nanoindentation, and homogenization models. Cem. Concr. Res.Concr. Res. **78**, 252–262 (2015)

The manufacturer's authorised representative in the EU is Springer Nature Customer Service Centre GmbH, Europaplatz 3, 69115 Heidelberg, Germany. If you have any concerns regarding our products, please contact ProductSafety@springernature.com

Printed and bound by CPI Group (UK) Ltd, Croydon, CR0 4YY
01/10/2026
02225937-0010